# the Library
www.imperial.ac.uk/library

# THREE WEEK LOAN (STANDARD)

Please return or renew by the due date.
Fines may be charged on items returned late.

# Electronic Device Architectures

### for the

# Nano-CMOS Era

## From Ultimate CMOS Scaling to Beyond CMOS Devices

# Electronic Device Architectures
### for the
# Nano-CMOS Era
## From Ultimate CMOS Scaling
## to Beyond CMOS Devices

Editor

## Simon Deleonibus
CEA-LETI, France

PAN STANFORD PUBLISHING

*Published by*

Pan Stanford Publishing Pte. Ltd.
5 Toh Tuck Link
Singapore 596224

*Distributed by*

World Scientific Publishing Co. Pte. Ltd.
5 Toh Tuck Link, Singapore 596224
*USA office:* 27 Warren Street, Suite 401-402, Hackensack, NJ 07601
*UK office:* 57 Shelton Street, Covent Garden, London WC2H 9HE

**British Library Cataloguing-in-Publication Data**
A catalogue record for this book is available from the British Library.

**ELECTRONIC DEVICE ARCHITECTURES FOR THE NANO-CMOS ERA**
**From Ultimate CMOS Scaling to Beyond CMOS Devices**

ISBN-13 978-981-4241-28-1
ISBN-10 981-4241-28-8

*Typeset by* Research Publishing Services
E-mail: enquiries@rpsonline.com.sg

Printed in Singapore by Mainland Press Pte Ltd.

# Acknowledgments

I wish to congratulate all contributors and their peers, all of whom are world-renowned researchers from top universities, institutions and organisations, for the results of their research. Their convictions and efforts were key elements for the success of this enterprise.

I wish to specially acknowledge Professor Hiroshi Iwai of Tokyo Institute of Technology (Yokohama, Japan) and former IEEE Electron Device Society President, for his advice, chapter contribution and personal encouragement.

The support of Professors Jean-Pierre Colinge (Tyndall, Cork, Ireland), Cor Claeys (IMEC, Leuven, Belgium), the present IEEE Electron Device Society President, Masataka Hirose (AIST, Tsukuba, Japan), and Jim Hutchby (SRC, Durham-NC, USA), to the promotion of the book is also appreciated. Their influence in the field of Nanoelectronics, Nanotechnology and Nanoscience is a reflection of the high scientific level of the different contributions.

I have special thanks to address to Mr. Stanford Chong, Mr. Rhaimie Wahap and staff members of Pan Stanford Publishing for their responsiveness and *immense patience* demonstrated throughout the whole process of the book's publishing.

Finally, none of this would have been possible without the support of CEA-LETI. The moral support and attention from my wife, Geneviève and my son Tristan, have been of utmost importance to me. I wish to dedicate this work to them.

Simon Deleonibus
*CEA-LETI/MINATEC*
*CEA-Grenoble, 17 rue des Martyrs 38054*
*Grenoble Cedex 09, France*
*sdeleonibus@cea.fr*

# Contents

# Introduction

# Electronic Devices Architectures for the NANO-CMOS Era — From Ultimate CMOS Scaling to Beyond CMOS devices

Since the invention of the first calculation machines, miniaturization has been a constant challenge to increase speed and complexity. Electronic devices have brought, and will bring in the future, a far increasing number of new functions to the basic computing systems such as fast data computing, telecommunication, several kinds of actuations,...which are collectively fabricated on the same physical object named solid state circuit[1], integrated circuit or "chip". Electronic devices are so small, that billions of basic functions are accessible in a hand held system. Moreover, their unit cost has been divided by more than a factor of 100 millions over the past 30 years! The collective fabrication of electronic devices coupled with the increase of their speed has given a tremendous success, which is unique in the history of mankind, to Micro and Nanoelectronics by continuously introducing innovations in the fabrication process (Fig. 1). Linear scaling of devices dimensions to a quasi-nanometer level allows to build complex systems integrated on a chip (Fig. 1) which reduce drastically their volume and power consumption per function, whilst tremendously increasing their speed. In the future, opportunities will appear to build sytems in a molecule. Nanoscience and Nanotechnology researchers join their efforts to Nanoelectronics actors in order to offer mankind possibilities of pervasion of their knowledge into the construction of nanosystems.

*Electronic Devices Architectures for the NANO-CMOS Era,* is a review for the use of Nanoelectronics, Nanoscience and Nanotechnology researchers and engineers, in which we address:

(1) the options to linearly scale down logic CMOS or memories;
(2) the possible competing breakthrough architectures allowing to relax on the linear scaling challenges;
(3) the new paths for integrated electronics.

The pending alternatives are two ways:

(1) try to continue the scaling of *Ultimate CMOS* requesting new materials or

(2) introduce new devices, systems architectures or paradygms *Beyond CMOS*. These questions are very much linked to the progress law that microelectronics has been following since the 1960's.[2]

In the 1960's, Gordon Moore[2] first reported a progress law of microelectronics by asserting that the number of transistors on a chip will increase by a factor of 2 every year. Electrostatics and power dissipation weighed versus the efficiency/speed of devices, required scaling rules which Robert Dennard, Giorgio Baccarani and co authors[3,4] expressed in the 1970's and 1980's. Since then, linear scaling of silicon devices has been dominating the microelectronics world due to the success of miniaturization techniques through collective fabrication, even though bipolar transistors have been replaced by CMOS. Today, the most advanced production integrated circuits are built on CMOS devices with minimum feature sizes of 40 nm. Scientists and engineers are facing, for the first time, new challenges dealing with ultimate scaling of CMOS devices. For example, a high dielectric constant (HiK) material is introduced to replace $SiO_2$, because the scaling

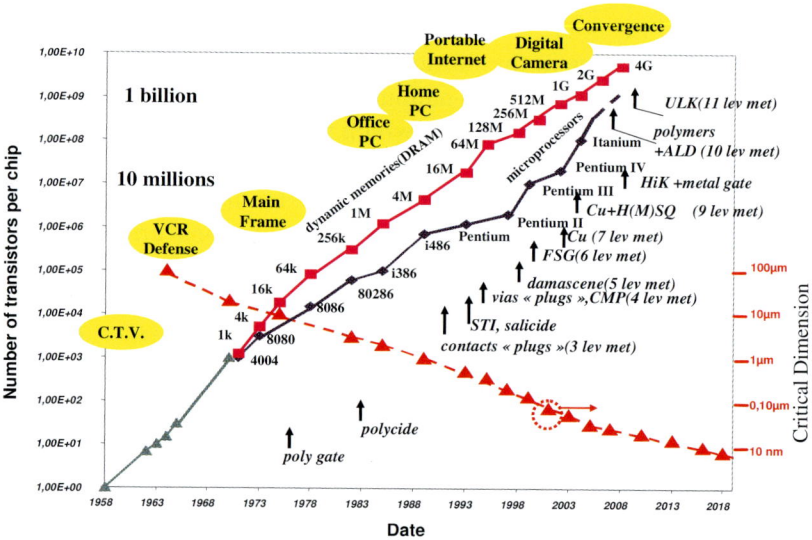

Fig. 1. Evolution of microelectronics devices since the invention of integrated circuits in 1958. On the double Y-axis, the number of transistors per chip (on the *left hand side*) and their critical dimension (gate length) (*right hand side*) are reported. Fabrication technology (arrows) and System (bubble) innovations are indicated.

of CMOS gate oxide cannot satisfy anymore the power dissipation specifications required to design practical and usable chips for the increasing Nomadic market needs. Other roadblocks appeared in microelectronics history in the 90's such as the whole interconnect system functionality and density which was enabled by the introduction of the plug concept technology and copper interconnect.

Device physicists and microelectronics engineers have been investigating various paths to continue the integration race through linear scaling down of silicon devices and searching new devices architectures or new state variables and why not new information processing paradygms.

We first overview the possible technological boosters that will allow CMOS nanoelectronics to reach the end of the roadmap in section 1. The challenges for Core CMOS and memory devices architectures scaling are addressed in sub sections 1 and 2. The various architectures and the physics of ultimate MOSFETs require to benchmark integration limits and transport in ultra small devices. These aspects are overlooked in Chapters 1 and 2 by S. Deleonibus *et al.* and T. Poiroux, G. Lecarval respectively. Possible materials alternatives are compared for channel, gate stack and source and drain engineering. What strain can bring to transport properties is reviewed by S. Takagi *et al.* for SOI or GeOI condensed channels in Chapter 3. A major breakthrough that has been expected for more than 10 years has finally been announced for manufacturing of large scale devices: high dielectric constant materials (HiK) are now used as gate dielectrics in combination with metal gates. In Chapter 4, H. Wong *et al.* address the issue of keeping high channel mobility together with low dielectric leakage current. The properties of rare earth oxides, promising for the realization of the HiK and the future scaling, are reviewed and benchmarked. Acces resistance becomes a severe issue whenever shallow junctions are scaled down as far as bulk Si or SOI devices are concerned. In Chapter 5, B. Mizuno highlights the promising potential of new doping techniques such as plasma doping combined with laser thermal processing or fast thermal processing to activate the dopants.

In the next decade, active devices architectures will need some breakthroughs whereas interconnect architectures went through the same issues in the 1990s. In Chapter 6, S. Laval *et al.* stress on the eventual use of optical interconnect and interfaces in Nanoelectronics chips to replace Copper. How can this paradygm help in reducing the power consumption and increase speed? After exploiting interchip solutions at the level of a system, intra chip solutions are the major research subjects today.

The challenges for memory devices are numerous. Achieving low writing and acces times combined with high retention time is still the Holy Graal searched for high density memory devices. In Chapter 7, K. Kim and G. Jeong review the main challenges in the different served applications to improve memory power consumption, speed and density evolving towards versatile devices properties.

FeRAM and MRAM have been considered as good candidates for fast operation of highly non volatile memory: they are very seductive to microelectronics engineers because these devices can be as fast as DRAM and demonstrate high retention times. In Chapter 8, Y. Arimoto reviews, their potentalities after recalling their principles based on remanent polarization of Ferroelelectric insulators capacitors for FeRAMs or magnetic tunnel junctions in MRAMs.

Current flash memories based on floating gate electron charging will be potentially limited by retention issues beyond the 32 nm node, whenever a reduced number of electrons will be used for switching or charge storage operation. In Chapter 9, B. de Salvo and G. Molas review the potentiality of discrete traps storage nodes to recover high retention: Silicon nanocrystals or molecules used in different conformations, or oxido-reduction states in self organized or cross bar matrices are likely to be considered for future high density low cost memories.

If the above mentioned solutions to proceed on the CMOS roadmap are not efficient or fully operating, we will need to consider new paths to propose alternatives or explore new paradygms bringing added value to circuit designs. Section 2 is devoted to the exploration of New Concepts for Nanoelectronics. CMOS operation at nanometer range dimensions or molecules will use a reduced number of electrons. In Chapter 10, J. Gautier *et al.* address the question on the operation of single electron devices based on Coulomb blockade. If theses devices cannot replace CMOS straightforwardly, they could be associated in a hybrid architecture for niche type of applications due to their very high charge sensitivity, or offer increased functionalities if an extra control gate is added.

In the nanoscale range, the operation of functions by using molecules is of interest due to their potential compacity. In Chapter 11, D. Vuillaume describes the electronic properties of organic monolayers and molecular devices. Hopefully, tunnel barriers, molecular wires, rectifying and NDR diodes, bistable and memories devices have been demonstrated possible with extension to cross bar architectures of highest density.

Carbon nanotubes (CNTs) have demonstrated very exciting characteristics on the thermal and electrical sides whereas their band structure can allow to build semiconductor or metal based devices. In Chapter 12, V. Derycke *et al.* achieve an overview from the materials electronics properties to the building of field effect transitors (FETs) demonstrating high carrier velocity and long carrier mean free path. The placement of CNTs and sorting their chirality are still issues to solve if one wishes to build circuits.

The ITRS teaches us that it is quite difficult to achieve the lowest power consumption together with high performance with electron charge based devices. Could we transfer state variables other than electron charge to address low power and high performance devices architectures? One of the alternatives could be based on spin transfer and detecting it selectively through so called spin valves. In Chapter 13, Kyung-Jin Lee and Sang Ho Lim give an historical review of spin electronics through the use of magnetoresistance in memory devices to the latest attempts to realize so called spin-FETs.

Searching alternative ways to enhance the efficiency of computing that contribute to the improvement of power/speed systems figures of merit is a permanent challenge for design. Can quantum wave functions be used for computing, allowing thus an infinite number of states per bit and compete with binary type operation based algorithms? In Chapter 14, P. Jorrand addresses the basic principles of quantum information processing and communication. The success of quantum algorithms has been proven in speeding up integer factoring or unordered search.

The authors of this review are well-recognized researchers in their field and have give then best to realize this review of the research on the state of the art of NanoCMOS architectures and beyond. They came from well-recognized universities, institutes and microelectronics companies worldwide to deliver tremendous efforts to develop devices and systems using nanotechnologies that make our daily life objects complex functions possible.

**Simon Deleonibus**

*CEA-LETI/MINATEC CEA-Grenoble,*
*17 rue des Martyrs 38054 Grenoble Cedex 09 France*
*sdeleonibus@cea.fr*

# References

1.  J. Kilby and E. Keonjian, IEDM Proceedings of Technical Digest, pp. 76–78, Washington (DC), Oct 29–30, (1959).
2.  G. Moore, *Electronics, Volume 38*, **8**, April 19, 1965.
3.  R. H. Dennard, F. H. Gaensslen, H. N. Yu, V. L. Rideout, Bassous E and A. R LeBlanc, *IEEE J Solid-State Circ*, **9**(5) ,256–68, 1974.
4.  G. Baccarani, M. R. Wordeman and R. H. Dennard, *IEEE Trans Electron Devices*, **31**(4), 452–62, 1984.

# Section 1

..........................................

# CMOS Nanoelectronics. Reaching the End of the Roadmap

# Sub-section 1.1

..........................................

# Core CMOS

# 1

# Physical and Technological Limitations of NanoCMOS Devices to the End of the Roadmap and Beyond

Simon Deleonibus*, Olivier Faynot, Barbara de Salvo, Thomas Ernst, Cyrille Le Royer, Thierry Poiroux and Maud Vinet

CEA-LETI/MINATEC CEA-Grenoble, 17 rue des Martyrs 38054 Grenoble Cedex 09 France.

*sdeleonibus@cea.fr

· · · · · · · · · · · · · · · · · · · · · · · · · · · · ·

Since the end of the 1990s, the microelectronics industry has been facing new challenges as far as CMOS devices scaling is concerned. Linear scaling will be possible in the future if new materials are introduced in CMOS device structures or if new device architectures are implemented. Innovations in the electronics history have been possible because of the strong association between devices and materials research. The demand for low voltage, low power and high performance are the great challenges for the engineering of sub 50 nm gate length CMOS devices because of the increasing interest and necessities of Nomadic Electronic Systems. Functional CMOS devices in the range of 5 nm channel length have been demonstrated. In this chapter, alternative architectures that allow increase to devices' drivability and reduce power consumption are reviewed such as multigate, multichannel architectures and nanowires. The issues in the field of gate stack, channel, substrate, as well as source and drain engineering are addressed. HiK gate dielectric and metal gate are among the most strategic options to implement for power consumption and low supply voltage management. By introducing new materials (Ge, Carbon based materials, III–V semiconductors,

5

HiK, …), Si based CMOS will be scaled beyond the ITRS as the future System-on-Chip Platform integrating also new disruptive devices. For these devices, the low parasitics required to obtain high performance circuits, makes competition against logic CMOS extremely challenging.

# 1. International Technology Roadmap of Semiconductors Acceleration and Issues

Since 1994, the International Technology Roadmap for Semiconductor (ITRS)[1] (Fig. 1) has accelerated the scaling of CMOS devices to lower dimensions continuously despite the difficulties that appear in device optimization.

However, technical roadblocks in lithography principally, economics and physical limitations have slowed down the evolution. Also, for the first time, since the introduction of poly gate in CMOS devices process, showstoppers other than lithography appear to be attracting special attention and require some breakthrough or evolution if we want to continue scaling at the same rate. Design will also be affected by this evolution.

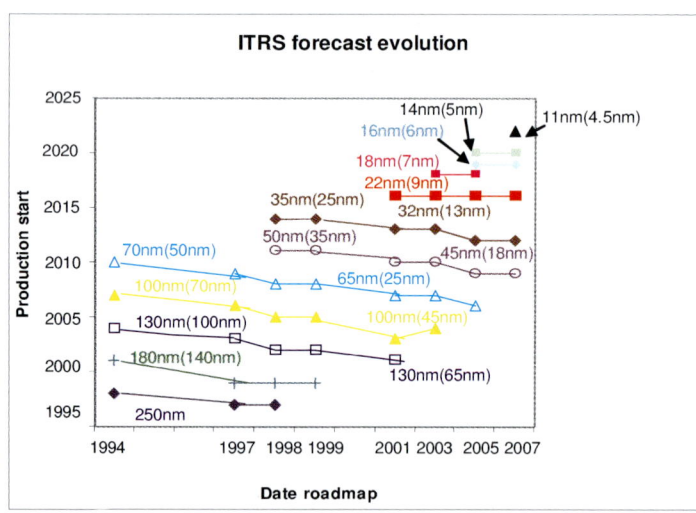

Fig. 1. ITRS forecast evolution since 1994 for MPU devices (HP devices).[1] The half pitch (technology node) appears as a parameter. The minimum physical gate length is given in brackets.

Which are the main showstoppers for CMOS scaling? In this paper, we focus on the possible solutions to investigate and guidelines for research in the next years in order to propose solutions to enhance CMOS performance before we need to skip to alternative devices. In other words, how can we offer a second life to CMOS?

To that respect, the roadmap distinguishes today three types of products: High Performance (HP) (Fig. 1), Low Operating Power (LOP) and Low Standby Power (LSTP) devices. In the HP case, a historical fact will happen by the 32 nm node: the contribution of static power dissipation will become higher than the dynamic power contribution to the total power consumption! This main fact could affect the MOSFET saturation current as can be observed on historical trends of smallest gate length devices.[2] Multigate devices could improve somewhat this evolution (see Section 4.2.2.) by improving the ratio between saturation current and leakage current. In this paper, we will analyze the various mechanisms giving rise to leakage current in a MOS device and that can impact consumption of final devices. Gate leakage current is already a concern. A High Dielectric Constant (HiK) gate insulator will be needed in order to limit static consumption (see Section 4.2).

In Section 2 of this review, we will first analyze the main limitations and showstoppers affecting bulk CMOS scaling. In Section 3, the issues in lowering supply voltage to reduce power dissipation are identified. In Section 4, the limitations to scaling must be taken into account in the device optimization in terms of gate stack, channel and source and drain engineering as well as new devices architectures (FDSOI or multigate devices). The alternative possibilities offered by new materials for enhancement of device transport properties or power dissipation are reviewed in Sections 5 and 6. Finally, in Section 7, we review the applications demonstrated by single or few electronics in the field of memories or possible alternatives to CMOS.

## 2. Limitations and Showstoppers Coming from CMOS Scaling

CMOS device engineering consist of minimizing leakage current together with maximizing the output current. In sub 100 nm CMOS devices, non stationary transport gains more importance as compared to diffusive transport.

## 2.1.  *Origin of leakage current in CMOS devices*

Several mechanisms can generate devices leakage in ultra small MOSFETs, which can be sorted in two categories:

a) Classical type.

- Drain Induced Barrier Lowering (DIBL) is due to the capacitive coupling between source and drain.
- Short Channel Effect (SCE) due to the charge sharing in the channel in the short channel devices at low $V_{ds}$.
- Punch-Through between source and drain due to the extension of source space charge to the drain.

b) Tunneling currents

- Direct tunneling through the gate dielectric.
- Field assisted tunneling at the drain to channel edge. This effect occurs if electric field is high and tunneling is enhanced through the thinnest part of the barrier.
- Direct tunneling from source to drain. This effect will occur in silicon for a thicker barrier than on $SiO_2$ because the maximum barrier height is lower (1.15 eV in Si versus 3.2 eV in $SiO_2$).

## 2.2.  *Issues related to non stationary transport*

Velocity overshoot and ballistic transport are the mechanisms that will enhance drivability in sub 50 nm channel lengths devices. However, the impact of Coulomb scattering by dopants on transport is non negligible even in the 5 nm range channel lengths.[3,4] Superhalo doping is efficient to improve SCE and DIBL in 16 nm finished gate length (Fig. 2)[5] but will degrade the channel transport properties[5] by dopant Coulomb scattering (Fig. 3(a)) and high transverse electric field.

The degradation of transport properties can be observed on short channel mobility measurement by using a specific method with direct $L_{eff}$ measurement[6] (Fig. 3(b)). A mobility degradation of a factor 2 to 3 or more can be measured on the most aggressive nano-scaled bulk technologies. The ITRS target of a transconductance increase by a factor $2^1$ is still very challenging on such gate length even if an enhancement is reported on long channels. Furthermore, for such gate lengths access resistance due to extension scaling is an issue (Fig. 3(a)).[4]

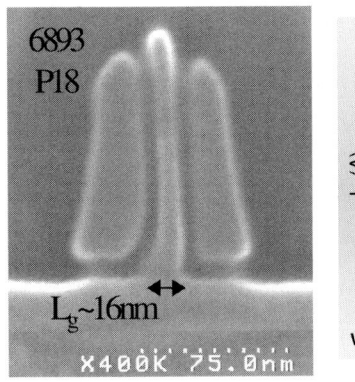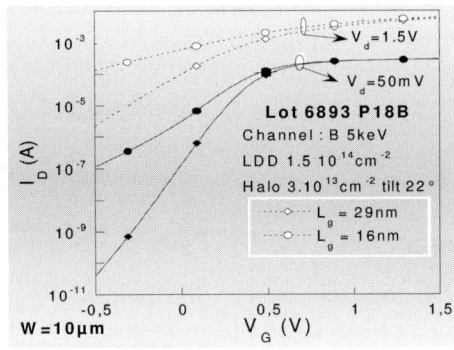

Fig. 2. Functional finished gate length 16 nm bulk n-MOSFET sub threshold characteristics. Gate oxide thickness is 1.2 nm.[4] Isat is 600 $\mu$A/$\mu$m.

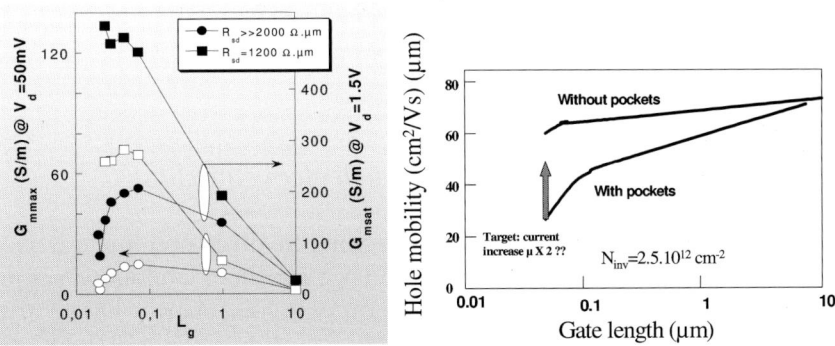

Fig. 3. (a) Effect of halo doping on nMOSFET short channel saturation and linear transconductance (Lg as low as 16 nm). The role of access resistance through extension doping is also investigated[4]; (b) Typical measured p channel mobility loss when gate length is down-scaled due to halo/pockets doping.[6]

## 3.     Issues in Supply Voltage Down Scaling

In the future, the electronics market will require portable objects used in daily life and consequently low standby power dissipation and low active power consumption will be needed. Scaling down of supply voltage is an essential leverage to decrease power dissipation. However, it raises several questions about the possible lower limits.

The power dissipation $P$ of a MOSFET is due to static and dynamic contributions expressed by:

$$P = P_{stat} + P_{dyn} \tag{1}$$

$$P_{stat} = V_{dd} \times I_{off} \tag{2.1}$$

and

$$P_{dyn} = CV_{dd^2} f \tag{2.2}$$

$P$ is the total power dissipation; $P_{stat}$ and $P_{dyn}$ are the static and the dynamic power dissipations respectively. The strong impact of supply voltage on power dissipation appearing in (1), (2.1) and (2.2), will also preclude a strategy of threshold voltage value adjustment depending on the application.

Information theory and statistical mechanics as well as the electrostatics of the device will set the limits of switching of binary devices. Moreover, dopant fluctuations will affect the control of device characteristics substantially: that is why low doping of CMOS channel will help in the down scaling of supply voltage.

## 3.1. Fundamental limits of binary devices switching

Quantum mechanics illustrates that switching involves non linear devices that would demonstrate a gain. That could occur with or without wavefunction phase changing. The Quantum limit on switching energy will be given by the Heisenberg's uncertainty principle:

$E \geq \frac{\hbar}{\tau}$ which gives a minimum switching energy of $E_{min} = 10^{-5} aJ$ considering $\tau = 10 \, ps$, $h = 2\pi\hbar$ is Planck's constant equal to $6.34 \times 10^{-34}$ J.s.

The second principle of thermodynamics imposes the maximization of entropy at temperature T. Applied to information theory this has a consequence on the minimal energy that a system, based on binary states of each bit of information, will require to switch from one state to the other: $E \geq kTLn \,(2)$ with entropy $S = kLn \,(2)$ linked the quantity of information available in such a system. Thus:

$$E \geq 3 \times 10^{-3} aJ \text{ at } T = 300 \, K$$

If the system has a large number of gates N, with a response time $\tau$ that could switch at an average rate time $\tau_{mbf}$, then the mean time between failures (MTBF) is given by the expression: $\tau_{mbf} = \frac{\tau}{N}\frac{1}{P} = \frac{\tau}{N}e^{\frac{E}{kT}}$

$P = e^{-\left[\frac{E}{kT}\right]}$ is the switching probability of a single gate. We can demonstrate that the minimum switching energy is given by:

$$E \geq kTLn\left(\frac{N.\tau_{mbf}}{\tau}\right).$$

If we consider $N = 10^9, \tau = 10\,ps$ and MTBF $= 1000\,h$ (i.e. $3.6 \times 10^6 s$), then we get: $E \geq 0.25\,aJ$.

Among the three limitations mentioned above, the latter is the largest one.

In order to estimate the associated minimal switching voltage $V_{min}$ one must consider the capacitive load $C_L$ associated to a switching gate. We will then extract $V_{min}$ from the following relation:

$$kTLn\left(\frac{N.\tau_{mbf}}{\tau}\right) = C_L V_{min^2}$$

and get

$$V_{min} = \left(\frac{kTLn\left(\frac{N.\tau_{mbf}}{\tau}\right)}{C_L}\right)^{1/2}$$

At $T = 300\,K$, $V_{min} = 10\,mV$ will be the limit if the load capacitance is in the range 0.4 fF (corresponding to 1 nm gate oxide thickness).

## 3.2.   Issues related with decananometer gate length devices

In the decananometer range (less than 100 nm), besides classical 2 dimensional electrostatic effects, tunneling currents will contribute significantly to MOSFET leakage. In the following, we review the principal parasitic effects that could limit ultimate MOSFETs operation.

**3.2.1.   *Direct tunneling through SiO$_2$ gate dielectric*** is significant for a thickness less than 2.5 nm. It contributes to the leakage component of power consumption. Less than 1.4 nm thin SiO$_2$ is usable without affecting devices reliability.[3,7−9]

**3.2.2.   *High doping levels in the channel*** reaching more than $5\times10^{18}$ cm$^{-3}$ enhances Fowler-Nordheim field assisted tunneling reverse current in sources and drains up to values of 1 A/cm$^2$ (under 1 V).[10]

**3.2.3.** *Direct tunneling from source to drain* is easily measurable for very short channel lengths[4,5] lower than 10 nm. It will affect subthreshold leakage substantially at room temperature for channel lengths less than 5 nm.

**3.2.4.** *Classical small dimension effects* are more severe than the fundamental limits of switching (quantum fluctuations, energy equipartition, or thermal fluctuations). A minimum value is required for threshold voltage due to:

- *subthreshold inversion.* For ideal fully-depleted SOI(FDSOI) 59.87 mV/dec subthreshold swing can be obtained at 300 K. The limit $V_T$ value is 180 mV precluding a supply voltage $V_S$ lower than 0.50 V. Impact Ionization MOS (I-MOS) would allow reducing subthreshold swing to 5 mV/dec. However, performance and reliability remain issues.[11]
- *short channel effect* due to the charge sharing along the transistor channel following the relation:

$$\Delta V_T = -4\varphi_F \frac{C_w}{C_{\mathrm{ox}}} \frac{x_j}{L} \left[ \left( 1 + 2\frac{W}{x_j} \right)^{1/2} - 1 \right]$$

$$= -4\varphi_F \frac{\varepsilon}{\varepsilon_{\mathrm{ox}}} \frac{t_{\mathrm{ox}}}{L} \frac{x_j}{W} \left[ \left( 1 + 2\frac{W}{x_j} \right)^{1/2} - 1 \right] \tag{3}$$

Here $V_T$ is expressed by:

$$V_T = V_{FB} + 2\varphi_F - \frac{Q_B}{C_{\mathrm{ox}}} \tag{4}$$

where

$$V_{FB} = \varphi_{MS} - \frac{Q_{\mathrm{ox}}}{C_{\mathrm{ox}}} \tag{5}$$

and

$$C_{\mathrm{ox}} = \frac{\varepsilon_{\mathrm{ox}}}{t_{\mathrm{ox}}}; \quad \varphi_{MS} = \varphi_M - \varphi_s \tag{5.1}$$

$\Delta V_T$ is the threshold voltage decay; $t_{\mathrm{ox}}$ is the gate dielectric thickness; $\varepsilon$ and $\varepsilon_{\mathrm{ox}}$ are the silicon and gate dielectric constant respectively; L is the channel length; $X_j$ is the drain or source junction depth; W is the space charge region depth; $V_T$ is the threshold voltage; $V_{FB}$ the flatband voltage; $\varphi_F$ the distance from Fermi level to the intrinsic Fermi level; $Q_B$ the gate controlled charge; $C_{\mathrm{ox}}$ is the unit area capacitance of the gate insulator. $\varphi_{MS}$

is the difference between the workfunctions of the gate and the semiconductor; $Q_{ox}$ is the oxide charge density; $\varphi_M$ and $\varphi_S$ are the metal and the semiconductor workfunction.

Gate depletion and quantum confinement in the inversion layer will play an important role on short channel effect by adding their contribution to the gate to channel capacitance $C_G$. SCE is the main limitation to minimal design rule. For low $V_T$ values it can be of the order of $V_T$. In order to maintain inverter delay degradation to less than 30%, we must observe the condition $V_T = -\frac{V_{DD}}{3}$.[12] $V_{DD}$ is the supply voltage.

• *Drain Induced Barrier Lowering (DIBL)*

Classically, DIBL is due to the capacitive coupling between drain and source resulting in a barrier lowering on the source side. An eased charge injection from the source allows an increased control of the channel charge by the source and drain electrodes and reduces the threshold voltage. This effect (thus $\Delta V_T$) increases with increasing Vds and decreasing L. A simple model shows that:

$$\Delta V_T = -\gamma \frac{Vds}{L^2} \quad (\gamma \text{ is in the range of } 0.01 \ \mu m^2)$$

### 3.3. Variability from statistical dopant fluctuations and Line Edge Roughness

The effect of dopant fluctuations has already been considered by Shockley in 1961.[13] Recently, special attention has been paid to this subject because the number of dopants in the channel of a MOSFET tends to decrease with scaling of devices geometry.[14,15] The random placement of dopants in the MOSFETs channel by ion implantation will affect devices characteristics for geometries lower than 50 nm. The discrete nature of dopant distribution can give rise to asymmetrical device characteristics[15] which will impact seriously the building of a complete integrated system with a large number of devices.

Dopant fluctuations and Fowler Nordheim limitation of leakage at high electric fields will encourage the use of low doped thin SOI.

Atomistic, *ab initio* approaches are used to simulate the contribution of the discrete number of dopants to the parameter variability as well as the Line Edge Roughness[14] which becomes an important source of dispersion brought by ultimate lithography resist or the underlying gate material roughness. These contributions will be added to the films interface rough-ness and thickness fluctuations to affect transport properties or noise figures at the level of a device or a complete integrated system.

## 4.    Technological Options to MOSFET Optimization

In Sub Sections 4.1, 4.3, the possible solutions to overcome the physical limitations encountered in classical scaling are reviewed through gate stack and channel/substrate engineering as well as source and drain engineering. Mastering and improvement of transport properties by strained channels and substrate engineering will be of primary importance in the future and not only limited to threshold voltage adjustment as it was the case in the past. The gate stack will also be reviewed on the electrical properties side as well as on the defect density view point. Source and drain engineering has to be addressed not only on the dopant activation side but also on the architecture side: access resistance to the channel can drastically reduce any advantage brought from channel transport properties optimization.

In Sub Section 4.2, we review the alternative architecture candidates to replace bulk devices by leveraging the trade off between performance and power consumption. Power dissipation limitation will be the hardest challenge to face in the future whereas portable devices and systems will drive the market in the nanoelectronics era. That is why thin films and Multigate architectures are major alternative approaches to extend CMOS life to the end of the roadmap and possibly beyond.

### 4.1.    Gate stack and channel/substrate engineering

Threshold voltage management issues in classical bulk MOSFET will guide its scaling.

Gate and channel engineering must be optimized together because both physical characteristics affect the nominal $V_T$ value of expression (4) which can be written as:

$$V_T = V_{FB} + 2\varphi_F - Q_B/C_G \tag{6}$$

(gate depletion and channel quantum effects are taken into account).

Low $V_T$ values will result from:

- *Tuning surface doping concentration (see Section 4.1.1)*
- *Strained channel engineering (see Section 4.1.2)*

- *Choosing the gate material (see Section 4.1.3)*
- *Adjusting gate insulator thickness (see Section 4.1.4)*

*4.1.1. Tuning surface doping concentration* as low as possible. Excellent localization of the dopant profile is needed to minimize junction parasitic capacitance and body effect. Selective Si epitaxy of the channel has also been demonstrated to achieve almost ideal retrograde profiles.[16] Selective epitaxial Si:C acts as a Boron diffusion barrier and thus help to improve drastically short channel effect[17] (Fig. 4(a)) as well as low field mobility. Multibarrier channels, using an alternated Si/SiGeC epitaxial channel structure, have been proven to be efficient in optimizing short channel effects immunity compatible with high devices drivability[18] (Fig. 4(b)). These solutions can give a longer breath to bulk CMOS devices scaling.

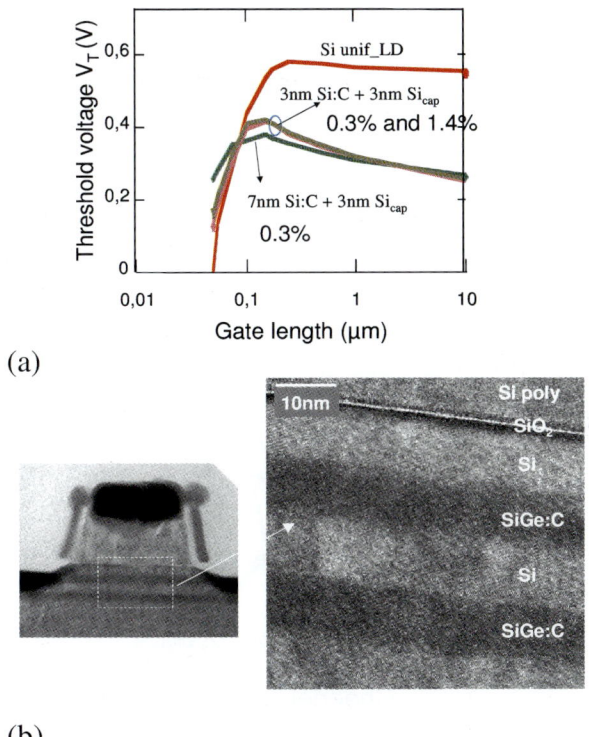

(a)

(b)

Fig. 4. Introduction of Carbonated silicon in MOSFET channel: (a) Influence on short channel effect17; (b) Optimization by a multibarrier channel.[18]

## 4.1.2. Strained channel engineering

### 4.1.2.1. Global strain

Strained SiGe,[19] SiGe$_x$C$_y$ based alloys or strained Si epitaxy have been studied to increase the channel mobility[17,20] by introducing compressive or tensile strain to enhance hole or electron effective mass respectively. In order to achieve such channel architectures, bulk relaxed SiGe pseudo substrates obtained by graded SiGe buffer were intensively developed during the last decades.[21,22] High-quality pseudomorphic silicon layer with very high biaxial-strain values (typically 1.2–1.5 MPa or more) can be grown on those substrates. The resulting degeneracy leverage on the conduction bands leads to effective electron mass reduction and mobility increase up to around 80%.

The quality of those substrates has been spectacularly improved. Independently of possible remaining defects (dislocation pile ups, stacking faults, etch pits[23]) a major limitation remains: the reported gain in current enhancement decreases with gate length reduction[24] (Fig. 5). This $I_{ON}$ gain decrease with $L$ was attributed to self heating (monitored pulse drain

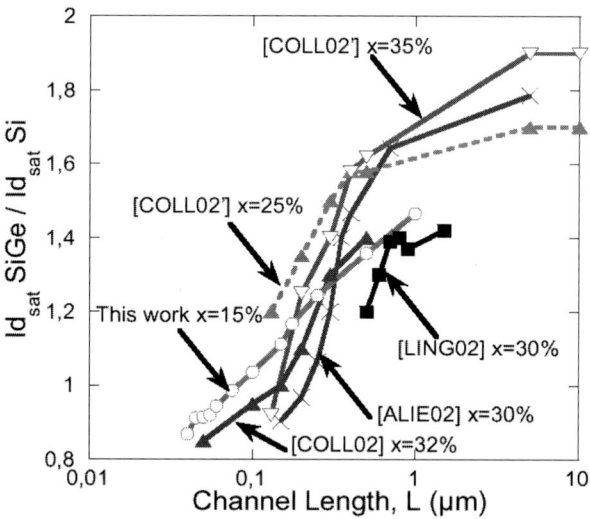

Fig. 5. Gain in drain current vs. gate lengths at VGT = VDS = −1.3 V for [ALIE98];[25] at VGT = −0.5 V VDS = −2 V for [LING02][26] and at VGT = −1 V VDS = −1.5 V for [COLL02];[27] VGT = (VG.− VT) for [COLL02'] (see Refs. 28 and 24).

current measurement) due to low thermal conductivity of SiGe.[29] But some authors have pointed out than even at low drain voltage (insensitive to self heating) the gain current loss is still relevant. Both possible S/D implantation damages[30] and lateral strain S/D relaxations[31] may explain the loss on mobility increase on those short channel strained devices.

However, high quality gate insulator and subthreshold characteristics optimization require a Si cap layer on top of the channel and low thermal budget.[15] Ultimately, a HiK gate insulator is needed in these architectures.[32,33]

In parallel, high quality strained silicon on insulator substrate, with or without SiGe for dual channel operation has been developed.[34,35] SiGe condensation technique can lead to high quality SiGe on Insulator (SGOI) whereas high quality SGOI and sSOI substrated by Smartcut® were reported.

## 4.1.2.2. Process induced strain

Process induced strain is the most mature option for today's IC and is proposed in the 65 nm and 45 nm platforms.[36] In those technologies, external strain, mostly uni-axial, is applied by various means. The most currently used approach is the compressive or tensile contact etch stop layer to obtain respectively tensile channel nMOS or compressive channel pMOS. Recent studies quantify by direct measurements the mobility enhancement on short channels with process induced strain[37] showing a direct correlation between low and high $V_d$ regime.

## 4.1.2.3. Other substrate solutions

Unstrained solutions may use the chemical composition of the substrate or the crystalline surface or transport orientation.

Changing surface silicon orientation or transport orientation can lead to mobility improvement by a factor 2 or more.[38] The (110) surface orientation lead to an improvement for hole. Dual channel with (100) orientation for electrons and (110) orientation for holes was reported.[39] Germanium and Germanium-on-insulator were proposed as unstrained substrates. One of the higher channel mobility improvement by using column IV elements is compressive Germanium with more than a factor 10 of hole inversion charge mobility improvement[40] which could bring a solution for dual channel optimization.

### 4.1.3. Choosing the gate material

Ideal transfer CMOS inverters characteristics requires symmetry of threshold voltage for $n$ and $p$ channel devices (i.e. $V_{TP} = -V_{TN}$). Several alternatives have been envisaged:

- *The use of n+ poly gate for nMOSFET and p+ poly gate for pMOSFET.* This solution suffers from Boron penetration into $SiO_2$ coming from the $p+$ doped gate. Nitrided $SiO_2$ *limits* this effect *without avoiding it*: trapping centers are created near or at the $SiO_2/Si$ interface decreasing carrier mobility.
- *The use of metal gate material.* No gate depletion is observed in this case. The use of midgap gate (TiN for example) on bulk silicon or partially depleted SOI will be dedicated to supply voltages higher than 1 V. Workfunction engineering for dual metal gates is challenging: the highest CMOS performance/lowest leakage current trade off can be obtained. It is mandatory on low doped FDSOI.

Several approaches have been proposed for metal gate integration. The classical process integration, so called direct gate, requires the protection of the metal gate material from ion implantation as well as from oxidation during the dopant activation anneal. TiN has often been chosen as a gate material[41] because it is available as a standard in the industry. Alternatives such as the damascene gate (Fig. 6)[42,43] have been achieved in order to avoid the issue of source and drain activation temperature. It is noteworthy that, thanks to the damascene architecture, High Frequency and Multi threshold devices could be embedded in Systems On Chip. Complete silicidation of polysilicon gate has been demonstrated to lead to metallic behavior of both $n$ and $p$ gates.[44−46] However, integration with HiK dielectrics gives rise to the so called Fermi level pinning similar to what is obtained with polysilicon gates.[47]

### 4.1.4. Gate dielectric engineering

The gate leakage due to direct tunneling in standard $SiO_2$ or $SiO_xN_y$ is one major show stopper.[1] It will impact directly the static power dissipation $P_{stat}$ according to relation (2.1) Let us consider a circuit with active area of the order of $1 \, cm^2$ and gate oxide $SiO_2$ $t_{ox} = 1.2 \, nm$. Considering the contribution of gate leakage to Ioff under the condition $V_{dd}=0.5 \, V$, then $P_{stat}(0.5 \, V)= 5 \, W$. We would get $P_{stat} (1.5 \, V) = 750 \, W$ if $V_{dd} =1.5 \, V$!! This

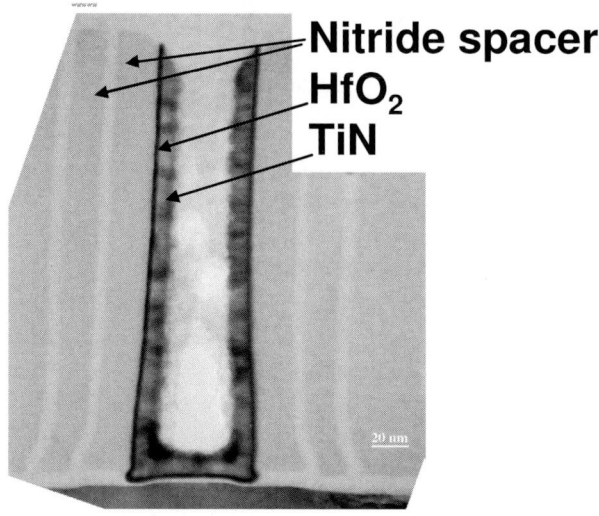

Fig. 6. TEM cross section of TiN/HfO$_2$ Damascene gate stacks.[43]

results as a major show stopper for scaling of CMOS technology. That is why High K will be urgently needed in the near future. Besides affecting static power, gate leakage also impacts negatively delay time[48] and affects the functionality of logic circuits.

## 4.1.4.1. From SiO$_2$ to High K gate dielectrics

A decrease of devices performance has been reported if SiO$_2$ thickness is lower than 1.3 nm[49] suggesting a surface roughness limited mobility process due to the proximity of sub-oxide. The strong band bending due to quantum mechanical corrections affects the lower limit of supply voltage in the constant field scaling approach.[50] Solutions compatible with silicon gate are also investigated to keep compatibility with a standard CMOS process flow: HfSiO$_x$, ZrSiO$_x$ are given much attention as good candidates.[51] These solutions are *dielectric thickness budget* consuming (SiO$_x$ interface) and Fermi level pinning occurs at the HiK/poly gate interface.[47]

Very low leakage current has been reported by using HfO$_2$ of 1.3 nm Equivalent Oxide Thickness (EOT) combined with a TiN gate integrated on 45 nm CMOS by a damascene process[43] (Fig. 6). Electron mobility degradation is reported compared to SiO$_2$ gate dielectric[43] attributed to stress induced phonon scattering (Fig. 7(a)). These materials have a smaller bandgap than SiO$_2$: thus trapping is a strong reliability issue.[5] That is why

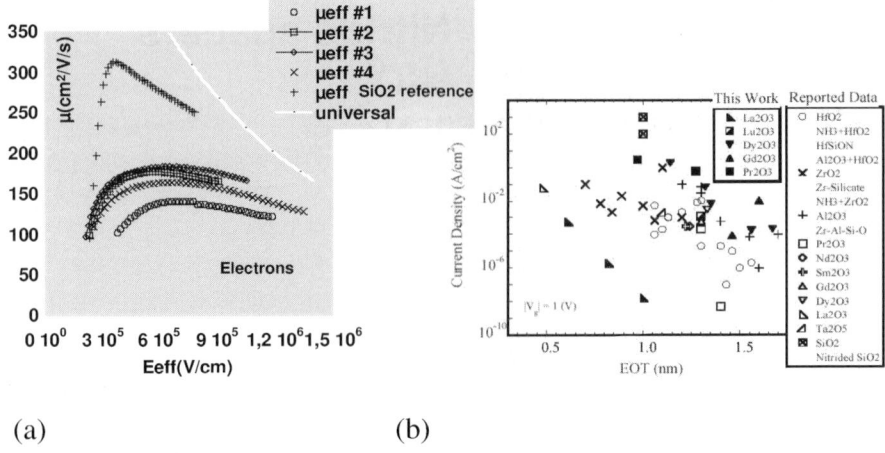

(a)                                              (b)

Fig. 7. (a) Degradation of electron mobility with $HfO_2/Si$[43]; (b) Leakage current as a function of EOT for various HiK materials reported from Ref. 52.

a SiON interface could be helpful to reduce the leakage current thanks to the higher bandgap of SiON.

$La_2O_3$ films with EOT as thin as $= 0.61$ nm have been proven to demonstrate very low leakage current as low as $J = 5.5 \times 10^{-4} A.cm^{-2}$ [52] compatible with high interface quality and acceptable mobility values (Fig. 7(b)). These results are obtained on low temperature end of process and aluminum gate. Integration into a direct gate process is still an issue.

### 4.1.4.2. Combining gate stack and channel workfunction engineering

Specific technological optimization may be necessary to maximize the transport gain in short channels. In particular, maintaining the high stress of 1.2 or more GPa in a nanoscaled device and reducing ion implantation damages are among the main challenges. Meanwhile, the combination of strained Si and SiGe channel can be a promising solution for future applications. For instance, it was shown that both surface conduction and hole mobility enhancement (65% at high transverse electric field) could be achieved by using selective SiGe for PMOS coupled with high-k and metal gate[33,53] (Fig. 8).

Even in the case of low gain in short channel $I_{ON}$ values,[33] it is possible to adjust $V_T$ by locally strained layers by using a mid gap metal gate.

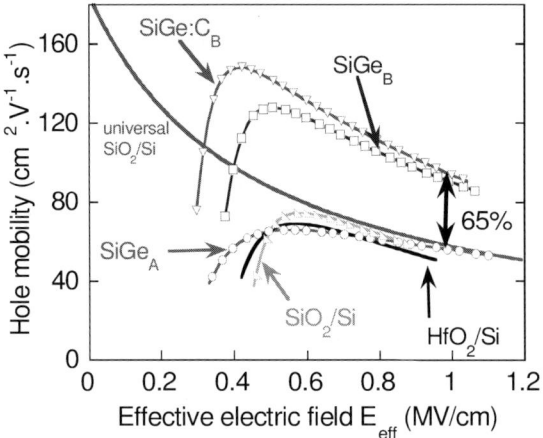

Fig. 8. Effective hole mobility versus effective field for the various channel-gate dielectric stacks.[53]

## 4.2. Architecture alternatives to improve CMOS performances and integration

### 4.2.1. Fully depleted SOI devices

In order to obtain the lowest subthreshold slope (60 mv/dec) and acceptable DIBL on FDSOI a practical rule is used: $T_{Si} \leq L_{gate}/4$.[54] The spreading of potential into the buried oxide, due to the coupling with the top gate, increases the coupling between source and drain and thus DIBL. Ultra-low SOI films thickness is difficult to control. That is why partially depleted SOI has been proposed.[54,55] Because of complete isolation of the SOI devices as well as lower junction capacitance, improved figures of merit are obtained as compared to bulk.[54] The threshold voltage is dependent on Si film thickness whenever the film thickness becomes lower than the space charge region. VT is then expressed as[54]:

$$V_T = V_{FB} + 2\varphi_F + \frac{qN_AT_{Si}}{2C_{\text{ox}}} \tag{7.1}$$

In the case of a low doped channel, expression (7.1) can be simplified as the well known relation:

$$V_T = \left(\varphi_M - \frac{E_i}{q}\right) + \frac{kT}{q}\ln\left(\frac{2.Cox.kT}{q^2n_iT_{Si}}\right) \tag{7.}$$

$N_A$ is the acceptor concentration; $T_{Si}$ is the silicon thickness; $C_{ox}$ is the gate insulator capacitance; $E_i$ is the semiconductor intrinsic Fermi level energy; $n_i$ is the intrinsic carrier concentration.

Scaling of FD devices encounters some limitations due to the quantum confinement of carriers in ultra thin films and its incidence on the threshold voltage value[56]: the increase of the fundamental level of the conduction band will increase flat band voltage and $V_T$ consequently.

The functionality of ultra small 6 nm gate length devices on 7 nm thin Si film was demonstrated.[57] However, the electrical performances of these devices are extremely sensitive to the SOI film thickness variations due to the fact that a compromise must be found between series resistance minimization and DIBL.[58]

Combination of strained channels and SOI could result in optimized trade off between short channel effects reduction and enhanced transport properties. A Si and SiGe Dual strained channels on insulator architecture has been demonstrated functional down to gate lengths of 15 nm (Fig. 9).[34,37]

For sub 100 nm range channel lengths and widths, the strain induced by the environing thin films affects devices characteristics. The loss of global strain observed in short channels is recovered by the lateral strain induced on the narrow active areas (Fig. 10(a)).[34,59,60] This effect has been evidenced quite clearly on FDSOI films[34,59] where the biaxial and uniaxial strain are additive effects which balance the loss of strain that could be induced by

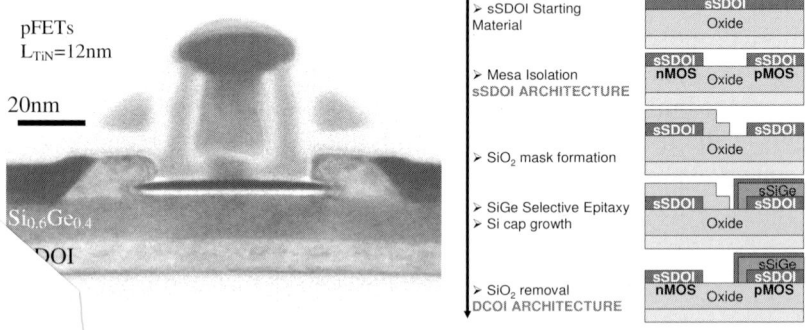

Cross sectional TEM pictures of the co-integrated dual channels MOSFETs on h a HfO₂/TiN/Poly/NiSi gate stack.[34,37]; (b) Strained Dual channels CMOS .[34]

s et al.

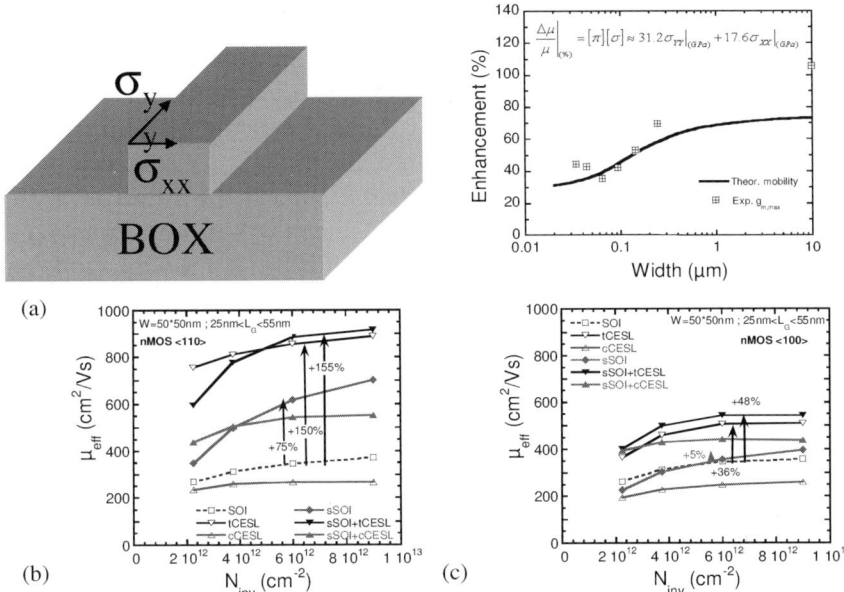

(a)

(b)                                (c)

Fig. 10. A piezoelectric model is applied to describe the effects induced by strain on the MOSFET electrical behaviour of: (a) short and narrow devices on SOI. Experimental gm, max enhancement vs. device width is compared to the piezoelectric model. Inset: Approximation of the used piezo-electric model.[34] Short and narrow n-channel electron mobility vs. inversion charge along orientations: (b) $\langle 110 \rangle$; (c) $\langle 100 \rangle$.[59,60]

source and drain and the process steps to implent contacts architecture. For electrons, these effects are more pronounced on $\langle 110 \rangle$ than on $\langle 100 \rangle$ (Figs. 10(b) and 10(c)).[60]

### 4.2.2. Multigate devices

SOI material should allow to realize attractive devices like multi gated MOSFETs[61] that will extend further scaling of FD devices which are limited by the quantum confinement and splitting of allowed energy bands as well as DIBL via the coupling of the gate with buried oxide[56] (Fig. 11(a)). With multi gate devices (Fig. 11(b)), short channel effects and leakage current can be drastically reduced because 60 mV/dec subthreshold swing and high drivability can be obtained. In the saturation regime, transport occurs by volume inversion due to the coupling of both gates. The conditions for controling short channel can be relaxed compared to single gate FD devices.[56,62−66] Nevertheless, the control of thin SOI and design of high density circuits with these devices have to be demonstrated.

Another main feature of these devices is to bring a solution to the channel dopant fluctuation issue in small volume. Reducing the film thickness to the minimum, allows using nearly intrinsic Si films because bulk punchthrough is no more a problem. Adjusting $V_T$ to match the overdrive defined by $(V_s - V_T)$ with a low supply voltage $V_S$ index will require adjusting the gate workfunction $\varphi_M$ according to relation (5.1). That is why, workfunction engineering on metal gate and HiK stacks is mandatory for low $V_S$ applications.

Among the various studies published on multi-gate devices,[67−69] many architectures have been proposed in which the channel is controlled by two or more gates.

In planar architectures, the structure can be non self-aligned, i.e. fabricated with one photo-lithography step for each gate, or self-aligned, using only one lithography step to define both gates. The non self-aligned architecture by wafer bonding is the most straightforward approach to fabricate planar double gate. The success of this approach depends on the lithography capability to align very short gates one to the other. Figure 11(b) shows a 10 nm non self-aligned planar double gate transistor, fabricated thanks to the use of wafer bonding and e-beam lithography.[70−73] Notice that a quasi-perfect gate alignment, with an accuracy of a few nanometers, could

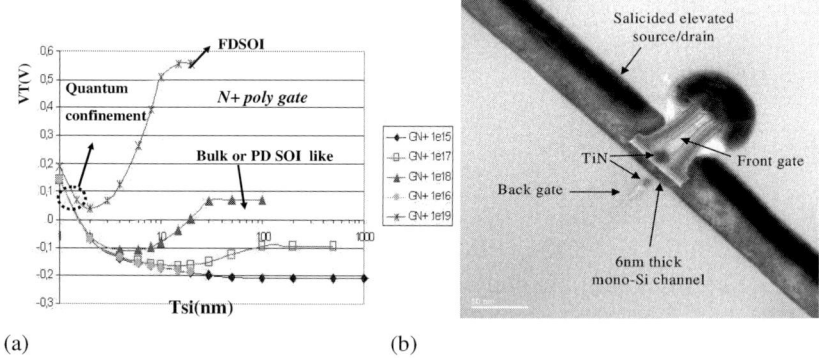

(a)                                                (b)

Fig. 11. (a) Threshold voltage dependence of SOI devices as a function of SOI thickness for different values of channel doping;[56] (b) TEM cross-section of a 10 nm planar bonded double gate transistor with TiN metal gate.[70]

be achieved thanks to the self-aligned regeneration of the alignment marks after the bonding step.[74]

Several approaches have been proposed to fabricate self-aligned planar double gate MOSFETs. The first one consisted in patterning a narrow silicon active area on a SOI substrate, etching a localized cavity under this active area into the buried oxide, and its filling by the gate material.[75] After gate patterning, the silicon active area is surrounded by the gate. Another gate-all-around (GAA) architecture, based on the silicon-on-nothing (SON) process, has been proposed more recently[76] and demonstrated down to very short gate lengths. This approach relies on successive epitaxial growth of crystalline SiGe and Si layers. The SiGe layer is then selectively etched to form a tunnel below the silicon film, and this tunnel is filled by the gate material.

In the PAGODA architecture,[77] the unpatterned back gate stack is deposited and encapsulated before wafer bonding. After initial substrate removal, the front gate is patterned and silicon spacers recrystallized from the channel are formed and silicided. These silicided spacers are used as a hard-mask for back gate etching and undercut.

The process flow proposed in[78] starts also from back gate stack deposition and wafer bonding. The whole stack, comprising the front gate, the channel and the back gate is then patterned. Insulated layers are formed beside the gates by use of oxidation rate difference between the gate and the channel materials. Source/drain regions are then regenerated by lateral epitaxial regrowth from the channel edges.

The key technological issues of the planar architectures are the precise controls of the very thin film thickness and of the back gate dimension, since the back gate is not directly accessible from the top of the wafer. However, with the planar bonded architectures it is possible to bias the front and back gate independently[74] (Figs. 12(a) and (b)). That allows the use of different transistors families with several threshold voltages values available on the same chip by using one single type of device. The electrical characteristics of the devices can fulfill the specifications of the 3 families of devices proposed in the ITRS[1], so-called High Performance (HP), Low Operating Power (LOP) and Low Standby Power (LSTP)[74] (Fig. 12(b)). Moreover, the planar bonded Double Gate devices are co integratable with single gate FDSOI and allow a metallic Ground plane by using the backside gate. The planar bonded architecture approach brings a unique innovative option to future Systems On Chip.[79]

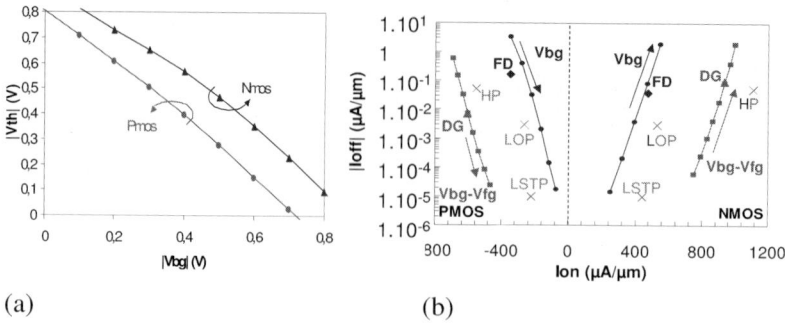

(a)                                        (b)

Fig. 12. (a) Tunable threshold voltage of the devices as a function of back gate voltage; (b) $I_{off}$ vs. $I_{on}$ of tunable DG MOS (adjustable $V_{bg}-V_{fg}$) and tunable DG MOS operating in FD mode (adjustable $V_{bg}$) from Low-stand-by-power (LSTP) to High-performance (HP) –90 nm node.[70]

On the other hand, structures with fingered vertical channel, such as FinFET[80] (Fig. 13(a)), Trigate[81] (Fig. 13(b)), $\Omega$-FET[82] (Fig. 14(a)), $\Pi$-Gate[83] and nanowire-FET[84] have been extensively studied. Fabrication of FinFETs relies on high aspect ratio fin definition and short gate patterning on this topography (Fig. 13(a)). Conversely to planar devices, the conduction takes place on the vertical sidewalls of the fin. The conduction width is thus twice the fin height ($h_{fin}$). As the fin height is limited to typically 50 to 100 nm, FinFETs are usually designed as multifinger transistors, with a conduction width quantified by $2.h_{fin}$. In order to obtain the same drive current per silicon area as planar double gate transistors, the spacing between the fingers has to be lower than the fin height.

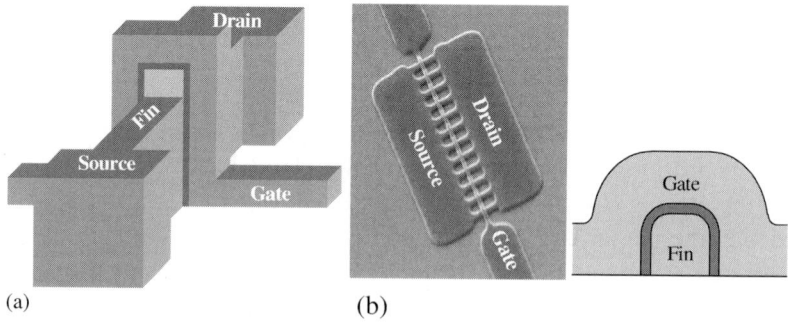

(a)                                        (b)

Fig. 13. (a) Schematic of a FinFET device. (b) Left: SEM top-view of a 20 nm gate length multifinger Trigate device. Right: Schematic cross-section of one Trigate fin.

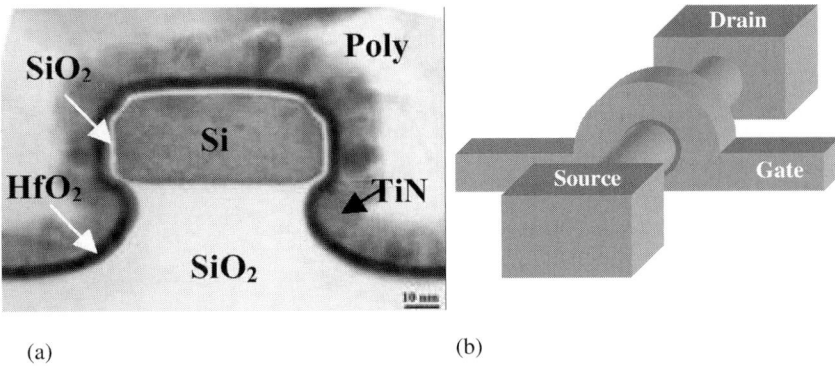

(a)                                                   (b)

Fig. 14. (a) $\Omega$- shaped FET. Functional devices with gate length as low as 10 nm are obtained.[86] (b) Schematic of a cylindrical surrounding-gate device.[84]

Thus, one key technological issue lies in the multi-fin definition. Dense array of narrow fins have to be patterned, with a good control of the fin width and shape. The use of spacers as hard-mask for fin patterning seems unavoidable, as it allows to double the fin density and to design sub-10 nm wide fins.[85]

Another approach consists in designing the fin with roughly a square cross-section (Fig. 13(b)). In that case, the channel is controlled by the gate on three sides. This device, so called Trigate,[81] has a conduction width given by twice the fin height plus the fin width. Trigate is still a multifinger device, and the spacing between fins has to be lower than $h_{fin} + w_{fin}/2$ to obtain higher drive currents per silicon area than with planar devices. This limit is far more strict for Trigate than for FinFET, since the fin height must be as low as the fin width in order to operate in trigate mode, and comparable to the gate length to benefit from a good electrostatic channel control.

The $\Omega$-FET[86] and $\Pi$-Gate architectures are basically similar to Trigate, but their channel control is close to that of a quadruple-gate device, thanks to the extension of the gate below the fin into the buried oxide.[87] The best electrostatic control can be achieved theoretically in a cylindrical channel completely surrounded by the gate (Fig. 14(b)). The most advanced practical realization of such a device is the 5 nm gate length nanowire-FET.[84]

Thanks to their better electrostatics control, multiple gate transistors are likely to allow a triple drive current with respect to single gate transistors at a given off-state current.[73,88]

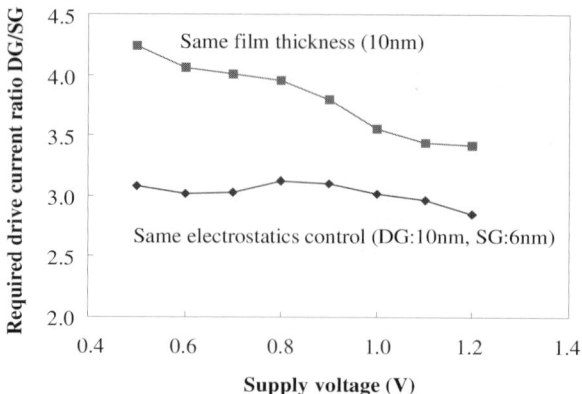

Fig. 15. Experimental drive current ratio between a 20 nm double gate and two 20 nm single gate devices as a function of the supply voltage.[73]

To illustrate this, we have plotted on Fig. 15 the ratio of the drive currents obtained experimentally on 20 nm co-integrated single gate and double gate devices. The drive current of the double gate transistor is $1230 \, \mu A/\mu m$ for an off-state current of $1 \, \mu A/\mu m$ at $V_{dd}$=1.2 V, which can be considered as a high performance device.

Two cases can be considered:

(1) Both devices have the same film thickness of 10 nm. The single gate transistor suffers from much more electrostatic control loss and the drive current ratio at Ioff = $1 \mu A/\mu m$ is between 3.4 and 4.0.

(2) Both devices exhibit roughly the same electrostatic control (subthreshold swing and DIBL respectively lower than 100 mV/dec and 250 mV/V). The film thickness is reduced to 6 nm for the single gate transistor. The current ratio is still around 3, because of the increased access resistances due to a thinner film for the single gate device.

Furthermore, if we consider loading capacitances (for example wires and junctions) in addition to intrinsic gate capacitance in the previous discussion, the multiple gate device advantage over single gate is further increased, because of the higher drive currents delivered by the multiple gate architectures.

Finally, since each added gate allows a better device scalability,[79,87,89] the advantage of multiple gate devices is more and more evident as the gate length is reduced.

Several critical issues are associated with the use of thin film or narrow fin devices. An intrinsic limitation is the mobility reduction observed for film thickness below 5 to 7 nm.[90] This effect is partly due to an increased phonon scattering mechanisms on thin films[91] and can be further accentuated by a more pronounced impact of the surface roughness.

In addition, devices with ultra-thin films are sensitive to thickness fluctuations through short channel effects variations. The scaling length $\lambda$ derived in[92] for low-doped double gate transistors is given by the expression:

$$\lambda = \frac{t_{Si}}{2}\sqrt{\frac{1}{2} + \frac{2.C_{Si}}{C_{ox}}} \qquad (8)$$

For an EOT of 1 nm, $\delta\lambda/\lambda$ is about 70% of $\delta t_{Si}/t_{Si}$. As short channel effects depend on $L/\lambda$, a fluctuation of 1 nm on a film thickness of 7 nm is equivalent to a gate length variation of 10%.

### 4.2.3. Multichannels Multigated devices for improved output current and integration density. Paving the way to the use of Nanowires

The increase of devices drivability could be obtained by multiplying the number of channels. Increasing the drivability capabilities while keeping high integration density is possible by stacking devices in parallel. The exploitation of the third dimension is an elegant and efficient way to achieve such a goal. Several teams have recently published results on multichannel architectures.[93−96] Figure 16 shows a 3-level CMOS Nanobeams stack of 30 to 70 nm widths: these devices demonstrate up to $3 \times I_{ON}$ increase compared to 1 level trigate.[95,96] A high current density/surface is obtained thanks to 3D integration. Starting from a SOI substrate, a (Si/SiGe) super-lattice is grown.[95] After the silicon nitride deposition, the superlattices are etched anisotropically in order to pattern stacked fins. Then the SiGe is selectively removed between the Si nanowires isotropically.

If the channel width reaches nanometer range dimensions, *the quantized width, imposed by the nanowires structure,* may reduce significantly the driving current and/or the design flexibility compared to planar architectures. This limitation can be overcome by 3D approaches. The 3D Gate-All-Around (GAA) architecture requires some specific integration strategy:[95,96] 3D Nano-Wire-GAA architectures (NWG) can be integrated by a damascene-gate FinFET to obtain suspended nanowires with GAA HO2/TiN/Poly gate.

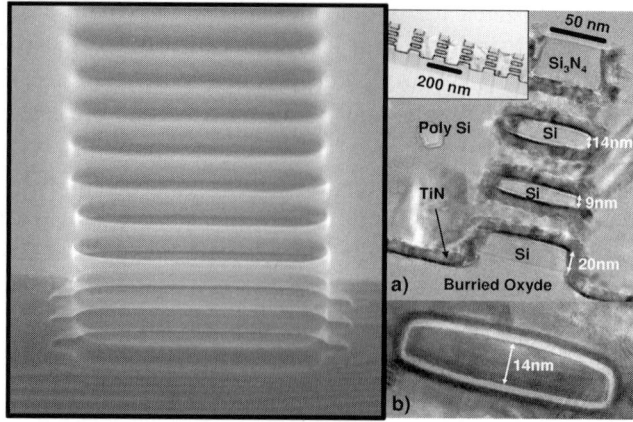

Fig. 16. Left: Three stacked levels nanobeam matrix after the Fin etch and the SiGe removal. Right: Cross sectional TEM pictures perpendicular to the beams a) of one stacked Si channels, Inset: $3 \times 50 = 150$ beams b) of one Si channel: excellent Si crystalline quality is obtained; $HfO_2$, TiN and Poly-Si conformity is achieved.[95]

Photo-resist trimming and optimized hydrogen annealing are employed to obtain rounded and continuous suspended nanowires:[96] hydrogen annealing was used intentionally for 3-D profile transformation by rounding sharp corners while diminishing surface roughness[97] which improves electrical characteristics of FinFETs.[98] In Fig. 17 an example of stack made of up to

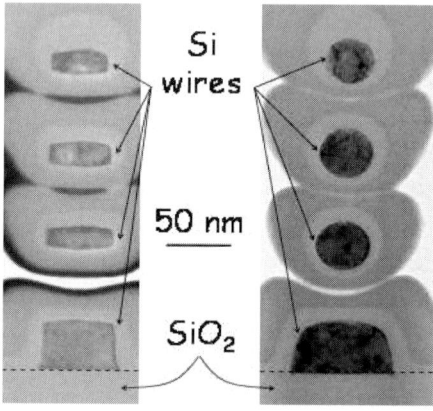

Fig. 17. TEM cross section of the multilayers nanowires. (a) before annealing — not rounded nanowire (b) annealed at $850°C$ — rounded nanowires. The lower Si nanowires are on $SiO_2$. Every wire is capped with $SiO_2$, $Si_3N_4$ and W for TEM imaging convenience.[96]

4 Nanobeams is shown: subsequent resist trimming and hydrogen anneal at 850°C gives a rounded shape to the Nanobeams which will turn out to behave as nanowires.[96]

Zipping between beams appears as a basic limit when we increase the wire density. This phenomenon is related to the smaller distance between beams when the number of beams is increased. In order to avoid strain relaxations (and thus misfit dislocations) in the initially grown super-lattice, the SiGe thickness between Si layers is decreased for an increasing number of beams. Capillary forces can induce sticking of the beams duuring the wet surface preparation step prior to the $HfO_2$ deposition. We showed that a shorter beam length avoids zipping when increasing the beams density.[95]

## 4.3. Source and drain engineering

Low energy ($<1$ keV)[49] and heavy molecules ($BF3$,[99] $B10H14$,[100] ...) have been extensively studied to replace Boron to achieve $p+$ shallow junctions. Plasma doping is investigated as an alternative to obtain as implanted $p+$ junction depths lower than 10 nm.[101,102] Transient Enhanced Diffusion (TED) is still the limiting process to reach the specified final junction depths (Fig. 18). Fast ramp up and down — so called spike or Flash annealing[102] — must be combined with Low Energy Ion Implantation[102] to reduce TED

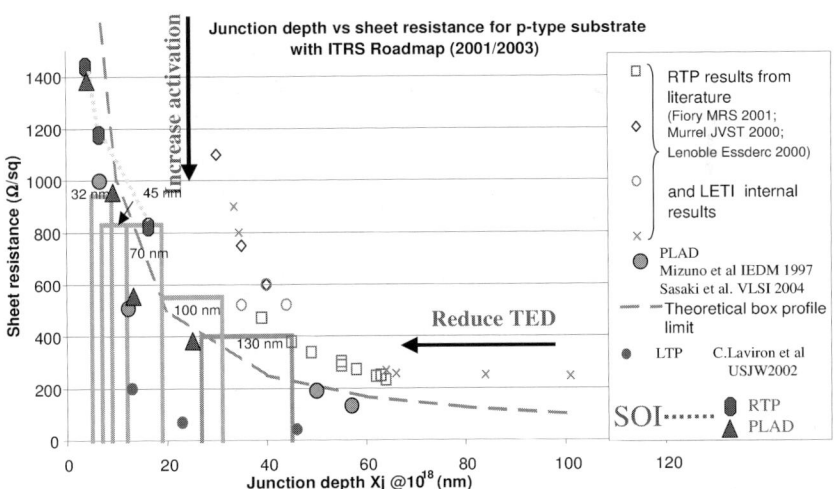

Fig. 18. $P+$ Sheet resistance as a function of junction depth on bulk or Si thickness for SOI.[101–104]

as much as possible, by reducing the role played by extended and dopant defects. Excimer Laser Anneal (Fig. 18)[103,104] has demonstrated the best trade off between low sheet resistance and junction depth shallowness: highest solid solubility combined with fast processing can be achieved. Low sheet resistance combined with low silicon consumption can be obtained with monosilicides (NiSi, PtSi) instead of disilicides (TiSi2, CoSi2).[105]

The same behavior will apply to SOI as well as bulk substrates (Fig. 18). However, on SOI films, several issues are linked with the access resistance optimization. As the film thickness decreases, achieving silicon doping becomes more and more challenging, because on one hand the square resistance of the silicon film increases in $1/t_{Si}$ as shown on Fig. 18. On the other hand, increasing dose and/or energy leads to surface silicon amorphization[73]: as long as the whole layer is not damaged, activation annealing allows the recrystallization of the film giving thus an active doping process window which is very narrow for a 5 nm thick silicon film. The surface species diffusion velocity during high thermal processes being strongly dependent on temperature and silicon thickness, the film becomes very sensitive to high temperature treatments[73,106] as silicon thickness decreases.

Devices on thin SOI will require raised sources and drains by epitaxial growth to facilitate further silicidation: pre-anneal before epitaxial growth can lead to a destabilization which dramatically transforms the continuous silicon film into silicon solid droplets on the buried oxide as shown on Fig. 19(a). Therefore selective epitaxy of raised source/drain requires technological developments such as temperature optimization, modulation of the interface energy between silicon and buried oxide to ensure that the silicon film will keep its integrity during the whole fabrication process. Figure 19(b) illustrates results obtained when the temperature of the pre-anneal is lowered (down to 650°C).

Silicidation process also requires technological optimization. Indeed diffusive metals have been introduced to suppress the voiding that occurs in the silicon films when silicon diffuses into the silicide. One way to overcome these technological difficulties could be to design MOS transistors with metallic source and drain either based on Schottky barriers[107] or modified Schottky barrier.[108] In both cases, selective epitaxy can be suppressed as source and drain are made out of metal. The key issue in this option is to find metals for N and PMOS with adjusted work function to design either adequate Schottky barrier or low specific resistance ohmic contacts.

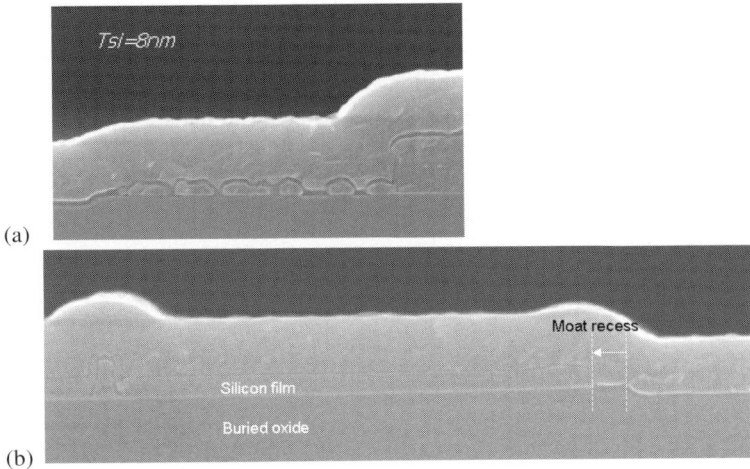

Fig. 19. (a) SEM cross-section- After $H_2$ anneal, silicon agglomeration is observed for thin films. (b) Lowering the anneal temperature leads to less dramatic consequences of silicon agglomeration as in this case, only moat recess is observed.[73]

## 5. Exploiting Non-Stationary Transport or CMOS on Semiconductors other than Silicon?

The introduction of strained channels is limited by saturation velocity values at high electric fields. Under these conditions, non stationary transport can occur for very short channels and devices performances can benefit from velocity overshoot. Unless transport is limited by surface roughness or impurity scattering[4,109,110] ballistic transport can offer a new degree of freedom to the increase of devices performance in sub 100 nm Si channel length devices. If the low field mobility is high, then the mean free path of carriers becomes comparable to or higher than the channel length: ballistic transport is likely to be taken into account.[49,111−113] These transport properties can be enhanced whenever undoped or nearly undoped channels can be used. Architectures based on ultra thin bodies like Fully Depleted SOI or Multigate devices can ease the exploitation of these phenomena due to the fact that short channel doping can be minimized while keeping low short channel leakage. Reduction of channel length and supply voltage poses the issue of new scaling paradigms through the exploitation of non stationary effects. Germanium and GaAs for example have low field carrier drift velocities higher than in silicon. However, at high electric fields the reverse situation occurs. Still the energy relaxation time is higher in Germanium than it is in silicon thus velocity overshoot may occur for less aggressive channel lengths. Limitations will however come from integration of the new materials which could request new gate dielectrics. Typically, High K materials are needed to fabricate Ge based CMOS devices due to the Ge oxides instabilities. In these devices, hole mobility has been reported to be improved whereas electron mobility enhancement is still an issue (see Section 6.2). Germanium offers the unique possibility for low temperature dopant activation.[114,115]

## 6. Optimization of Carrier Transport and Power Dissipation

### 6.1. Electrostatics, transport and self heating issues

The best choice to maximize the CMOS integration density is obtained under the condition $\mu_n = \mu_p$ ($\mu_n$ and $\mu_p$ are respectively the n-channel and p channel mobilities). Dual channels obtained from strained epitaxial layers could be a possible approach[40] (see Section 4.1.3). As far as a monolithic solution can be found, this unique condition occurs in the case of C-diamond (Table 1). However, $n$ dopant activation in this material is still limited[116] whereas, recently progress has been made for $p$ doping.[117] However, ohmic contacts of metal to diamond need to be optimized. Moreover, C-diamond is far the highest thermal conducting material (10 times the thermal conductivity of silicon or 50 times the thermal conductivity of $Al_2O_3$) and could be integrated as a buried layer to limit self heating in future Semiconductor On Insulator substrates. The dielectric constant of

Table 1. Electrons, holes bulk mobilities and saturation velocities (at 300 K) of mostly used semiconductor materials.

| Material | $\mu_n$ $(cm^2V^{-1}s^{-1})$ | $\mu_p$ $(cm^2V^{-1}s^{-1})$ | $V_{sat}$ $(10^7$ cm/s) |
|---|---|---|---|
| Si | 1400 | 500 | 0,86 |
| Ge | 3900 | 1900 | 0,60 |
| GaAs | 8900 | 400 | 0,72 |
| C Diamond | 1800 | 1800 | 2,7 |
| 4HSiC | 900 | 120 | 2,0 |
| InSb | 78000 | 750 | 5,0 |

Table 2. Electrons affinity, bandgap, maximum valence band level, thermal conductivity and dielectric constant for various pertinent mostly used semiconductors and High K materials.

| Material | Electron Affinity (V) | Gap (V) | Ev (V) | Thermal Conductivity $\sigma$th (W/m/K) | Dielectric constant $K$ |
|---|---|---|---|---|---|
| *Si* | 4.05 | 1,12 | 5,17 | **141** | **11.9** |
| *Ge* | 4.13 | 0,66 | 4,79 | 59.9 | 16 |
| *GaAs* | 4,07 | 1,42 | 5,49 | 46 | 12.5 |
| *C diamond* | 0 | 5,47 | 5,47 | **>2000** | **5.7** |
| *4HSiC* | 3,55 | 3,00 | 6,55 | 500 | 6.52 |
| *InSb* | 4,59 | 0,16 | 4,75 | | 16.0 |
| *SiO2* | 1,10 | 9,00 | 10,1 | 1.38 | 3.9 |
| *Si3N4* | 2,00 | 5,00 | 7,00 | 30.1 | 7.5 |
| *Al2O3* | 1,92 | 6,2 | 8,12 | 25.1 | 10 |
| *HfO2* | 2,07 | 5,6 | 7,67 | 11.4 | 24 |
| *ZrO2* | 2,07 | 5,5 | 7,57 | 1.30 | 24 |
| *AlN* | 2.00 | 6,2 | 8.20 | **175** | **8.9** |
| *BeO* | 2.00 | 10,6 | 12.6 | **260** | **6.7** |

C-diamond ($K_C = 5.7$) offers the best compromise between HiK and SiO$_2$ to control short channel effect according to relation (3).

However, the isolation on the valence band side is difficult (Table 2): the C/Si barrier height is far less than the SiO$_2$/Si barrier height (0.30 eV for C/Si instead of 4.93 eV for SiO$_2$/Si!). That is why a HiK insulator is needed. Among the best candidates, BeO or AlN offer a good compromise in terms of short channel effect ($K_{Beo} = 6.7$ or $K_{AlN} = 8.9$) and thermal conductivity (Table 2). Furthermore, their valence band is at least at $-6.2$ or $-10.6$ eV from vacuum. Thus a good isolation is obtained for holes whereas for C-diamond by itself would not be a good insulator on the valence band side.

Thus the integration of C-diamond has to be combined with HiK buried insulators if we wish to integrate it on silicon as a possible solution to limit power dissipation and suppress self-heating of CMOS devices (Fig. 20).[118]

## 6.2. Germanium on insulator: a second life for germanium?

Germanium was initially used to fabricate microelectronics through the realization of the first transistor. Many interesting properties can be

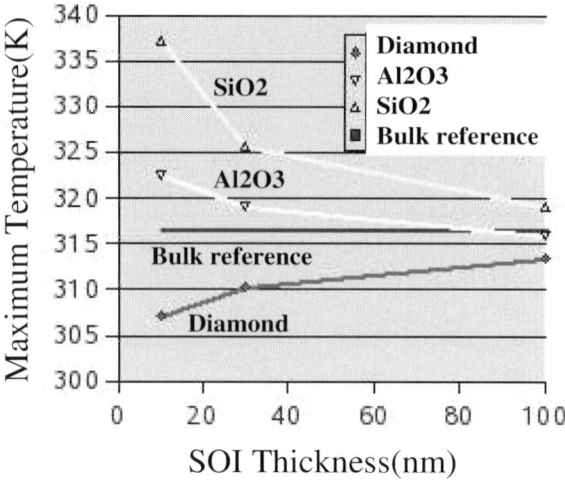

Fig. 20. Maximum channel temperature in Lg = 50 nm FDSOI transistors with different Buried Insulators as a function of SOI thickness. VDD = 1.2 V.[118]

accounted to Ge: larger low electric field mobility values than in Si as well as smaller $\mu_n/\mu_p$ ratio (see Table 1), despite lower saturation velocity at high fields. However, Ge has a higher energy relaxation time which potentially relaxes linear gate length scaling constraint to gain performance as compared to Si.

Due to its compatibility with silicon processing and its availability in many fabs, Ge has recently been given much interest again as a promising candidate for high performance MOSFETs. Thanks to High-K materials, the non stable native Ge oxide is not a limitation anymore for the use of Ge in the CMOS technology. Low band gap materials show high diode leakage current. The impact of this leakage on MOS characteristics (IOFF, bulk leakage) is a severe limitation for the use of bulk Ge for CMOS devices. Thus, a more realistic use of Ge for CMOS is Germanium On Insulator(GeOI) Fully Depleted MOSFETs since the bulk leakage is suppressed by the BOX and S/D leakage can be reduced by using ultra thin Germanium in a device operating in the Fully Depleted regime. We have realized Fully Depleted deep sub-micron (gate length down to 0.25 $\mu$m) Ge p-MOSFETs on Ultra Thin Germanium-On-Insulator (GeOI) wafers.[119] The Ge layer obtained by hetero-epitaxy on Si wafers is transferred using the Smart-Cut$^{TM}$ process to fabricate 200 mm GeOI wafers with Ge thickness down to 60 nm (Fig. 21).

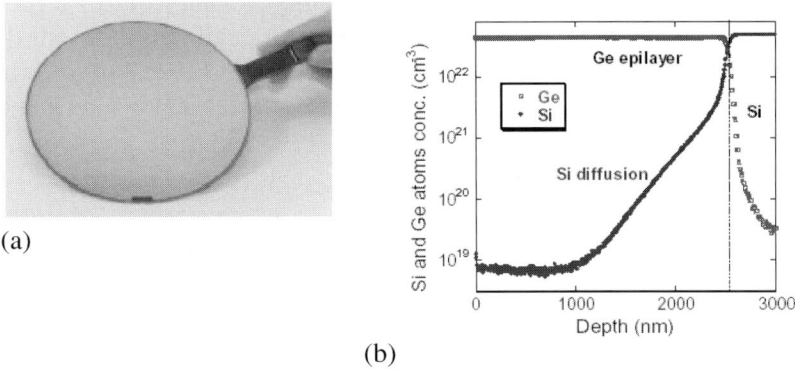

(a)

(b)

Fig. 21. Features of GeOI using epitaxial Ge on Si.[119] (a) Top view photograph of a final GeOI wafer 200 mm in diameter (TGe = 60 nm, TBOx = 400 nm). The donor wafer is a 200 mm epiwafer. (b) SIMS depth profile of the Si and Ge atoms inside a 2.5 $\mu$m thick Ge layer grown on Si(001) that has subsequently submitted to *in situ* anneals.

A full CMOS compatible p-MOSFET process was implemented with HfO$_2$/TiN gate stack. An ION/IOFF ratio higher than 10$^3$ and a 300 mV/decade sub-threshold slope are measured. These results suggest that both the quality of the Ge layer and the gate stack have to be improved. Nevertheless ION vs. LG state-of the-art values reported in Fig. 22 for Ge and GeOI devices illustrate the excellent performances of our devices.[115,120−122] We have also performed TCAD simulations of GeOI

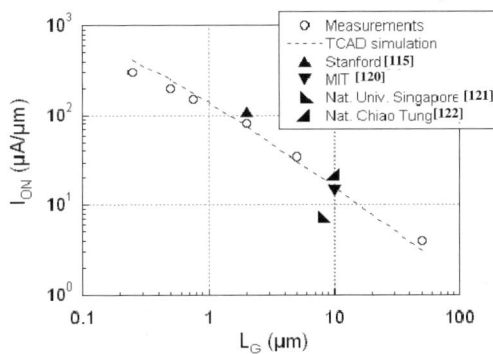

Fig. 22. Comparison of the ION performance of our GeOI P-MOSFETs (LGmin = 0.25 $\mu$m) with literature. The ON current is measured for VDS = −1.5 V, VGS-VT = −2 V. TCAD simulations of GeOI devices show good agreement with the electrical results.[113]

MOSFET structures using a Ge CVT mobility model. The CVT parameters were theoretically calculated or adapted by calibration. From these simulations the ION current values for LG down to 0.25 $\mu$m have been extracted, and show a good agreement with our electrical results and also with literature data.[115,120-122]

## 7.     Alternative CMOS or Alternative to CMOS on Silicon?

Many research teams are making efforts on Single Electron Transistors (SET) operation based on the Coulomb blockade principle. Demonstration of CMOS inverter operation at 27 K has been achieved by using a Vertical Pattern Dependent Oxidation (V-PADOX) process.[123] No solution has been found that could compete with CMOS devices. Some possibilities to achieve memory functional devices by using single electron trapping by a Coulomb blockade effect for DRAM,[124] or Non Volatile applications[125-127] have been pointed out. This effect supposes that the Coulomb energy: $e^2/2C$ (9) is larger than the thermal energy of electrons kT (e is the electron charge; C is the capacitance of the quantum box). This energy is necessary to localize the electrons in a Coulomb box provided that tunneling is the limiting process: implicitly, one has to use very low capacitance and sufficiently high tunneling resistance. However, the Coulomb blockade process will be self limiting due to charge repulsion which reduces the speed of the charge transfer. Non Volatile Memory (NVM) applications can be envisaged by using trapping in nanometer size Si Nanocrystals (SiNc)[126]: Al2O3 has been chosen as the tunnel insulator due to the increased dot density as compared to other materials (in the range of $10^{12}$cm$^{-2}$), with reasonable interface states density (less than $10^{11}$ cm$^{-2}$). Whether the involved writing or erase mechanisms are due or not to single electron transfer has been a controversial debate. In large area devices, with a large amount of randomly distributed SiNc, it is very difficult to identify whether the single electron transfer is occurring or not, due to the large distribution of dot sizes and consequently of Coulomb energies. It is thus very important to use a device of the smallest size possible, containing only one dot or a low number of dots, to get a high sensitivity to single electron transfer. Such a result has been obtained at room temperature on 20 nm $\times$ 20 nm Non Volatile Memory Silicon wire based on Silicon quantum dots (Fig. 23(a))[128]: current spikes on the writing or erasing characteristics have been identified as single electron trapping or detrapping respectively. Coulomb blockade oscillations can be observed

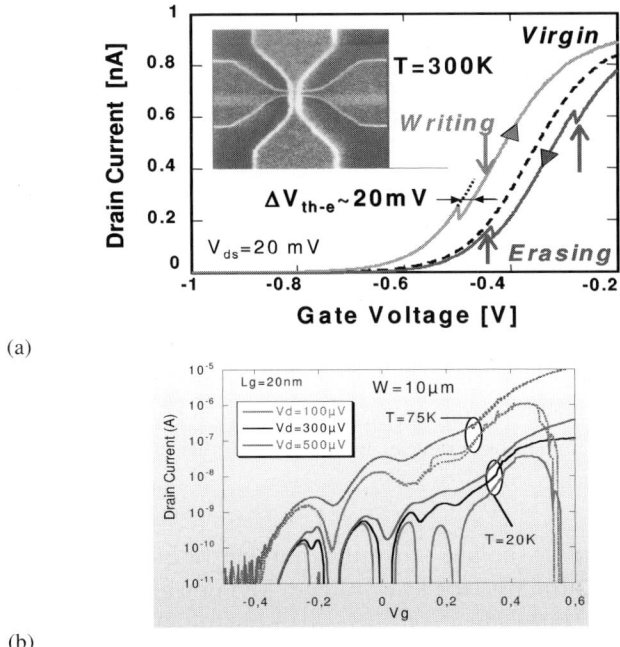

(a)

(b)

Fig. 23. Devices characteristics evidencing Single Electron phenomena. (a) Writing and erase characteristics of 20 nm×20 nm(W×L) devices at room temperature. Top view of 20 nm×20 nm nanowire[128] inserted. (b) Drain current oscillations in a Lg = 20 nm MOSFET at 75 and 20 K, demonstrating that Coulomb blockade is possible in such devices.[5]

if the series access resistance with the quantum well is high enough compared to the resistance quantum:[129] $(e^2/h)^{-1}(10)$. This effect has already been reported on 50 nm gate length N channel MOS transistors at 4.2 K[130] making CMOS transistors attractive as single electron devices candidates. As gate length is scaled down to 20 nm, access resistance becomes larger and channel conductance oscillations appear at higher temperatures (here 75 K) (Fig. 23(b)).[4]

The Si-Nc technology (Fig. 24(a)) offers new scaling possibilities to Flash memories in the sub-90 nm nodes (Fig. 24(b))[127] because of superior Stress Induced Leakage(SILC) immunity of the tunnel oxide. Thus NOR type architectures show a larger tolerance to threshold voltage fluctuations than NAND type devices[127]: if one considers a Si-Nc density of $10^{12}$cm$^{-2}$, NOR type can be scaled down to the 35 nm node whereas NAND type would reach the 65 nm node (Fig. 24(b)). The stored charge discreteness makes these devices much sensitive to stochastic fluctuations of writing and

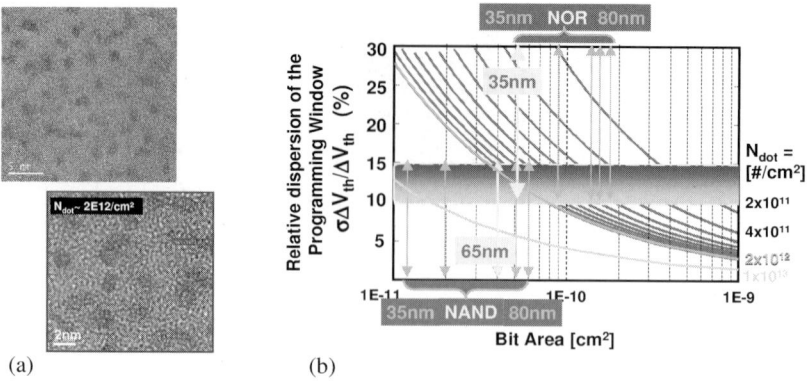

(a)                                                                    (b)

Fig. 24. Si-nc based Flash memories use (a) $2 \times 10^{12} \mathrm{cm}^{-2}$ CVD density of nanometers size Si dots; (b) the scaling of the devices will depend on their architecture and thus on their programming scheme.[127]

retention times[131]: the use of limited number of electrons makes the Si-nc devices more attractive for low voltage, low power operation (Fig. 25).[131] Double bit operation has also been demonstrated.[127,132] This solution is compatible with high standard retention times and endurance cycles,[127] down to gate lengths of 35 nm.[132] The use of High K as a coupling dielectric between the control gate and the SiNc will enhance the coupling ratio and thus allows their integration in NAND architectures.[133]

More generally, discrete traps memories are of interest to address the scaling of NVM via the SONOS architectures[134] for embedded architectures (see also Chapters 7 to 9 of this book). These architectures are challenged by an increasing interest of Resistor Phase Change memories devices (Chapter 7).

## 8.    Conclusions

By the end and beyond the end of the roadmap, power consumption will be the greatest issue whatever the application. We reviewed the physical limitations of MOSFET that will be encountered in the optimization of the performance versus leakage trade off and screened the different possibilities on the architecture or material sides. Multigate devices using strained channels will be widely used for high performance CMOS. Si based alloys or compatible semiconductors will be introduced to enhance the possibilities of future Systems on Chip. New materials including HiK dielectrics, Ge

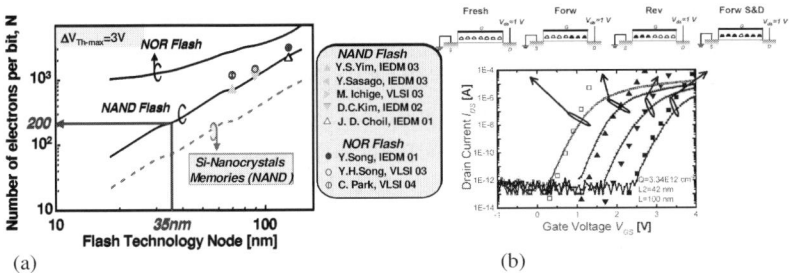

(a)  (b)

Fig. 25. Si-nc allow: (a) lower number of electrons per bit for programming: that reduces the programming voltages and power consumption.[131] (b) Double bit operation: transfer characteristics of a scaled SOI device charged consecutively on drain, source and on both sides with the same stressing conditions. Four clear states are apparent also if the two pockets of charge are very close to one another.[132]

and C-based materials could be integrated to optimize integration density of logic circuits as well as for limitation of short channel effects and power dissipation. New devices architectures requiring a low number of electrons for operation have good potentials in low power, low voltage Flash memories applications by the use of silicon nanocrystals. Single electronics will be a major study subject to optimize the use of ultra small devices.

## 9.    Acknowledgments

We wish to warmly thank the members of the LETI — Electronics Nanodevices Laboratory and Nanotechnologies Division for their various contributions to this chapter as well as the LETI Silicon Technologies Platform for wafer processing. Many of these studies were carried out thanks to the funding by industrial collaborations with STMicroelectronics, Freescale, NXP, Texas Instruments, ATMEL, as well as in the frame of European Commission programs in FP4, FP5 and FP6, MEDEA+, Basic Research French National Programs RTB, ANR and LETI-Carnot Institute labels.

## References

1.    The International Technology Roadmap for Semiconductors (2007 edition).

2.    H. Iwai, *IEDM Tech. Digest 2004*, pp 11–14, San Francisco (CA), 2004.

3.    S. Deleonibus, C. Caillat, G. Guegan, M. Heitzmann, M. E. Nier, S. Tedesco, B. Dal'zotto, F. Martin, P. Mur, A. M. Papon, G. Lecarval, S. Biswas and D. Souil, IEEE Electron Dev. Letters, pp 173–175, April 2000.

4.    G. Baccarani, M. R. Wordeman and R. H. Dennard, *IEEE Trans Electron Devices*, **31**(4), 452–462, 1984.

5.    G. Bertrand, S. Deleonibus, B. Previtali, G. Guegan, X. Jehl, M. Sanquer and F. Balestra, *Solid State Electron.*, **48** (4), pp 505–509, 2004.

6.    K. Romanjek, F. Andrieu, T. Ernst and G. Ghibaudo, *IEEE Electron Dev. Lett.*, **25**(8) pp 583–585, (2004).

7.    S. Deleonibus, C. Caillat, J. N. Gautier, G. Guegan, M. Heitzmann, F. Martin and S. Tedesco, *ESSDERC Tech. Digest 1999*, pp 119–126, Leuven, 1999.

8.    H. Iwai and H. S. Momose, *IEDM Tech. Digest 1998*, pp 163–166, San Francisco (CA), 1998.

9.    C. Caillat, S. Deleonibus, G. Guegan, S. Tedesco, B. Dal'zotto, M. Heitzmann, F. Martin, P. Mur, B. Marchand and F. Balestra, *VLSI Tech. Symp. Tech. Digest 1999*, pp 89–90, Kyoto (Japan), 1999.

10.   S.-H. Lo, D. A. Buchanan, Y. Taur and W. Wang, *IEEE Electron Device Lett*, **18**(1), pp 209–221, 1997.

11.   K. Gopalakrishnan, P. B. Griffin and J. D. Plummer, *IEDM2002 Tech. Digest*, pp 289–291, San Francisco (CA), 2002.

12.   H. Oyamatsu, M. Kinugawa and M. Kakumu, *VLSI Tech. Symp. Tech. Digest*, pp 89–90, Kyoto (Japan), 1993.

13.   W. Schockley, *Solid State Electron.*, 2, pp 35–67. 1961.

14.   A. Asenov, *VLSI Tech. Symp. Tech. Digest 2007*, pp 86–87, Kyoto (Japan), 2007.

15.   H.-S. Wong and Y. Taur, *IEDM Tech. Digest 1993*, pp 705–708, Washington (DC), 1993.

16.   T. Ohguro, N. Sugiyama, S. Imai, K. Usuda, M. Saito, T. Yoshitomi, M. Ono, H. Kimijima, H. S. Momose, Y. Katsumata and H. Iwai, *IEEE Trans. Electron Dev.*, **45**(3), pp 710–716, 1998.

17.   T. Ernst, F. Ducroquet, J.-M. Hartmann, O. Weber, V. Loup, R. Truche, A. M. Papon, P. Holliger, L. Brevard, A. Toffoli, J. L. Di Maria and S. Deleonibus, *VLSI Tech. Symp. Tech. Digest*, pp 51–52, Kyoto (Japan), 2003.

18.   F. Ducroquet, T. Ernst, J.-M. Hartmann, O. Weber, F. Andrieu, P. Holliger, F. Laugier, P. Rivallin, G. Guégan, D. Lafond, C. Laviron,

V. Carron, L. Brévard, C. Tabone, D. Bouchu, A. Toffoli, J. Cluzel and S. Deleonibus, *IEDM Tech. Digest 2004*, pp 437–440, San Francisco (CA), 2004.

19. M. Carroll, T. Ivanov, S. Kuehne, J. Chu, C. King, M. Frei, M. Mastrapasqua, R. Johnson, K. Ng, S. Moinian, S. Martin, C. Huang, T. Hsu, D. Nguyen, R. Singh, L. Fritzinger, T. Esry, W. Moller, B. Kane, G. Abeln, D. Hwang, D. Orphee, S. Lytle, M. Roby, D. Vitkavage, D. Chesire, R. Ashton, D. Shuttleworth, M. Thoma, S. Choi, S. Lewellen, P. Mason, T. Lai, H. Hsieh, D. Dennis, E. Harris, S. Thomas, R. Gregor, P. Sana and W. Wu, *IEDM Tech. Digest 2000*, pp 145–148, San Francisco (CA), 2000.

20. K. Rim, J. L. Hoyt and J. F. Gibbons, *IEDM Tech. Digest 1998*, pp 707–710, San Francisco (CA), 1998.

21. J. M. Hartmann, Y. Bogumilowicz, P. Holliger, F. Laugier, R. Truche, G. Rolland, M. N. Séméria, V. Renard, E. Olshanetsky, O. Estibal, D. Kvon, J. C. Portal, L. Vincent and A. Claverie, *Semicond. Sci. Technol.* **19**(3) 311–318, 2004.

22. E. A. Fitzgerald, M. T. Currie, S. B. Samavedam, T. A. Langdo, G. Taraschi, V. Yang, C. W. Leitz and M. T. Bulsara, *Phys. Status Solidi (a)*, 171–227, 1999.

23. L. Jinggang, Z. Renhua, G. Rozgonyi , E. Yakimov, N. Yarykin and M. Seacrist, *Proc. Electrochem. Soc. of Symp. Silicon Materials Science and Technology*, **19** (2), pp 569–577, Denver (CO), 2006.

24. F. Andrieu, T. Ernst, K. Romanjek, O. Weber, C. Renard, J.-M. Hartmann, A. Toffoli, A.-M. Papon, R. Truche, P. Holliger, L. Brévard, G. Ghibaudo and S. Deleonibus, *ESSDERC 2003 Proc.*, pp 267–270, Estoril (Portugal), 2003.

25. J. Alieu, P. Bouillon, R. Gwoziecki, D. Moi, G. Bremond and T. Skotincki, *ESSDERC*, p 144, Bordeaux (France), 1998.

26. A. C. Lindgren, P. E. Hellberg, M. von Haartman, D. Wu, C. Menon, S. Zhang and M. Östling, *ESSDERC* 2002, p 175, Firenze (Italy), 2002.

27. N. Collaert, P. Verheyen, K. De Meyer, R. Loo and M. Caymax, *Silicon Nanoelectronics Workshop Digest 2002*, 15–16, Honolulu (HI) and *IEEE Trans. of Nanotechnology*, **1**(4), pp 190–4, 2002.

28. N. Collaert, P. Verheyen, K. De Meyer, R. Loo and M. Caymax *ESSDERC 02*, p 263, Firenze (Italy), 2002.

29. K. A. Jenkins and K. Rim, *IEEE Electron. Dev. Let.*, **23**(6), 2002.

30. G. Xia, H. M. Nayfeh, M. J. Lee, E. A. Fitzgerald, D. A. Antoniadis, J. L. Hoyt, J. Li, D. H. Anjum and R. Hull, *IEEE Trans. on Electron Devices*, **51**(12), 2004.

31. H. Kawasaki, K. Ohuchi, A. Oishi, O. Fujii, H. Tsujii, T. Ishida, K. Kasai,Y. Okayama, K. Kojima, K. Adachi, N. Aoki, T. Kanemura, D. Hagishima,M. Fujiwara, S. Inaba, K. Ishimaru, N. Nagashima and H. Ishiuchi, *IEDM 2004 Tech. Digest*, p 169–172, San Francisco (CA), 2004.

32. K. Rim, E. P. Gusev, C. D'Emic, T. Kanarsky, H. Chen, J. Chu, J. Ott, K. Chan, D. Boyd, V. Mazzeo, B. H. Lee, A. Mocuta, J. Welser, S. L. Cohen, M. Ieong and H.-S. Wong, *VLSI Tech. Symp. Digest 2002*, pp 12–13, Honolulu (HI), 2002.

33. O. Weber, F. Ducroquet, T. Ernst, F. Andrieu, J.-F. Damlencourt, J.-M. Hartmann, B. Guillaumot, A.-M. Papon, H. Dansas, L. Brevard, A. Toffoli, P. Besson, F. Martin, Y. Morand and S. Deleonibus, *VLSI Tech. Symp. Digest 2004*, pp 42–43, Honolulu (HI), 2004.

34. F. Andrieu, C. Dupré, F. Rochette, O. Faynot, L. Tosti, C. Buj, E. Rouchouze, M. Cassé, B. Ghyselen, I. Cayrefourcq, L. Brévard, F. Allain, J. C. Barbé, J. Cluzel, A. Vandooren, S. Denorme, T. Ernst, C. Fenouillet-Béranger, C. Jahan, D. Lafond, H. Dansas, B. Previtali, J. P. Colonna, H. Grampeix, P. Gaud, C. Mazuré and S. Deleonibus, *VLSI Tech. Symp. Digest 2006*, 168–169, Honolulu (HI), 2006 and F. Andrieu, T. Ernst, O. Faynot, Y. Bogumilowicz, J.-M. Hartmann, J. Eymery, D. Lafond, Y.-M. Levaillant, C. Dupré, R. Powers, F. Fournel, C. Fenouillet-Beranger, A. Vandooren, B. Ghyselen, C. Mazure, N. Kernevez, G. Ghibaudo and S. Deleonibus, *IEEE Intern. SOI Conf. Digest*, pp 223, Honolulu (HI), 2005.

35. T. Tezuka, N. Sugiyama, T. Mizuno and S. Takagi, *VLSI Tech. Symp. Digest*, pp 96–97, Honolulu (HI), 2002.

36. K. Mistry, C. Allen, C. Auth, B. Beattie, D. Bergstrom, M. Bost, M. Brazier, M. Buehler, A. Cappellani, R. Chau, C.-H. Choi, G. Ding, K. Fischer, T. Ghani, R. Grover, W. Han, D. Hanken, M. Hattendorf, J. He, J. Hicks, R. Huessner, D. Ingerly, P. Jain, R. James, L. Jong, S. Joshi, C. Kenyon, K. Kuhn, K. Lee, H. Liu, J. Maiz, B. McIntyre, P. Moon, J. Neirynck, S. Pae, C. Parker, D. Parsons, C. Prasad, L. Pipes, M. Prince, P. Ranade, T. Reynolds, J. Sandford, L. Shifren, J. Sebastian, J. Seiple, D. Simon, S. Sivakumar, P. Smith, C. Thomas, T. Troeger, P. Vandervoorn, S. Williams and K. Zawadzki, *IEDM 2007 Digest*, pp. 247–250, Dec 2007, Washington (DC).

37. F. Andrieu, T. Ernst, F. Lime, F. Rochette, K. Romanjek, S. Barraud, C. Ravit, F. Boeuf, M. Jurczak, M. Casse, O. Weber, L. Brévard, G. Reimbold, G. Ghibaudo and S. Deleonibus, *2005 VLSI Tech. Symp. Digest*, pp 176–177, Kyoto (Japan), 2005.

38. T. Mizuno, N. Sugiyama, T. Tezuka, Y. Moriyama, S. Nakaharai and S. Takagi, *VLSI Techn. Symp. Digest 2003*, p 97–98, Kyoto (Japan), 2003.

39. M. Yang, V. Chan, S. H. Ku, M. Ieong, L. Shi, K. K. Chan, C. S. Murthy, R. T. Mo, H. S. Yang, E. A. Lehner, Y. Surpris+, F. F. Jamin, P. Oldiges, Y. Zhang, B. N. To, J. R. Holt, S. E. Steen, M. P. Chudzik, D. M. Fried, K. Bernstein, H. Zhu, C. Y. Sung, J. A. Ott, D. C. Boyd and N. Rovedo, *VLSI Tech. Symp 2004*. pp 160–161, Honolulu (HI), 2004.

40. M. L. Lee and E. A. Fitzgerald, *IEDM Tech. Digest 2003*, pp 429–131, Washington (DC), 2003.

41. A. Chatterjee, R. A. Chapman, K. Joyner, M. Otobe, S. Hattangady, M. Bevan, G. A. Brown, H. Yang, Q. He, D. Rogers, S. J. Fang, R. Kraft, A. L. P. Rotondaro, M. Terry, K. Brennan, S.-W. Aur, J. C. Hu, H.-L. Tsai, P. Jones, G. Wilk, M. Aoki, M. Rodder and L. C. Chen, *IEDM Tech. Digest 1998*, pp 777–780, San Francisco (CA), 1998.

42. A. Yagashita, T. Saito, K. Nakajima, S. Inumiya, Y. Akasaka, Y. Ozawa, G. Minamihaba, H. Yano, K. Hieda, K. Suguro, T. Arikado and K. Okumura, *IEDM Tech. Digest 1998*, pp 785–788, San Francisco (CA), 1998.

43. B. Guillaumot, X. Garros, F. Lime, K. Oshima, B. Tavel, J. A. Chroboczek, P. Masson, R. Truche, A. M. Papon, F. Martin, J. F. Damlencourt, S. Maitrejean, M. Rivoire, C. Leroux, S. Cristoloveanu, G. Ghibaudo, J. L. Autran, T. Skotnicki and S. Deleonibus, *IEDM Tech. Digest 2002*, pp 335–338, San Francisco (CA), 2002.

44. B. Tavel, T. Skotnicki, G. Pares, N. Carrière, M. Rivoire, F. Leverd, C. Julien, J. Torres and R. Pantel, *IEDM 2001 Digest*, p 825–828, Washington (DC), 2001.

45. J. Kedzierski, E. Nowak, T. Kanarsky, Y. Zhang, D. Boyd, R. Carruthers, C. Cabral, R. Amos, C. Lavoie, R. Roy, J. Newbury, E. Sullivan, J. Benedict, P. Saunders, K. Wong, D. Canaperi, M. Krishnan, K.-L. Lee, B. A. Rainey, D. Fried, Peter Cottrell, H.-S. P. Wong, M. Ieong and W. Haensch, *IEDM 2002 Digest*, pp 247–250, San Francisco (CA), 2002.

46.    W. P. Maszara, Z. Krivokapic, P. King, J.-S. Goo and M.-R. Lin, *IEDM Tech. Digest 2002*, pp 367–370, San Francisco (CA), 2002.

47.    C. Hobbs, L. Fonseca, V. Dhandapani, S. Samavedam, B. Taylor, J. Grant, L. Dip, D. Triyoso, R. Hegde, D. Gilmer, R. Garcia, D. Roan, L. Lovejoy, R. Rai, L. Hebert, H. Tseng, B. White and P. Tobin, *VLSI Tech. Symp. 2003 Tech. Digest*, pp 9–10, Kyoto (Japan), 2003.

48.    D. Souil, G. Guégan, T. Poiroux, O. Faynot, S. Deleonibus and G. Ghibaudo, *3rd ULIS Workshop 2002*, pp 139–142, Munich (FRG), 2002.

49.    G. Timp, K. K. Bourdelle, J. F. Bower, F. H. Baumann, T. Boone, R. Cirelli, K. Evans-Lutterodt, J. Garno, A. Ghetti, H. Gossmann, M. Green, D. Jacobson, Y. Kim, R. Kleiman, A. Kornblit, C. Lochstampfor, W. Mansfield, S. Moccio, D. A. Muller, L. F. Ocola, M. L. O'Malley, J. Rosamilia, J. Sapjeta, P. Silverman, T. Sorsch, D. M. Tenant, W. Timp and B. F. Weir, *IEDM Tech. Digest 1998*, pp 615–618, San Francisco (CA), 1998.

50.    S. Takagi, M. Takayanagi-Takagi and A. Toriumi, *IEDM Tech. Digest 1998*, pp 619–622, San Francisco (CA), 1998.

51.    B. H. Lee, L. Kang, W.-J. Qi, R. Nieh, Y. Jeon, K. Onishi and J. C. Lee, *IEDM Tech. Digest 1999*, pp 133–136, Washington (DC), 1999.

52.    H. Iwai, S. Ohmi, S. Akama, C. Ohshima, A. Kikuchi, I. Kashiwagi, J. Taguchi, H. Yamamoto, J. Tonotani, Y. Kim, I. Ueda, A. Kuriyama and Y. Yoshihara, *IEDM Tech. Digest 2002*, pp 625–627, San Francisco (CA), 2002.

53.    O. Weber, F. Andrieu, M. Cassé, T. Ernst, J. Mitard, F. Ducroquet, J.-F. Damlencourt, J.-M. Hartmann, D. Lafond, A.-M. Papon, L. Militaru, L. Thevenod, K. Romanjek,C. Leroux, F. Martin, B. Guillaumot, G. Ghibaudo and S. Deleonibus, *IEDM 2004*, pp 867–670, San Francisco (CA), 2004.

54.    J.-L. Pelloie, *ISSCC Tech. Digest 1999*, p 428, San Francisco (CA), 1999.

55.    E. Leobandung, M. Sherony, J. Sleight, R. Bolam, F. Assatleraghi, S. Wu, D. Schepis, A. Ajmera, W. Rausch, B. Davari and G. Shahidi, *IEDM Tech. Digest 1998*, pp 403–407, San Francisco (CA), 1998.

56.    J. Lolivier, S. Deleonibus and F. Balestra, *ECS Spring 2003 Proc.*, pp 379, Paris (France), 2003.

57.    B. Doris, M. Ieong, T. Kanarsky, Y. Zhang, R. A. Roy, O. Dokumaci, Z. Ren, F.-F. Jamin, L. Shi, W. Natzle, H.-J. Huang, J. Mezzapelle,

A. Mocuta, S. Womack, M. Gribelyuk, E. C. Jones, R; J. Miller, H.-S. P. Wong and W. Haensch, *IEDM 2002 Tech. Digest*, pp 267–270, San Francisco (CA), 2002.

58. J. Lolivier, M. Vinet, T. Poiroux, Q. Rafhay, B. Previtali, T. Chevolleau, J. M Hartmann, O. Faynot, A.-M. Papon, R. Truche, F. Balestra and S. Deleonibus, *SOI Conference 2004*, pp 17–18, Charleston (SC), 2004.

59. F. Andrieu, O. Faynot, F. Rochette, J.-C. Barbé, C. Buj, Y. Bogumilowicz, F. Allain, V. Delaye, D. Lafond, F. Aussenac, S. Feruglio, J. Eymery, T. Akatsu, P. Maury, L. Brévard, L. Tosti, H. Dansas, E. Rouchouze, J.-M. Hartmann, L. Vandroux, M. Cassé, F. Boeuf, C. Fenouillet-Béranger, F. Brunier, I. Cayrefourcq, C. Mazuré, G. Ghibaudo and S. Deleonibus, *VLSI Symposium 2007*, pp 50–51, Kyoto (Japan), 2007.

60. F. Andrieu, F. Allain, C. Buj-Dufournet, O. Faynot, F. Rochette, M. Cassé, V. Delaye, F. Aussenac, L. Tosti, P. Maury, L. Vandroux, N. Daval, I. Cayrefourcq and S. Deleonibus, *SSDM 2007*, pp 888–889, Tokyo (Japan), 2007.

61. H.-S. P. Wong, Kevin K. Chan and Y. Taur, *IEDM Tech. Digest*, pp 427–430, Washington (DC), 1997.

62. F. Allibert, A. Zaslavsky, J. Pretet and S. Cristoloveanu, *ESSDERC 2001*, pp 267–270, Nurnberg (FRG), 2001.

63. B. Yu, L. Chang, S. Ahmed, H. Wang, S. Bell, C.-Y. Yang, C. Tabery, C. Ho, Q. Xiang, T.-J. King, J. Bokor, C. Hu, M.-R. Lin and D. Kyser, *IEDM Tech. Digest*, pp 251–253, San Francisco (CA), 2002.

64. J. Kedzierski, E. Nowak, T. Kanarsky, Y. Zhang, D. Boyd, R. Carruthers, C. Cabral, R. Amos, C. Lavoie, R. Roy, J. Newbury, E. Sullivan, J. Benedict, P. Saunders, K. Wong, D. Canaperi, M. Krishnan, K.-L. Lee, B. A. Rainey, D. Fried, P. Cottrell, H.-S. P. Wong, M. Ieong and W. Haensch, *IEDM Tech. Digest*, pp 247–250, San Francisco (CA), 2002.

65. F.-L. Yang, H.-Y. Chen, F.-C. Chen, Y.-L. Chan, K.-N. Yang, C.-J. Chen, H.-J. Tao, Y.-K. Choi, M.-S. Liang and C. Hu, *VLSI Tech. Symp. 2002 Digest*, pp 109–110, Honolulu (HI), 2002.

66. L. Chang, S. Tang, T.-J. King, J. Bokor and C. Hu, *IEDM Tech. Digest*, pp 719–722, San Francisco (CA), 2000.

67. T. Sekigawa and Y. Hayashi, *Solid State Elec.* **27**, 827–828, 1984.

68. D. Hisamoto, W.-C. Lee, J. Kedzierski, E. Anderson, H. Takeuchi, K. Asano, T.-J. King, J. Bokor and C. Hu, *Technical Digest of IEDM*, 833–836, 1998.

69. F. Balestra, S. Cristoloveanu, M. Benachir, J. Brini and T. Elewa, *IEEE Elec. Dev. Lett.* **8**, 410–412, 1987.

70. M. Vinet, T. Poiroux, J. Widiez, J. Lolivier, B. Previtali, C. Vizioz, B. Guillaumot, Y. Le Tiec, P. Besson, B. Biasse,F. Allain, M. Cassé, D. Lafond, J.-M. Hartmann, Y. Morand, J. Chiaroni and S. Deleonibus, *IEEE Electron Dev. Letters*, **26**(5), pp 317–319, 2005.

71. M. Vinet, J. Widiez, B. Biasse, T. Poiroux, B. Previtali, J. Lolivier and S. Deleonibus, *ECS Spring meeting 2005 Proc.*, pp 285–296, Québec (CA), 2005.

72. T. Poiroux, M. Vinet, O. Faynot, J. Widiez, J. Lolivier, B. Previtali, T. Ernst and S. Deleonibus, *ULIS 2005 Proc.*, pp 71–74, Bologna (Italy), 2005.

73. T. Poiroux, M. Vinet, O. Faynot, J. Widiez, J. Lolivier, B. Previtali, T. Ernst and S. Deleonibus, *Microelectronics Engineering*, 80, pp 378–385, 2005.

74. M. Vinet, T. Poiroux, J. Widiez, J. Lolivier, B. Previtali, C. Vizioz, B. Guillaumot, P. Besson, J. Simon, F. Martin, S. Maitrejean, P. Holliger, B. Biasse, M. Cassé, F. Allain, A. Toffoli, D. Lafond, J. M. Hartmann, R. Truche, V. Carron, F. Laugier, A. Roman, Y. Morand, D. Renaud, M. Mouis and S. Deleonibus, *Int. Conf. SSDM 2004 Proc.*, pp 768–769, Tokyo (Japan), 2004.

75. J. P. Colinge, M. H. Gao, A. Romano-Rodriguez, H. Maes and C. Claeys, *IEDM 1990 Digest*, pp 595–598, San Francisco (CA), 1990.

76. S. Harrison, P. Coronel, F. Leverd, R. Cerutti, R. Palla, D. Delille, S. Borel, S. Jullian, R. Pantel, S. Descombes, D. Dutartre, Y. Morand, M. P. Samson, D. Lenoble, A. Talbot, A. Villaret, S. Monfray, P. Mazoyer, J. Bustos, H. Brut, A. Cros, D. Munteanu, J.-L. Autran and T. Skotnicki, *IEDM 2003 Digest*, pp 449–452, Washington (DC), 2003.

77. K. W. Guarini, P. M. Solomon, Y. Zhang, K. K. Chan, E. C. Jones, G. M. Cohen, A. Krasnoperova, M. Ronay, O. Dokumaci, J. J. Buc-chignano, C. Cabral Jr., C. Lavoie, V. Ku, D. C. Boyd, K. S. Petrarca, I. V. Babich, J. Treichler, P. M. Kozlowski, J. S. Newbury, C. P. D'Emic, R. M. Sicina and H.-S. Wong, *IEDM 2001 Digest*, pp 425–428, Washington (DC), 2001.

78. J.-H. Lee, G. Taraschi, A. Wei, T. A. Langdo, E. A. Fitzgerald and D. A. Antoniadis, *IEDM 1999 Digest*, pp 71–74, Washington (DC), 1999.

79.  J. Lolivier, J. Widiez, M. Vinet, T. Poiroux, F. Daugé, B. Previtali, M. Mouis, J. Jommah, F. Balestra and S. Deleonibus, *ESSDERC 2004 Proc.*, pp 177–180, Leuven (Belgium), 2004.

80.  X. Huang, W.-C. Lee, C. Kuo, D. Hisamoto, L. Chang, J. Kedzierski, E. Anderson, H. Takeuchi, Y.-K. Choi, K. Asano, V. Subramanian, T.-J. King, J. Bokor and C. Hu, *IEDM 1999 Digest*, pp 67–70, Washington (DC), 1999.

81.  B. Doyle, B. Boyanov, S. Datta, M. Doczy, S. Hareland, B. Jin, J. Kavalieros, T. Linton, R. Rios and R. Chau, *VLSI Tech. Symp. 2003 Digest*, pp 133–134, Kyoto (Japan), 2003.

82.  F.-L. Yang, H.-Y. Chen, F.-C. Chen, C.-C. Huang, C.-Y. Chang, H.-K. Chiu, C.-C. Lee,C.-C. Chen, H.-T. Huang, C.-J. Chen, H.-J. Tao, Y.-C. Yeo, M.-S. Liang and C. Hu, *IEDM 2002 Tech. Digest*, pp 255–258, San Francisco (CA), 2002

83.  J.-T. Park, J.-P. Colinge and C. H. Diaz, *IEEE Electron Device Letters* **22**, pp 405–406, 2001.

84.  F.-L. Yang, D.-H. Lee, H.-Y. Chen, C.-Y. Chang, S.-D. Liu, C.-C. Huang, T.-X. Chung, H.-W. Chen, C.-C. Huang, Y.-H. Liu, C.-C. Wu, C.-C. Chen, S.-C. Chen, Y.-T. Chen, Y.-H. Chen, C.-J. Chen, B.-W. Chan, P.-F. Hsu, J.-H. Shieh, H.-J. Tao, Y.-C. Yeo, Y. Li, J.-W. Lee, P. Chen, M.-S. Liang and C. Hu, *VLSI Tech. Symp. 2004 Digest*, pp 196–197, Honolulu (HI), 2004.

85.  Y. K. Choi, T. J. King and C. Hu, *Solid State Elec.*, **46**, 1595–1601, 2002.

86.  C. Jahan, O. Faynot, M. Cassé, R. Ritzenthaler, L. Brévard, L. Tosti, X. Garros, C. Vizioz, F. Allain, A. M. Papon, H. Dansas, F. Martin, M. Vinet, B. Guillaumot, A. Toffoli, B. Giffard and S. Deleonibus, *IEEE VLSI Tech. Symp.*, pp 112–113, Kyoto (Japan), 2005.

87.  J. P. Colinge, *Solid State Elec.*, **48**, 897–905, 2004.

88.  J. G. Fossum, L. Ge and M.-H. Chiang, *IEEE Trans. on Elec. Dev.* **49**, pp 808–811, 2002.

89.  H.-S. P. Wong, D. J. Frank and P. M. Solomon, *IEDM 1998 Digest*, pp 407–410, San Francisco (CA), 1998.

90.  T. Ernst, S. Cristoloveanu, G. Ghibaudo, T. Ouisse, S. Horiguchi, Y. Ono, Y. Takahashi and K. Murase, in *IEEE Trans. on Elec. Dev.*, **50**, 830–838, 2003.

91.  F. Gamiz, J.-B. Roldan, A. Godoy, P. Cartujo-Cassinello and J.-E. Carceller, *Journal of Appl. Phys.*, **94**, 5732–5741, 2003.

92.  K. Suzuki, T. Tanaka, Y. Tosaka, H. Horie and Y. Arimoto, *IEEE Trans. on Elec. Dev.*, **40**, 2326–2329, 1993.

93. S.-Y. Lee, E.-J. Yoon, S.-M. Kim, C. W. Oh, M. Li, J.-D. Choi, K.-H. Yeo, M.-S. Kim, H.-J. Cho, S.-H. Kim, D.-W. Kim, D. Park and K. Kim, *VLSI Tech. Symp*, p 200–201, 2004.

94. S. M. Kim, E. J. Yoon, M. S. Kim, S. D. Suk, M. Li, L. Jun, C. W. Oh, K. H. Yeo, S. H. Kim, S. Y. Lee, Y. L. Choi, N.-Y. Kim, Y.-Y. Yeoh, H.-B. Park, C. S. Kim, H.-M. Kim, D.-C. Kim, H. S. Park, H. D. Kim, Y. M. Lee, D.-W. Kim, D. Park and B.-I. Ryu, *VLSI T. Symp.*, p 84–85, 2006.

95. T. Ernst, C. Dupré,C. Isheden, E. Bernard, R. Ritzenthaler, V. Maffini-Alvaro, J.-C. Barbé, F. De Crecy, A. Toffoli, C. Vizioz, S. Borel, F. Andrieu, V. Delaye, D. Lafond, G. Rabillé, J.-M. Hartmann, M. Rivoire, B. Guillaumot, A. Suhm, P. Rivallin, O. Faynot, G. Ghibaudo and S. Deleonibus, *IEDM 2006 Tech. Digest*, pp 997–1000, San Francisco (CA), 2006.

96. T. Ernst, C. Dupré, E. Dornel, J. C. Barbé, S. Bécu, C. Vizioz, V. Delaye, F. Andrieu, J-M. Hartmann, S. Barnola, T. Poiroux, O. Faynot, G. Ghibaudo and S. Deleonibus, *SSDM 2007 Proc.*, pp 200–201, Tokyo (Japan), 2007.

97. M.-C. M. Lee and M. C. Wu, *J. MEMS* **15**(2), 338 (2006).

98. W. Xiong, G. Gebara, J. Zaman, M. Gostkowski, B. Nguyen, G. Smith, D. Lewis, C. Rinn Cleavelin, R. Wise, S. Yu, M. Pas, T.-J. King and J.-P. Colinge, *IEEE Electr. Dev. Lett.* **25** (8), 541 (2004).

99. J. M. Ha, J. W. Park, W. S. Kim, S. P. Kim, W. S. Song, H. S. Kim, H. J. Song, K. Fujihara, H. K. Kang, M. Y. Lee, S. Felch, U. Jeong, M. Goeckner, K. H. Shim, H. J. Kim, H. T. Cho, Y. K. Kim, D. H. Ko and G. C. Lee, *IEDM Tech. Digest 1998*, pp 639–642, San Francisco (CA), 1998.

100. K. Goto, J. Matsuo, Y. Tada, T. Tanaka, Y. Momiyama, T. Sugii and I. Yamada, *IEDM Tech. Digest 1997*, pp 471–474, Washington (DC), 1997.

101. M. Takase, K. Yamashita, A. Hori and B. Mizuno, *IEDM Tech. Digest 1997*, pp 475–478, Washington (DC), 1997.

102. Y. Sasaki, C. G. Jin, H. Tamura, B. Mizuno, R. Higaki, T. Satoh, K. Majima, H. Sauddin, K. Takagi, S. Ohmi, K. Tsutsui and H. Iwai, *VLSI Techn. Symp. 2004 Tech. Digest*, pp 180–181, Honolulu (HI), 2004.

103. T. Noguchi, *J. of the Korean Physical Society*, **34**, pp 265–267, 1999.

104.  C. Laviron, M. N. Semeria, D. Zahorski, M. Stehlé, M. Hernandez, J. Boulmer, D. Débarre and G. Kerrien, 2nd IWJT, *IEEE-Cat. No. 01EX541C*, 91–4., Tokyo (Japan), 2001.

105.  T. Ohguro, *ECS Sympon ULSI 1997*, p 275, Montreal (CA), 1997.

106.  R. Nuryadi, Y. Ishikawa, M. Tabe and Y. Ono, *Sci. Tech.* B20(1) (2002), 167.

107.  E. Dubois and G. Larrieu, *Solid State Elec.*, 997, 2002.

108.  B. Y. Tsui and C. P. Lin, *IEEE Elec. Dev. Lett.*, 430–433, 2004.

109.  G. Niu, J. D. Cressler, S. J. Mathew and S. Subbanna, *IEEE Trans Electron Dev.*, **46**, pp 1912–1914, 1999..

110.  K. Chen, C. Hu, P. Fang and A. Gupta, *Solid State Electronics*, **39**, pp 1515–1518, 1996.

111.  S. Datta, F. Assad and M. S. Lundstrom, *Superlattices and Microstructures*, **23**(3)(4), pp 771–780, 1998.

112.  F. Assad, Z. Ren, D. Vasileska, S. Datta and M. Lundstrom, *IEEE Trans. on Electron Devices*, **47**(1), pp 232–240, 2000.

113.  F. Assad, Z. Ren, S. Datta and M. Lundstrom, *IEDM Tech. Digest*, pp 547–550, Washington (DC), 1999.

114.  V. M. Gusev, M. I. Guseva, E. S. Ionova, A. N. Mansurova and C. V. Starinin, *Phys. Stat. Sol. (a)*, **21**, pp 413–418, 1974.

115.  C. O. Chui, H. Kim, D. Chi, B. B. Triplett, P. C. McIntyre and K. C. Saraswat, *IEDM Techn. Digest*, pp 437–440, San Francisco (CA), 2002.

116.  M. Nishitami-Gamo, E. Yasu, X. Changyong, Y. Kikuchi, K. Ushizawa, I. Sakaguchi, T. Suzuki and T. Ando, *Diamond and Related Materials*, **9**, pp 941–947, 2000.

117.  J.-P. Lagrange, A. Deneuville and E. Gheeraert, *Carbon*, **37**, pp 807–810, 1999.

118.  S. Deleonibus, B. de Salvo, T. Ernst, O. Faynot, T. Poiroux, P. Scheiblin and M. Vinet, *Int. J. High Speed Electron. and Syst.*, **16**(1), pp 193–219, 2006.

119.  L. Clavelier, C. Le Royer, C. Tabone, J. M. Hartmann, C. Deguet, V. Loup, C. Ducruet, C. Vizioz, M. Pala, T. Billon, S. Deleonibus, F. Letertre, C. Arvet, Y. Campidelli, V. Cosnier and Y. Morand, *2005 Silicon Nanolectronics Workshop*, pp 18–19, Kyoto (Japan), 2005.

120.  A. Ritenour, S. Yu, M. L. Lee, N. Lu, W. Bai, A. Pitera, E. A. Fitzgerald, D. L. Kwong and D. A. Antoniadis, *IEDM Tech. Digest*, pp 433–436, Washington (DC), 2003.

121.  Z. Shiyang, L. Rui, S. J. Lee, M. F. Li, A. Du, J. Singh, Z. Chunxiang, A. Chin and D. L. Kwong, *IEEE Elec. Dev. Lett.*, **26**(2), pp 81–83, 2005.
122.  D. S. Yu, C. H. Huang, A. Chin, C. Zhu, M. F. Li, B. J. Cho and D.-L. Kwong, *IEEE EDL* **25**(3), pp 138–140, 2004.
123.  Y. Ono, Y. Takahashi, K. Yamazaki, M. Nagase, H. Namatsu, K. Kurihara and K. Murase, *IEDM 1998 Tech. Digest*, pp 367–370, San Francisco (CA), 1998.
124.  S. Tiwari, F. Rana, H. Hanafi, A. Hartstein, E. Crabbé and K. Chan, *Appl. Phys. Lett.*, **68**, 1377, 1996.
125.  K. Yano, T. Ishii, T. Sano, T. Mine, F. Murai, T. Kure and K. Seki, *IEDM Tech. Digest 1998*, pp 107–110, San Francisco (CA), 1998.
126.  A. Fernandes, B. DeSalvo, T. Baron, J. F. Damlencourt, A. M. Papon, D. Lafond, D. Mariolle, B. Guillaumot, P. Besson, P. Masson, G. Ghibaudo, G. Pananakakis, F. Martin and S. Haukka, *IEDM 2001 Tech. Digest*, pp 155–158, Washington (DC), 2001.
127.  B. DeSalvo, C. Gerardi, S. Lombardo, T. Baron, L. Perniola, D. Mariolle, P. Mur, A. Toffoli, M. Gely, M. N. Semeria, S. Deleonibus, G. Ammendola, V. Ancarani, M. Melanotte, R. Bez, L. Baldi, D. Corso, I. Crupi, R. A. Puglisi, G. Nicotra, E. Rimini, F. Mazen, G. Ghibaudo, G. Pananakakis, C. Monzio Compagnoni, D. Ielmini, A. Spinelli, A. Lacaita, Y. M. Wan and K. van der Jeugd, *IEDM Tech. Digest 2003*, pp 597–600, Washington (DC), 2003.
128.  G. Molas, B. De Salvo, D. Mariolle, D. Fraboulet, G. Ghibaudo, A. Toffoli, N. Buffet and S. Deleonibus, *WODIM 2002 Proc.*, pp 175–178, Grenoble (France), 2002.
129.  X. Jehl, M. Sanquer, G. Bertrand. G. Guégan, S. Deleonibus and D. Fraboulet, SNW 2003, pp 70–71 Kyoto (Japan), 2003.
130.  M. Specht, M. Sanquer, C. Caillat, G. Guegan and S. Deleonibus, *IEDM Tech. Digest 1999*, pp 383–341, Washington (DC), 1999.
131.  G. Molas, D. Deleruyelle, B. De Salvo, G. Ghibaudo, M. Gely, S. Jacob, D. Lafond and S. Deleonibus, *IEDM Tech. Digest 2004*, pp 877–880, San Francisco (CA), 2004.
132.  L. Perniola, G. Iannaccone, B. De Salvo, G. Ghibaudo, G. Molas, C. Gerardi and S. Deleonibus, *IEDM Tech. Digest 2005*, pp 877–880, Washington (DC), 2005.
133.  G. Molas, M. Bocquet, J. Buckley, J. P. Colonna, L. Masarotto, H. Grampeix, F. Martin, V. Vidal, A. Toffoli, P. Brianceau, L. Vermande, P. Scheiblin, M. Gély, A. M. Papon, G. Auvert, L. Perniola, C. Licitra,

T. Veyron, N. Rochat, C. Bongiorno, S. Lombardo, B. De Salvo and S. Deleonibus, *IEDM Tech. Digest 2007*, pp 453–456, Washington (DC), 2007.

134.  J. Buckley, M. Bocquet, G. Molas, M. Gely, P. Brianceau, N. Rochat, E. Martinez, F. Martin, H. Grampeix, J.-P. Colonna, A. Toffoli, V. Vidal, C. Leroux, G. Ghibaudo, G. Pananakakis, C. Bongiorno, D. Corso, S. Lombardo, B. DeSalvo and S. Deleonibus, *IEDM Tech. Digest 2006*, pp 251–254, San Francisco (CA), 2006.

# 2

# Advanced CMOS Devices on Bulk and SOI: Physics, Modeling and Characterization

Thierry Poiroux* and Gilles Le Carval

Department of Nanotechnology, CEA-LETI/Minatec,
17, rue des Martyrs, 38054 Grenoble, France.

*thierry.poiroux@cea.fr

..................................

The modeling and the characterization of decananometer MOSFETs require taking into account several effects that could be neglected on previous technological generations. In this chapter, we make an overview of the main physical effects that must be accounted for to properly describe the electrostatics and the carrier transport in modern transistors. We discuss about the underlying physics, and we indicate the appropriate tools and methods available for simulation and characterization purpose.

## 1. Introduction

For several decades, the performance improvement trend that makes the success of semiconductor industry relies on the dimension shrinking of the basic circuit component: the MOSFET. With gate lengths in the decananometer range, continuing with the same performance enhancement slope while keeping the power consumption under control requires the use of novel materials and novel device architectures. The dimension downscaling and these material and architecture changes are accompanied by some evolutions in the device modeling, simulation and characterization.

While the physics of the field-effect transistor is obviously unchanged from the beginning of the CMOS adventure, the dimension scaling down to the nanometer range requires taking into some physical effects that could be neglected in previous technology nodes, such as the quantum nature of the carriers. Furthermore, some averaged behaviors are no longer meaningful on very short transistor. For example, the MOSFET behavior can no longer be described assuming a large number of dopant atoms or a large number of interactions in the channel.

In this chapter, we describe the physical ingredients required for a proper modeling, simulation and characterization of advanced transistors. In the first part, dedicated to power consumption, we discuss about the device electrostatics of conventional and novel device architectures, as well as the parasitic currents which have to be controlled. The second part is dedicated to the transistor performance, and is mainly focused on the carrier transport and on the parasitic resistances.

## 2. Power Consumption

One of the main issues for advanced CMOS technology nodes is the control of circuit power consumption when the circuit is idle (static consumption) or active (dynamic consumption). At the transistor level, keeping a low static power consumption requires an excellent control of the off-state current of the MOSFET, and thus a very good electrostatic control of the transistor channel by the gate, as well as limited parasitic leakage currents, such as gate tunneling current, gate induced drain leakage (GIDL), junction leakage and direct source-drain tunneling. Dynamic power consumption has also to be considered at transistor level, where parasitic capacitances must be reduced as much as possible.

### 2.1. Electrostatic control

#### 2.1.1. Short channel effects

Bi-dimensional electrostatics of planar MOSFETs is characterized by the so-called short channel effects, including the threshold voltage roll-off (threshold voltage dependence with the gate length at low drain voltage), the drain induced barrier lowering (threshold voltage dependence with the drain bias), and the sub-threshold slope degradation (see Figure 1).

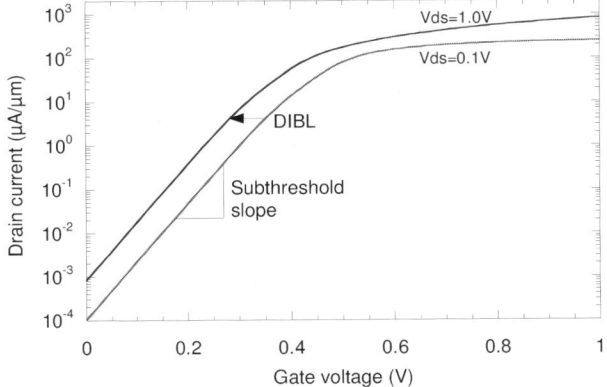

Fig. 1. Transfer characteristics of a 25 nm long transistor at low and high drain voltages, illustrating the drain induced barrier lowering (DIBL) and the subthreshold slope degradation.

These different manifestations of short channel effects have the same physical origin, which is the competition between the electrostatic influence of the source/drain electrodes on the channel with that of the gate electrode. Thus, in order to improve the electrostatic integrity of a transistor, one has to increase the gate to channel capacitive coupling relatively to the coupling between the channel and the source/drain. A metric of the transistor integrity is given by the ratio $C_{GC}/(C_{GC}+C_{DC}+C_{SC}+C_{BC})$, where $C_{GC}$, $C_{DC}$, $C_{SC}$ and $C_{BC}$ are respectively the gate, drain, source and bulk to channel capacitances. From these considerations, simple geometric analyses allow a rough estimate of the influence of the device design on its scalability.

For device optimization or compact modeling purpose, more sophisticated and accurate approaches of 2D-electrostatics modeling are based on pseudo-2D resolution of Poisson equation (see for example doping-voltage transformation[2]) or on superposition principle and development in series expansion of the bi-dimensional parts of the electrostatic potential in the channel.[3]

To minimize the short channel effect on bulk MOSFETs, pockets are implanted below the LDD areas. These additional implanted regions are of the same type as the channel and contribute to flatten the threshold voltage versus gate length curve by two distinct effects. First, they limit the penetration of the electrical fields induced by the source and the drain into the channel, thus improving the electrostatic control by the gate. Second, because of these quite heavily implanted pockets near the source and the drain, the average channel doping is higher on short gate length. This induces a higher potential barrier between the source and the channel and thus a higher threshold voltage on short devices. This effect, known as reverse short channel effect, can be modeled by introducing a non uniform doping along the channel in the 2D Poisson equation and by using potential and field continuity equations to obtain the potential profile from the source to the drain and a closed form for the threshold voltage shift.[4]

The other ways used to ensure the transistor electrostatic integrity are the reduction of the gate dielectric thickness, in order to increase the gate to channel coupling, and the use of ultra-shallow source/drain junctions, to limit the source/drain to channel capacitance.

### 2.1.2. Thin film and multiple gate devices

While bulk MOSFETs are approaching their limits in terms of electro-static control for gate lengths of about 25 nm, thin film fully depleted SOI (FDSOI) devices offer an opportunity to further scale down the transistors. Indeed, the use of ultra-thin silicon film on a buried oxide (BOX) allows a significant reduction of the capacitive coupling between the source/drain electrodes and the channel. The short channel effects are then mainly con-trolled by adjusting first the silicon film thickness, and second the gate dielectric thickness. Numerical simulations as well as analytical modeling of such ultra-thin body transistors show that the channel length over film thickness ratio ($L_{ch}/t_{Si}$) has to be higher than 4 or 5 in order to keep the DIBL below 100 mV/V and the subthreshold swing below 80 mV/dec.[5,6] In such devices, the buried insulator thickness plays also a role on the electrostatic control since electrical fields generated by the drain in the buried oxide can be curved into this BOX and take part to the potential barrier lowering at the channel entrance. This effect, known as fringing field effect, can be modeled thanks to series development of Laplace equation[7] or through the concept of drain-induced virtual substrate bias (DIVSB).[8] The latter approach con-sists in considering the drain as a virtual back gate that influences not only the back channel but also the front channel through the coupling between interfaces. The bi-dimensional potential deformation induced by the drain is derived from Schwarz-Cristoffel conformal mapping.

A reduction of the buried insulator thickness helps suppressing this detrimental effect, but is not sufficient. Indeed, for ultra-thin buried oxides, the fringing fields can go through the depleted region of the substrate beneath the BOX if the substrate surface is left undoped.[8] One way to further reduce the fringing field effect is to use a metallic or a heavily doped layer at the substrate surface. In that case, the fringing fields vanish in this conductive layer, called ground-plane, and the drain to channel coupling is limited to a characteristic length fixed by the BOX thickness.

Reducing strongly the buried insulator thickness down to a nanometer size and connecting the ground plane electrically with the gate leads to the well known concept of double-gate transistor.[9] In such a configuration, the gate to channel coupling is doubled, leading to significantly improved electrostatic control at a given channel thickness. The minimum ratio $L_{ch}/t_{Si}$ required to keep a correct control of the transistor by the gate is reduced from 4 for single gate devices to about 2 for double gate.

Several double gate architectures can be envisaged. Planar double gate transistors can be fabricated thanks to wafer bonding[10] (see Figure 2) or starting from the Silicon-On-Nothing approach.[11]

On the other hand, the double gate behavior can be achieved by etching very narrow silicon fins to form the channel and by patterning a gate that controls the channel from both sides of the fin.[12] These devices, called FinFETs, require the patterning of high aspect ratio fins in order to ensure a good layout density and an excellent control of the fin width.

If the fin height is comparable to its width, a gate control can be obtained on three sides of the channel. Depending on the exact shape of the gate, these devices are called Trigate,[13] Pi-gate[14] or $\Omega$-FETs[15,16] (see Figure 2).

Ultimately, the best electrostatic control is obtained with a cylindrical channel completely surrounded by the gate.[17]

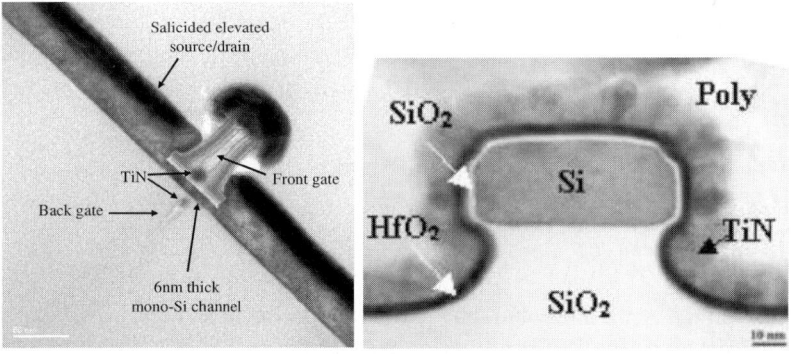

Fig. 2. Left: TEM cross-section of a 10 nm gate length planar double gate MOSFET.[10] Right: TEM cross-section of a 60 nm silicon finger $\Omega$-FET device.[16]

Table 1. Characteristic scale length expressions for various thin film device architectures calculated from 2D Poisson equation (from Refs. 18–21).

| Device Architecture | Surface Conduction Scale Length | Volume Conduction Scale Length |
|---|---|---|
| FDSOI Single gate | $\lambda = \sqrt{\dfrac{\varepsilon_{Si}}{\varepsilon_{ox}} t_{Si} t_{ox}}$ | $\lambda = \sqrt{\dfrac{\varepsilon_{Si}}{\varepsilon_{ox}} t_{Si} \left( t_{ox} + \dfrac{\varepsilon_{ox}}{\varepsilon_{Si}} \dfrac{t_{Si}}{2} \right)}$ |
| Double gate | $\lambda = \sqrt{\dfrac{\varepsilon_{Si}}{\varepsilon_{ox}} \dfrac{t_{Si}}{2} t_{ox}}$ | $\lambda = \sqrt{\dfrac{\varepsilon_{Si}}{\varepsilon_{ox}} \dfrac{t_{Si}}{2} \left( t_{ox} + \dfrac{\varepsilon_{ox}}{\varepsilon_{Si}} \dfrac{t_{Si}}{4} \right)}$ |
| Cylindrical channel | | $\lambda = \sqrt{\dfrac{\varepsilon_{Si}}{\varepsilon_{ox}} \dfrac{t_{Si}}{4} \left( \dfrac{t_{Si}}{2} \ln\left( 1 + \dfrac{2t_{ox}}{t_{Si}} \right) + \dfrac{\varepsilon_{ox}}{\varepsilon_{Si}} \dfrac{t_{Si}}{4} \right)}$ |

To compare the scaling potential of these various device architectures, it is very convenient to calculate their characteristics scale lengths from 2D Poisson equation and boundary conditions. The scale lengths obtained for several architectures and channel doping levels are given in Table 1 (sub-threshold volume conduction is obtained for low-doped channels while sub-threshold surface conduction corresponds to heavily doped channels).[18−21]

It should be noticed that in the case of low-doped channel devices, the subthreshold conduction is located in the middle of the film. Thus, reducing the film thickness leads also to an increase of the gate to channel capacitance, making the film thickness $t_{Si}$ more influent on electrostatics than the gate dielectric thickness $t_{ox}$.

The threshold roll-off, the DIBL and the sub-threshold swing degradation scale roughly as exp(-L/(2λ)) and the minimum channel length with acceptable short channel effects (DIBL<100 mV/V) is approximately 5λ.[5]

## 2.2. Parasitic currents

### 2.2.1. Gate leakage

In addition to the off-state current of the transistor, several parasitic currents can contribute to the static power consumption.

First, for nanometer size gate dielectrics, the extension of electron or hole wavefunctions through the gate oxide leads to a non negligible

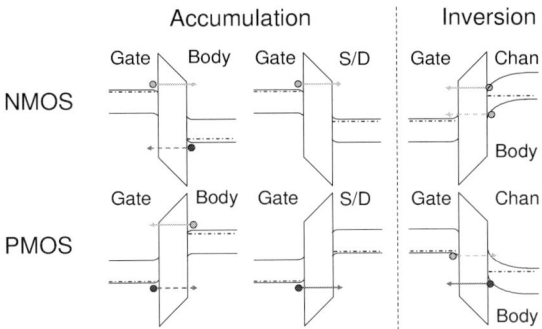

Fig. 3. A schematic illustration of MOSFET band structure and gate tunneling current components.

probability for carriers to tunnel between the gate and the channel. This gate tunneling current in conventional silicon dioxide or nitrided oxide gate dielectrics can contribute significantly to the total leakage current of advanced devices.[1] Depending on the device type (N or PMOSFET), on the biasing conditions and on the location of the leakage, several gate current components have to be considered[22] (see Figure 3).

In accumulation mode, a gate to body current is induced by conduction-band electron tunneling from the gate to the substrate (NMOSFETs) or from the substrate to the gate (PMOSFETs). A gate to source/drain current exists also in overlap regions, due to the tunneling of conduction-band electrons from the N-type gate in case of NMOSFETs, or valence-band holes from the P-type gate in case of PMOSFETs. In inversion mode, the gate current has two components: a gate to channel component due to the tunneling of electrons (resp. holes) from the inversion layer for NMOSFETs (resp. PMOSFETs) and, for gate bias close or higher than the semiconductor bandgap, a gate to body component corresponding to valence-band electron injection from the body (NMOSFETs) or from the gate (PMOSFETs).

Although this latter component is negligible with respect to the total gate current, it must be taken into account in SOI technologies since it can lead to gate-induced floating body effects (GIFBE).[23,24]

Modeling these gate current components implies to deal with a quantum problem with open boundaries. It has been shown that gate tunneling current can be modeled by calculating the carrier density at the injection side, their impact frequency against the barrier, and the tunneling probability through the barrier.[25] A compact modeling of the gate current can thus be obtained by using the transparency approach to calculate the tunneling probability, coupled with a variational approach to estimate the impact frequency.[26] Excellent agreement with numerical simulations and experimental results is obtained for a transparency calculated in the WKB approximation and accounting for wave reflections at the interfaces.[25]

An efficient way to reduce the gate leakage is to increase the physical dielectric thickness, while keeping a sufficient potential barrier height between the channel and the dielectric. In order to keep at the same time a large capacitive coupling between the gate and the channel, one has to use high-k materials as gate dielectrics. With hafnium-based materials, a gate current reduction of four decades can be achieved at a given gate to channel coupling.[27]

### 2.2.2. *Junction leakage and band-to-band tunneling*

The reduction of junction leakage current becomes a crucial issue for the next generations of devices. In addition to the source/drain to body current of the PN junction, largely reduced in SOI devices, a drain to body leakage can be induced when a high negative (resp. positive) gate to drain voltage is applied in NMOSFETs (resp. PMOSFETs). In the case of NMOSFETs, this gate induced drain leakage (GIDL) is mainly due to band to band tunneling of electrons between the conduction band in the drain region and the valence band in the accumulated region below the gate oxide.

Models of this band to band tunneling have to account for the effects of lateral and vertical electric fields near the drain to gate overlap region, for the drain doping profile, and rely generally on some approximations, such as the WKB approximation for transparency calculation.[28] More recent studies on this topic focus on the trap assisted tunneling GIDL observed at relatively low gate to drain voltage,[29] and on the development of Monte-Carlo tools to account for non-equilibrium transport in GIDL numerical simulations.[30]

### 2.2.3. *Source/drain direct tunneling*

Finally, for channel lengths below 10 nm, a significant amount of carriers can tunnel directly from source to drain through the barrier potential.[31] The simulation of this effect requires tools accounting for quantum effects in the transport direction, such as simulation tools based on tight-binding using the Green's function formalism.[32] Experimental evidence of this tunneling component can be obtained thanks to a study of the sub-threshold behavior of the transistor as a function of temperature,[33] since direct tunneling current is far less sensitive to temperature effects than the normal sub-threshold current of a MOSFET.

## 2.3. Variability

Variability is a major concern at circuit level in advanced CMOS technologies. Indeed, in addition to the deterministic variability induced by the technological layout-dependent dispersions of the device dimensions, aggressively scaled transistors will face some new sources of variability due to their reduced dimensions.

### 2.3.1. Channel doping fluctuations

For doped channel MOSFETs (bulk, partially-depleted SOI), the number of dopant atoms into the depleted region of the body is reduced to a few tens for the 32 and the 22 nanometer nodes. Consequently, the statistical fluctuations on this number of dopant atoms will be increased to more than 10%. Furthermore, if we consider also the random placement of these impurities, severe variations of the short channel effect amplitude will induce large fluctuations (several decades) on the off-state current of the MOSFETs.[34,35]

Three-dimensional numerical simulations are required to estimate the impact of this statistical variability. The simulation tools must allow a random placement of the dopant atoms and should also include quantum mechanical effects in order to avoid a too strong coulomb trapping of the mobile carriers near the ionized impurities.[36]

### 2.3.2. Thin film thickness control

For ultra-thin body devices with undoped channel, the film thickness control is of prime importance since it strongly conditions electrostatic control. From the scale lengths presented in paragraph 2.1, one can estimate that a 10% variation of the film thickness is equivalent to a 7% variation of the channel length. Furthermore, in such devices, a statistical fluctuation of the number and location of the dopant atoms in the extension regions can lead to additional variability and should be considered carefully.

From a circuit simulation point of view, these sources of variability have to be taken into account in the development of the library cells, and statistical simulations are required to explore the whole domain of parameter variations.[37]

# 3. Device Performance

As for power consumption, several physical effects that could be neglected in previous technology nodes have to be accounted for in order to estimate properly and to enhance the performance of advanced devices. In this part, we discuss about the physical ingredients required to model and characterize correctly the performance at the transistor level in both bulk and thin film SOI technologies.

Considering first device electrostatics, we describe the effect of quantum confinement and the impact of the gate material on the inversion charge in the channel. The second part, dedicated to carrier transport, presents the main collision mechanisms as a function of longitudinal and transverse electrical field, as well as the impact of quantum confinement and the possible ways to enhance the carrier mobility. The third part focuses on the characterization of access resistances, which is a first order parameter to be optimized in advanced devices.

## 3.1. Electrostatics

The on-state current of a transistor is given by the density of the mobile charge in inversion mode times the carrier velocity. As discussed in 2.1, a large gate to channel coupling is required to ensure a good electrostatic control of the channel, in order to reduce the off-state current of the transistor. This strong coupling is also required from a performance point of view, since a large gate to channel capacitance ensures a high inversion charge density at a given supply voltage.

### 3.1.1. Impact of quantum confinement

For a MOSFET in inversion mode, a potential well in which the mobile carriers are located is induced by the gate bias. As the carrier wave length is of the same order of magnitude as the size of this potential well (a few nanometers), charge confinement occurs in the channel (see Figure 4). The quantum (non local) nature of the carriers implies that the probability amplitude for finding them at the interface between the semiconductor and the gate dielectrics is low. The charge centroid is then located about one nanometer inside the semiconductor, which induces a so-called dark-space region at the gate dielectric interface.[38] The capacitance of this dark-space region is in series with the gate dielectric capacitance and is equivalent to

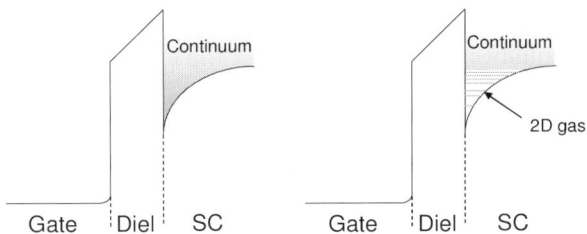

Fig. 4. A schematic illustration of NMOSFET band structure in inversion mode. In a classical picture (left), the conduction band of the semiconductor is considered as a quasi-continuum of energy sub-bands. The carrier gas is treated as a 3D electron gas. In a quantum mechanics picture (right), quantum confinement leads to a splitting of the energy sub-bands. Each sub-band is populated by a 2D electron gas according to the Fermi-Dirac statistics.

a few angströms (0.3–0.4 nm) of silicon dioxide. Thus, for gate dielectrics thicknesses aimed in advanced MOSFET technology nodes, equivalent to a $SiO_2$ thickness of about one nanometer,[1] this series capacitance can no longer be neglected.

In addition, because of quantum confinement, the sub-band energies are elevated (resp. lowered) with respect to the bottom (resp. top) of the conduction band (resp. valence band). This leads to a threshold voltage shift, which can be linked in a good approximation to the shift of the first energy sub-band.[39]

In ultra-thin film devices, an additional quantum confinement is induced by the potential well formed by the thin semiconductor film between the gate dielectrics and the buried oxide.

Taking into account quantum confinement with all sub-bands requires solving self-consistently Poisson and Schrödinger equations. This kind of simulation tools is required for a proper extraction of the physical gate dielectric thickness from capacitive and gate current measurements.[40] Quantum confinement can also be taken into account in TCAD simulation tools, through the so-called "density-gradient" formalism.[41] Density-gradient theory is a quantum mechanical macroscopic model obtained from the moments of the Wigner distribution function.[42] This generalization of the standard diffusion-drift transport incorporates lowest-order quantum effects by making the equations of state of the electron and hole gases depend not only on the gas densities but also on the gradients of their densities.

An analytical modeling of quantum confinement can be obtained from the variational method.[43] Such an approach starts with a trial envelop wavefunction of the carriers, depending on one parameter. The total energy per carrier is calculated by including this approximate wavefunction in the Schrödinger's equation, with a potential profile in the direction transverse to the transport plane calculated self-consistently. The wavefunction parameter is then found by minimizing the total energy per carrier. This approach, initially developed for bulk devices,[44] has been adapted to symmetrical double gate MOSFETs.[45]

### 3.1.2. Gate depletion and metal gates

The use of doped polysilicon as gate material induces another limitation of the gate to channel capacitive coupling. Indeed, if we consider a NMOSFET with an n-type polysilicon gate in inversion mode, the negative inversion charge into the channel has to be balanced by a positive charge in the gate. This positive charge is composed by the gate dopant atoms near the gate dielectric interface, region from which electrons are repelled by the transverse electric field. This depletion region has a thickness of about 1nm for an inversion charge of $10^{13}$ cm$^{-2}$ and an active dopant concentration in the gate of $10^{20}$ cm$^{-3}$. This gate depletion effect induces another additional capacitance in series with the gate dielectrics capacitance, equivalent to a 0.3–0.4 nm thick $SiO_2$ layer, and can be characterized by capacitive measurements in inversion mode.

In order to get rid of this detrimental effect, metal gates will be used for advanced technology nodes. In that case, the MOSFET threshold voltage depends on both the channel doping and the metal gate workfunction. Depending on the aimed application, suitable gate workfunctions for bulk technologies are in the 3.8–4.2 eV and 4.9–5.3 eV range respectively for N and PMOSFETs.[1]

In the case of thin film SOI devices (in the 10 nm range), the threshold voltage can no longer be adjusted thanks to channel doping, and has to be tuned by the use of appropriate gate workfunctions. This offers the opportunity to benefit from the enhanced transport properties of undoped channels, the device electrostatic integrity being ensured by the thinness of the film. Suitable gate workfunctions are then located around silicon midgap value, in the 4.4–4.8 eV range for N and PMOSFETs.[1]

### 3.2. Carrier transport

MOSFET performance is intimately correlated to the transport properties of the carriers in the channel, since the widely used CV/I metrics is in first approximation inversely proportional to the mean carrier velocity between the source and the drain.

### 3.2.1. *Carrier mobility*

At low longitudinal field ($\ll 10^4$ V/cm), carriers are in thermal equilibrium with the lattice. In that case, the Boltzmann Transport Equation (BTE) reduces to the well-known drift-diffusion equation for the current. In the drift component, that governs the carrier transport in a MOSFET in inversion mode at low drain voltage, the drift velocity is linked to the longitudinal field through the carrier mobility. This carrier mobility is the result of various elastic interaction mechanisms that occur in the channel. At low transverse electric field (in the subthreshold regime), carrier mobility is limited by Coulomb scattering induced by the presence of dopant atoms in the channel or in the source/drain, or by charges located at the gate dielectric interface or in the gate dielectrics. The latter effect is sometimes called remote Coulomb scattering in case of high-k dielectrics.[46] At higher electric field (in the moderate to strong inversion regime), Coulomb scattering is screened because of the high density of mobile charges in the channel, and the mobility is limited by acoustic phonon scattering. At high electric field (in the strong inversion regime), carriers are confined close to the gate dielectric interface and their mobility is governed by surface roughness scattering. The effective mobility plotted against the effective transverse electric field is a universal curve, found to be independent from channel doping[47] (Fig. 5).

A proper characterization of carrier mobility is mandatory to identify the main transport limiting mechanisms. In order to distinguish between these different contributions, the mobility dependence versus temperature and transverse field has to be analyzed. While the Coulomb-limited mobility is roughly proportional to temperature,[48] phonon scattering can be efficiently suppressed at low temperature. Low temperature measurements allow consequently the characterization of Coulomb-limited temperature at low transverse field and surface roughness scattering at high transverse field, while the mobility degradation as temperature increases is the signature of phonon scattering.

Several characterization techniques have been proposed to extract the carrier mobility from the transistor electrical characteristics. On long transistors, the most straightforward technique, named split-CV, is based on drain current measurement at low drain voltage coupled with capacitive

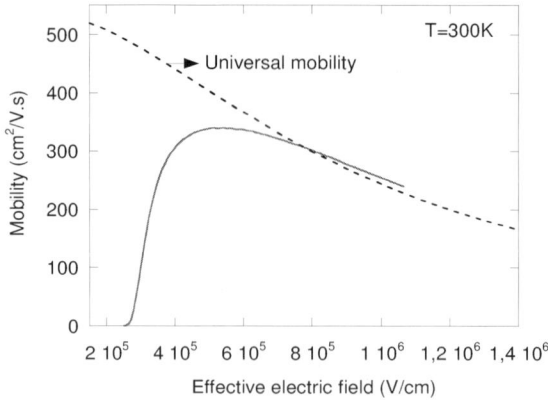

Fig. 5. Electron effective mobility as a function of the effective electric field. Phonon limited mobility and surface roughness limited mobility are observed respectively at moderate and high electric field.

measurements.[49] At a given gate voltage, inversion and depletion charges can be obtained respectively from the integration of the gate to channel and the gate to substrate capacitances. Dividing the drain current by the inversion charge gives the carrier drift velocity, and thus their mobility, while the effective field can be calculated from the inversion and the depletion charges. The mobility extraction on short channel transistors is more problematic, since the channel length is generally not precisely known, and since the drain current has to be corrected from the access resistance effect. Nevertheless, the split-CV technique has been adapted to short channel transistors.[50] This method is based on capacitive measurements on devices with various gate lengths. After the suppression of parasitic capacitances (independent from gate length), the channel length can be obtained by plotting the intrinsic gate to channel capacitance as a function of the gate length. Once the channel length is known, the drain current can be corrected from the access resistance effect and the mobility can be calculated. Some other available techniques for mobility extraction allowing also access resistance characterization are described in Section 3.3.

### 3.2.2.   High longitudinal field and non equilibrium transport

While the mobility concept discussed above is meaningful at low longitudinal field, the carrier drift velocity is no longer proportional to the electric field as this field reaches $10^4$ V/cm. Indeed, as the longitudinal field increases, the carrier energy increase leads to an efficient quantum emission of optical phonons with an energy relaxation time of the order of 0.1 ps in silicon at room temperature.[51] These interactions with optical phonons make the electron drift velocity saturate at about $10^7$ cm/s in silicon at 300 K. This velocity saturation effect can be accounted for in analytical modeling and in TCAD drift-diffusion simulation tools by introducing a dependence of the carrier mobility with the longitudinal field:

$$\mu_{eff} = \frac{\mu_{\text{low field}}}{\left[1 + \left(\frac{E_{//}}{E_{sat}}\right)^n\right]^{1/n}} \tag{1}$$

In Equation (1), $\mu_{\text{low field}}$ is the low longitudinal field mobility, $E_{//}$ is the longitudinal field, $E_{sat}$ is the saturation field, and the power n is about 1 to 2.[52]

In a MOSFET channel, carrier velocity is at its saturation value at the drain side when the longitudinal field is high enough ($> 10^4$ V/cm) and when carriers undergo a large number of inelastic interactions from the source to the drain. With an energy relaxation time of about 0.1 ps, the distance between two consecutive optical phonon emissions is a few tens of nanometers. Consequently, for channel length below 100 nm, the carrier transit time in the channel is comparable to the energy relaxation time. The carrier gas is then far from thermal equilibrium and the carrier velocity can exceed the saturation value. This velocity overshoot effect has been observed experimentally at low temperature.[53] This effect can be taken into account in TCAD simulation tools by solving the second moment of the Boltzmann Transport Equation (energy conservation), which is done in the so-called hydrodynamics[54] and energy-balance[55] approaches. Starting from the second moment of BTE and a closure relation on the energy flux, a simplified equation of the drain current can also be found, providing an analytical modeling of this non-static effect.[56]

If the channel length is further reduced, carrier transit time in the channel can be comparable also to the momentum relaxation time, which is the characteristic time between consecutive interactions (inelastic and elastic). In that case, the carriers undergo a small number of interactions between the source and the drain. This transport mode, called "quasi-ballistic", occurs for channel lengths below 20 to 30 nm in silicon MOSFETs. The simulation of quasi-ballistic transport requires tools dealing with a discrete number of interactions. This is the case of Monte-Carlo simulators, in which charged particles travel from the source to the drain with given (calculated) interaction frequencies.

In the field of analytical modeling, a ballistic MOSFET model has first been proposed,[57] in which the current is expressed as the product of the inversion charge at the channel entrance times the carrier injection velocity at this point. The carrier gas is assumed to be at thermal equilibrium with the carrier gas in the source reservoir. This model has been extended to quasi-ballistic operation, by introducing a probability for carriers to be backscattered towards the source,[58] leading to the following equation of the drain current:

$$I_{drain} = \sum_{sub-bands\,j} WQ_{inv,j} \frac{1-r}{1+r} v_{th,j} \frac{F_{1/2}(\eta_{F,j})}{F_0(\eta_{F,j})} \frac{1 - \dfrac{F_{1/2}(\eta_{F,j} - U_{ds})}{F_{1/2}(\eta_{F,j})}}{1 + \dfrac{1-r}{1+r} \dfrac{F_0(\eta_{F,j} - U_{ds})}{F_0(\eta_{F,j})}}$$

(2)

In Equation (2), W is the device width, $Q_{inv,j}$ is the inversion charge in the $j$th sub-band, r is the backscattering coefficient, $v_{th,j}$ the carrier thermal velocity in the $j$th sub-band, $\eta_{F,j}$ the position of the Fermi level with respect to the bottom of the $j$th sub-band (normalized to kT/q) and $U_{ds}$ the drain to source voltage normalized to kT/q. $F_0$ and $F_{1/2}$ are the integrals of Fermi function of order 0 and $\frac{1}{2}$ respectively.

A few methods have been proposed in order to characterize quasi-ballistic transport in very short MOSFETs,[59,60] all based on the quasi-ballistic current analytical model developed by Purdue University.[58] Since a proper characterization of ballisticity requires taking into account the population of each sub-band with its appropriate injection velocity, these methods rely on Poisson-Schrödinger simulations or analytical modeling of the sub-band energies.[61]

### 3.2.3. *Impact of quantum confinement*

In the conduction band of (001) silicon, one has to distinguish between the four energy valleys located in the transport plane, referred as Δ4 or primed bands, and the two valleys located along the direction normal to the transport plane, referred as Δ2 or unprimed bands. Indeed, electrons in the Δ4 valleys have an effective mass in the ⟨001⟩ direction equal to the transverse mass ($m_t = 0.192\,m_0$), while electrons in the Δ2 valleys have an effective mass equal to the longitudinal mass ($m_l = 0.918\,m_0$). For a MOSFET biased in inversion mode, the quantum confinement in the channel (described in 3.1.1) induces an energy band splitting, with lower energies for Δ2 valley sub-bands since their effective mass in the confinement direction is higher.

This energy band splitting is more pronounced in ultra-thin film transistors because of the additional confinement in the potential well formed by the thin film.[62] The relative population of unprimed sub-bands is thus increased with respect to the situation of bulk silicon. Thus, to describe the transport in silicon NMOSFETs (for example in the ⟨110⟩ direction), one has to deal with two bi-dimensional electron gases with different effective masses in the transport direction: $0.192\,m_0$ for $\Delta 2$ valleys and $0.371\,m_0$ for $\Delta 4$ valleys. The same discussion can be held for PMOSFETs, where heavy hole and light hole valleys must be distinguished.

Quantum confinement can be accounted for self-consistently in Monte-Carlo simulators by using an effective potential approach based on the concept of Bohm potential,[63,64] or by solving 1D Poisson-Schrödinger equation in the confinement direction in each slice of the channel.[65] Both approaches require the use of scattering frequencies calculated for a 2D carrier gas.[66]

### 3.2.4. Transport boosters

Several ways can be followed in order to improve the carrier transport in MOSFETs. The most straightforward way is to induce uniaxial or biaxial tensile (resp. uniaxial compressive) strain in NMOSFET (resp. PMOSFET) channels. For NMOSFETs, tensile strain induces an energy band splitting of the silicon conduction band rather similar to that induced by quantum confinement. $\Delta 2$ energy valleys, exhibiting a lower effective mass in the transport direction, are shifted towards lower energies, while $\Delta 4$ valleys are shifted towards higher energies.[67] The resulting re-population of the energy sub-bands, with a larger proportion of electrons in unprimed sub-bands, leads to a decrease of the overall effective mass in the transport direction, and thus, to higher mobilities. In addition, this energy band splitting is responsible for a strong reduction of phonon intervalley scattering.[67] The combination of both contributions leads to long channel mobility gains over unstrained transistors over 100% and to short channel saturation current gains over 20% on both bulk silicon and thin film technologies.[68–71] Biaxial strain can be induced by substrate engineering, for example from epitaxial growth of tensile silicon on relaxed SiGe layers,[70,71] and uniaxial tensile or compressive strain can be obtained from the optimization of Contact Etch Stop Layer (CESL).[72]

While hole mobility enhancement in ⟨110⟩ direction has been experimentally demonstrated with uniaxial compressive strain,[73] the explanation of this mobility gain is a bit more complex than for electrons. From band structure calculation based on k.p. formalism, it can be shown that this gain results mainly from a combination of band warping together with a quantum confinement sub-band splitting, leading to a reduction of the conductivity effective mass in the $\langle 110 \rangle$ direction.[74,75]

Another efficient way to improve the transport properties is to play with the crystalline orientation and the transport direction in order to further decrease the conductivity effective mass. This can be achieved on PMOSFETs by using $\langle 100 \rangle$ as the transport direction,[76] or by using silicon substrate with (110) surface, where large hole mobility gains with respect to (100) surface have been demonstrated in the $\langle 110 \rangle$ direction.[77]

Finally, other materials presenting higher electron and/or hole mobilities, such as compressive silicon-germanium[78] or germanium,[79] may also be required to further improve the transport in the channel.

### 3.3. Series resistance

In order to benefit from the performance enhancement, resistances in series with the transistor channel have to be reduced as much as possible. Indeed, for very short channel devices, the resistance of the source/drain and the extension regions can be comparable to that of the channel, making series resistance a first order parameter to be characterized and optimized. These series resistances have a gate bias independent component, composed by the contact resistance between the silicide and the doped source/drain region and by the sheet resistance of the doped source/drain, and a gate bias dependent component, composed of the spreading resistance at the channel entrance and the overlap resistance in the doped extension region overlapped by the gate.

Several characterization methods have been proposed to extract the series resistance, as well as the carrier mobility in short devices. First, the shift 'n' ratio method[80] assumes that the total MOSFET resistance is the sum of the channel resistance, obtained by the product of the effective channel length times a given function of the gate overdrive, and an extrinsic resistance independent from the gate bias. The derivative of this total resistance with respect to the gate bias is thus a function of the effective channel length (with a linear dependence) and of the gate overdrive. If we consider a long and a short channel transistors at the same gate overdrive (i.e. with an appropriate gate voltage shift to compensate the threshold voltage roll-off), the ratio between their respective derivatives gives the ratio between their effective channel lengths, allowing the extraction of the difference between the drawn gate length and the electrical channel length. Series resistance can then be extracted from the total resistance values, knowing the effective channel lengths. The main drawback of this method is that it assumes the same mobility for short and long channel transistors, which is generally not verified.

Another DC method is based on the so-called Y-function,[81] defined as the ratio between the drain current at low drain voltage over the square root of the transconductance. This function is in fact a transformation of the derivative of the total MOSFET resistance with respect to the gate bias. The slope of this Y-function versus the gate voltage gives a factor depending on the known drain bias, on the device geometry and on the low field mobility. Using this factor together with the slope of the X-function, defined as the inverse of the transconductance square root, versus the gate voltage for several gate lengths, one can extract the series resistance and the first order mobility degradation term. This method, initially developed with a simple dependence of the mobility with the gate bias, can be adapted to more complex mobility dependence as well as to gate bias dependent access resistance.[82]

RF measurements are also very useful for the extraction of parasitic resistance. A procedure has been proposed to extract properly the gate dependent and gate independent parts of the series resistance from S-measurements on appropriate RF test structures.[83]

## 4.     Conclusions and Outlook

In this chapter, we have described the material and device architecture changes needed to fulfill the performance requirements of the coming CMOS technology nodes, as well as the impact of these evolutions on the MOSFET physical description. Device electrostatics integrity will have to be improved. Thin film or multiple gate transistors are good candidates with that respect and offer the opportunity to get rid of channel dopant fluctuations by using undoped channels with appropriate metal gate workfunctions. High-k dielectrics are required to increase the gate to channel capacitive coupling while reducing gate tunneling currents. To improve the carrier transport, substrate and/or process induced strained channels are used. In addition, crystalline orientation and transport direction optimization, as well as new channel materials such as silicon-germanium or germanium can be envisaged to further increase the transistor performance.

All these evolutions lead to the need for complementary simulation tools. Tools for fundamental physics description are required, such as full-band calculations for strain and crystalline orientation optimization, or quantum transport simulation, in order to describe direct source/drain tunneling. Tools describing the transport with small numbers of interactions and impurities in advanced materials and including non-static effects, such as Monte-Carlo simulators, should include also quantum confinement effects. Finally, calibrated simplified tools accounting for these effects are needed for device optimization. Associated analytical models are also mandatory for device design and development of adapted characterization methodologies. Finally, physics-based compact models taking into account all these effects are strongly required for circuit simulation purpose.

## Acknowledgments

Authors would like to thank Thomas Ernst, Vincent Barral, Sylvain Barraud, Sébastien Soliveres, François Andrieu, Carine Jahan, Olivier Faynot and Simon Deleonibus for fruitful discussions about the content of this chapter.

## References

1.   *International Technology Roadmap for Semiconductors*, 2005 edition.
2.   T. Skotnicki, G. Merckel and T. Pedron, *IEEE Electron Device Lett.* **9**, 109 (1988).
3.   D. J. Frank, Y. Taur and H.-S. P. Wong, *IEEE Electron Device Lett.* **19**, 385 (1998).
4.   B. Yu, E. Nowak, K. Noda and C. Hu, *Symp. VLSI Technology*, **162** (1996).
5.   J. P. Colinge, *Solid State Electronics* **48**, 897 (2004).
6.   T. Poiroux, M. Vinet, O. Faynot, J. Widiez, J. Lolivier, B. Previtali, T. Ernst and S. Deleonibus, *Solid State Electronics* **50**, 18 (2006).
7.   J. S. Woo, *IEEE Trans. on Electron Devices* **37**, 1999 (1990).
8.   T. Ernst, C. Tinella, C. Raynaud and S. Cristoloveanu, *Solid State Electronics* **46**, 373 (2002).
9.   F. Balestra, S. Cristoloveanu, M. Benachir, J. Brini and T. Elewa, *IEEE Electron Device Lett.* **9**, 410 (1987).

10.  M. Vinet, T. Poiroux, J. Widiez, J. Lolivier, B. Previtali, C. Vizioz, B. Guillaumot, Y. Le Tiec, P. Besson, B. Biasse, F. Allain, M. Casse, D. Lafond, J. M. Hartmann, Y. Morand, J. Chiaroni and S. Deleonibus, *IEEE Electron Device Lett.* **26**, 317 (2005).
11.  S. Harrison, P. Coronel, F. Leverd, R. Cerutti, R. Palla and D. Delille, *IEEE Int. Electron Device Meeting (IEDM)*, 449 (2003).
12.  D. Hisamoto, T. Kaga, Y. Kawamoto and E. Takeda, *IEEE Int. Electron Device Meeting (IEDM)*, 833 (1989).
13.  B. Doyle, B. Boyanov, S. Datta, M. Doczy, J. Hareland, B. Jin, J. Kavalieros, T. Linton, R. Rios and R. Chau, *Symp. VLSI Technology*, 133 (2003).
14.  J. T. Park, J. P. Colinge and C. H. Diaz, *IEEE Electron Device Lett.* **22**, 405 (2001).
15.  F. L. Yang, H. Y. Chen, F. C. Chen, C. C. Huang, C. Y. Chang, H. K. Chiu, C. C. Lee, C. C. Chen, H. T. Huang, C. J. Chen, H. J. Tao, Y. C. Yeo, M. S. Liang and C. Hu, *IEEE Int. Electron Device Meeting (IEDM)*, 255 (2002).
16.  C. Jahan, O. Faynot, M. Cassé, R. Ritzenthaler, L. Brévard, L. Tosti, X. Garros, C. Vizioz, F. Allain, A. M. Papon, H. Dansas, F. Martin, M. Vinet, B. Guillaumot, A. Toffoli, B. Giffard and S. Deleonibus, *Symp. VLSI Technology*, 112 (2005).
17.  F. L. Yang, D. H. Lee, H. Y. Chen, C. Y. Chang, S. D. Liu, C. C. Huang, T. X. Chung, H. W. Chen, C. C. Huang, Y. H. Liu, C. C. Wu, C. C. Chen, S. C. Chen, Y. T. Chen, Y. H. Chen, C. J. Chen, B. W. Chan, P. F. Hsu, J. H. Shieh, H. J. Tao, Y. C. Yeo, Y. Li, J. W. Lee, P. Chen, M. S. Liang and C. Hu, *Symp. VLSI Technology*, 196 (2004).
18.  K. K. Young, *IEEE Trans. on Electron Devices* **36**, 504 (1989).
19.  R. H. Yan, A. Ourmazd and K. F. Lee, *IEEE Trans. on Electron Devices* **39**, 1704 (1992).
20.  K. Suzuki, T. Tanaka, Y. Tosaka, H. Horie and Y. Arimoto, *IEEE Trans. on Electron Devices* **40**, 2326 (1993).
21.  C. P. Auth and J. D. Plummer, *IEEE Electron Device Lett.* **18**, 74 (1997).
22.  W. C. Lee and C. Hu, *Symp. VLSI Technology*, 198 (2000).
23.  J. Prétet, T. Matsumoto, T. Poiroux, S. Cristoloveanu, R. Gwoziecki, C. Raynaud, A. Roveda and H. Brut, *Proc. European Solid State Device Research (ESSDERC)*, 515 (2002).
24.  T. Poiroux, O. Faynot, C. Tabone, H. Tigelaar, H. Mogul, N. Bresson and S. Cristoloveanu, *IEEE Int. SOI Conference*, 99 (2002).

25.  L. F. Register, E. Rosenbaum and K. Yang, *Applied Physics Lett.* **74**, 457 (1999).

26.  R. Clerc, P. O'Sullivan, K. G. McCarthy, G. Ghibaudo, G. Pananakakis and A. Mathewson, *Solid State Electronics* **45**, 1705 (2001).

27.  B. Guillaumot, X. Garros, F. Lime, K. Oshima, B. Tavel, J. A. Chroboczek, P. Masson, R. Truche, A. M. Papon, F. Martin, J. F. Damlencourt, S. Maitrejean, M. Rivoire, C. Leroux, S. Cristoloveanu, G. Ghibaudo, J. L. Autran, T. Skotnicki and S. Deleonibus, *IEEE Int. Electron Device Meeting (IEDM)*, 355 (2002).

28.  J. H. Chen, S. C. Wong and Y. H. Wang, *IEEE Trans. Electron Devices* **48**, 1400 (2001).

29.  D. Rideau, A. Dray, F. Gilibert, F. Agut, L. Giguerre, G. Gouget, M. Minondo and A. Juge, *Proc. Int. Conference on Microelectronic Test Structures (ICMTS)* **17**, 149 (2004).

30.  Z. Xia, G. Du, Y. Song, J. Wang and X. Liu, *Japanese Journal of Applied Physics* **46**, 2023 (2007).

31.  M. Städele, *Proc. European Solid State Device Research (ESS-DERC)*, 135 (2002).

32.  M. Bescond, J. L. Autran, D. Munteanu, N. Cavassilas and M. Lannoo, *Proc. European Solid State Device Research (ESSDERC)*, 395 (2003).

33.  J. Lolivier, M. Vinet, T. Poiroux, B. Previtali, T. Chevolleau, J. M. Hartmann, A. M. Papon, R. Truche, O. Faynot, F. Balestra and S. Deleonibus, *IEEE Int. SOI Conference*, 17 (2004).

34.  A. Asenov and S. Saini, *IEEE Trans. Electron Devices* **46**, 1718 (1999).

35.  D. J. Frank, *IBM Journal of Research and Development* **46**, 235 (2002).

36.  A. Asenov, *Short Course Ultimate Integration on Silicon Conf. (ULIS)* (2007).

37.  C. McAndrew, *Proc. Int. Symp. Quality Electronic Design (ISQED)*, 357 (2003).

38.  J. A. Lopez-Villanueva, P. Castujo-Casinello, J. Banqueri, F. Gamiz and S. Rodriguez, *IEEE Trans. Electron Devices* **44**, 1915 (1997).

39.  M. J. Van Dort, P. H. Woerlee and J. Walker, *Solid State Electronics* **37**, 411 (1994).

40.  C. Leroux, P. Mur, N. Rochat, D. Rouchon, R. Truche, G. Reimbold and G. Ghibaudo, *Microelectronics Engineering* **72**, 121 (2004).
41.  M. G. Ancona and H. F. Tiersten, *Phys. Rev. B* **35**, 7959 (1987).
42.  M. G. Ancona and G. J. Iafrate, *Phys. Rev. B* **39**, 9536 (1989).
43.  F. F. Fang and W. E. Howard, *Phys. Rev. Lett.* **16**, 797 (1966).
44.  F. Stern and W. E. Howard, *Phys. Rev.* **163**, 816 (1967).
45.  L. Ge and J. G. Fossum, *IEEE Trans. Electron Devices* **49**, 287 (2002).
46.  S. Saito, D. Hisamoto, S. Kimura and M. Hiratani, *IEEE Int. Electron Device Meeting (IEDM)*, 797 (2003).
47.  S. Takagi, A. Toriumi, M. Iwase and H. Tango, *IEEE Trans. Electron Devices* **41**, 2357 (1994).
48.  C. T. Sah, T. Ning and L. Tschopp, *Surface Science* **32**, 561 (1972).
49.  C. G. Sodini, T. Ekstedt and J. L. Moll, *Solid State Electronics* **25**, 833 (1982).
50.  K. Romanjek, F. Andrieu, T. Ernst and G. Ghibaudo, *IEEE Electron Device Lett.* **25**, 583 (2004).
51.  J. H. Chun, B. Kim, Y. Liu, O. Tornblad and R. W. Dutton, *Proc. Int. Conf. Simulation of Semiconductor Processes and Devices (SISPAD)*, 275 (2005).
52.  C. G. Sodini, P. Keung Ko and J. L. Moll, *IEEE Trans. Electron Devices* **31**, 1386 (1984).
53.  S. Y. Chou, D. A. Antoniadis and H. I. Smith, *IEEE Electron Device Lett.* **6**, 665 (1985).
54.  K. Blotekjaer, *IEEE Trans. Electron Devices* **17**, 38 (1970).
55.  R. Stratton, *Phys. Rev.* **126**, 2002 (1962).
56.  G. Baccarani and S. Reggiani, *IEEE Trans. Electron Devices* **46**, 1656 (1999).
57.  K. Natori, *Journal of Applied Physics* **76**, 4879 (1994).
58.  A. Rahman and M. S. Lundstrom, *IEEE Trans. Electron Devices* **49**, 481 (2002).
59.  M. J. Chen, H. T. Huang, Y. C. Chou, R. T. Chen, Y. T. Tseng, P. N. Chen and C. H. Diaz, *IEEE Trans. Electron Devices* **51**, 1409 (2004).
60.  V. Barral, T. Poiroux, F. Rochette, M. Vinet, S. Barraud, O. Faynot, L. Tosti, F. Andrieu, M. Casse, B. Previtali, R. Ritzenthaler, P. Grosgeorges, E. Bernard, G. Le Carval, D. Munteanu, J. L. Autran and S. Deleonibus, *Symp. VLSI Technology*, 128 (2007).
61.  M. Ferrier, R. Clerc, G. Ghibaudo, F. Bœuf and T. Skotnicki, *Solid State Electronics* **50**, 69 (2006).

62. S. Barraud, *Semiconductor Science and Technology* **22**, 413 (2007).

63. D. K. Ferry, R. Akis and D. Vasileska, *IEEE Int. Electron Device Meeting (IEDM)*, 287 (2000).

64. M. A. Jaud, S. Barraud, P. Dollfus, H. Jaouen and G. Le Carval, *Proc. Int. Conf. Simulation of Semiconductor Processes and Devices (SISPAD)*, 335 (2006).

65. J. Saint-Martin, A. Bournel, F. Monsef, C. Chassat and P. Dollfus, *Semiconductor Science and Technology* **21**, L29 (2006).

66. F. Monsef, P. Dollfus, S. Galdin-Retailleau, H. J. Herzog and T. Hackbarth, *Journal of Applied Physics* **95**, 3587 (2004).

67. S. Takagi, J. L. Hoyt, J. J. Welser and J. F. Gibbons, *Journal of Applied Physics* **80**, 1567 (1996).

68. I. Lauer, T. A. Langdo, Z. Y. Cheng, J. G. Fiorenza, G. Braithwaite, M. T. Currie, C. W. Leitz, A. Lochtefeld, H. Badawi, M. T. Bulsara, M. Somerville and D. A. Antoniadis, *IEEE Electron Device Lett.* **25**, 83 (2004).

69. J. R. Hwang, J. H. Ho, S. M. Ting, T. P. Chen, Y. S. Hsieh, C. C. Huang, Y. Y. Chiang, H. K. Lee, A. Liu, T. M. Shen, G. Braithwaite, M. Currie, N. Gerrish, R. Hammond, A. Lochtefeld, F. Singapore-wala, M. Bulsara, Q. Xiang, M. R. Lin, W. T. Shiau, Y. T. Loh, J. K. Chen, S. C. Chien and F. Wen, *Symp. VLSI Technology*, 103 (2003).

70. F. Andrieu, T. Ernst, O. Faynot, Y. Bogumilowicz, J. M. Hartmann, J. Eymery, D. Lafond, Y. M. Levaillant, C. Dupre, R. Powers, F. Fournel, C. Fenouillet-Beranger, A. Vandooren, B. Ghyselen, C. Mazure, N. Kernevez, G. Ghibaudo and S. Deleonibus, *IEEE Int. SOI Conference*, 223 (2005).

71. K. Rim, K. Chan, L. Shi, D. Boyd, J. Ott, N. Klymko, F. Cardone, L. Tai, S. Koester, M. Cobb, D. Canaperi, B. To, E. Duch, I. Babich, R. Carruthers, P. Saunders, G. Walker, Y. Zhang, M. Steen and M. Ieong, *IEEE Int. Electron Device Meeting (IEDM)*, 49 (2003).

72. H. S. Yang, R. Malik, S. Narasimha, Y. Li, R. Divakaruni, P. Agnello, S. Allen, A. Antreasyan, J. C. Arnold, K. Bandy, M. Belyansky, A. Bonnoit, G. Bronner, V. Chan, X. Chen, Z. Chen, D. Chidambar-rao, A. Chou, W. Clark, S. W. Crowder, B. Engel, H. Harifuchi, S. F. Huang, R. Jagannathan, F. F. Jamin, Y. Kohyama, H. Kuroda, C. W. Lai, H. K. Lee, W. H. Lee, E. H. Lim, W. Lai, A. Mallikarjunan, K. Matsumoto, A. McKnight, J. Nayak, H. Y. Ng, S. Panda, R. Ren-garajan, M. Steigerwalt, S. Subbanna, K. Subramanian, J. Sudijono,

G. Sudo, S. P. Sun, B. Tessier, Y. Toyoshima, P. Tran, R. Wise, R. Wong, I. Y. Yang, C. H. Wann, L. T. Su, M. Horstmann, T. Feudel, A. Wei, K. Frohberg, G. Burbach, M. Gerhardt, M. Lenski, R. Stephan, K. Wieczorek, M. Schaller, H. Salz, J. Hohage, H. Ruelke, J. Klais, P. Huebler, S. Luning, R. van Bentum, G. Grasshoff, C. Schwan, E. Ehrichs, S. Goad, J. Buller, S. Krishnan, D. Greenlaw, M. Raab and N. Kepler, *IEEE Int. Electron Device Meeting (IEDM)*, 1075 (2004).

73.  C. H. Ge, C. C. Lin, C. H. Ko, C. C. Huang, Y. C. Huang, B. W. Chan, B. C. Perng, C. C. Sheu, P. Y. Tsai, L. G. Yao, C. L. Wu, T. L. Lee, C. J. Chen, C. T. Wang, S. C. Lin, Y. C. Yeo and C. Hu, *IEEE Int. Electron Device Meeting (IEDM)*, 73 (2003).

74.  T. Guillaume and M. Mouis, *Solid State Electronics* **50**, 701 (2006).

75.  S. E. Thompson, G. Sun, Y. Sung Choi and T. Nishida, *IEEE Trans. Electron Devices* **53**, 1010 (2006).

76.  M. V. Fischetti, Z. Ren, P. M. Solomon, M. Yang and K. Rim, *Journal of Applied Physics* **94**, 1079 (2003).

77.  M. Yang, E. P. Gusev, M. Ieong, O. Gluschenkov, D. C. Boyd, K. K. Chan, P. M. Kozlowski, C. P. D'Emic, R. M. Sicina, P. C. Jamison and A. I. Chou, *IEEE Electron Device Lett.* **24**, 339 (2003).

78.  O. Weber, F. Ducroquet, T. Ernst, F. Andrieu, J. F. Damlencourt, J. M. Hartmann, B. Guillaumot, A. M. Papon, H. Dansas, L. Brevard, A. Toffoli, P. Besson, F. Martin, Y. Morand and S. Deleonibus, *Symp. VLSI Technology*, 42 (2004).

79.  O. Weber, Y. Bogumilowicz, T. Ernst, J. M. Hartmann, F. Ducroquet, F. Andrieu, C. Dupre, L. Clavelier, C. Le Royer, N. Cherkashin, M. Hytch, D. Rouchon, H. Dansas, A. M. Papon, V. Carron, C. Tabone and S. Deleonibus, *IEEE Int. Electron Device Meeting (IEDM)*, 143 (2005).

80.  Y. Taur, D. S. Zicherman, D. R. Lombardi, P. J. Restle, C. H. Hsu, H. I. Nanafi, M. R. Wordeman, B. Davari and G. G. Shahidi, *IEEE Electron Device Lett.* **13**, 267 (1992).

81.  G. Ghibaudo, *Electronic Lett.* **24**, 543 (1988).

82.  A. Cros, S. Harrison, R. Cerutti, P. Coronel, G. Ghibaudo and H. Brut, *Proc. Int. Conference on Microelectronic Test Structures (ICMTS)*, 69 (2005).

83.  E. Torres-Rios, R. Torres-Torres, G. Valdovinos-Fierro and E. A. Gutierrez, *IEEE Trans. Electron Devices* **53**, 571 (2006).

# 3

# Devices Structures and Carrier Transport Properties of Advanced CMOS Using High Mobility Channels

Shinichi Takagi*,†,§, Tsutomu Tezuka*, Toshifumi Irisawa*, Shu Nakaharai*, Toshinori Numata*, Koji Usuda*, Naoharu Sugiyama*, Masato Shichijo†, Ryosho Nakane† and Satoshi Sugahara‡

*MIRAI-AIST, 1 Komukai Toshiba-cho, Saiwai-ku, Kawasaki, 212-8582, Japan

†The University of Tokyo, 7-3-1 Hongo, Bunkyo-ku, Tokyo 113-8656, Japan

‡The Tokyo Institute of Technology, 4259 Nagatsuta, Midori-ku, Yokohama, 226-8503, Japan

§s-takagi@mirai.aist.go.jp, takagi@ee.t.u-tokyo.ac.jp

........................................

Mobility enhancement technologies have currently been recognized as mandatory for future scaled MOSFETs. In this paper, the recent mobility enhancement technologies including application of strain and new channel materials such as SiGe, Ge and III–V materials are reviewed. These carrier transport enhancement technologies can be classified into three categories; global enhancement techniques, local enhancement techniques and global/local-merged techniques. We present our recent results on MOSFETs using these three types of the technologies with an emphasis on the global strained-Si/SiGe/Ge substrates and the combination with the local techniques. Finally, issues on device structures merged with III–V materials are briefly described.

**Keywords:** MOSFET, Mobility, Strain, SiGe, Ge, III–V semiconductors.

# 1. Importance of Enhancement of Carrier Transport Properties

It has been well recognized that, under sub-100 nm regime, conventional device scaling concept has confronted with several physical and essential limitations. These limitations provide the trade-off relationships among on-current, power consumption or leakage current and short channel effects, shown in Fig. 1. The important device parameters in MOSFETs and the physical origins yielding these trade-off relationships are also shown here. Therefore, any new device engineering to overcome these difficulties and to realize advanced CMOS is strongly needed to dissolve or mitigate the constraints in the trade-off relations. A group of theses new device technologies including the introduction of new materials and new geometrical structures, which are shown as the solutions in Fig. 1, have recently been called the technology boosters in International Technology Roadmap for Semiconductors (ITRS).[1]

These technology boosters can be classified mainly into three categories, as schematically shown in Fig. 2. The first one is the gate stack engineering including high k gate insulators and metal gate electrodes for

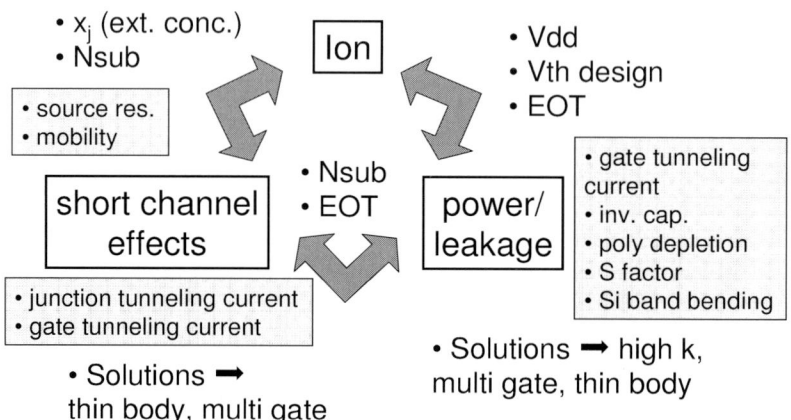

Fig. 1. Trade-off factors among on-current, power consumption/leakage current and short channel effects under simple device scaling and possible solutions to mitigate the relationship.

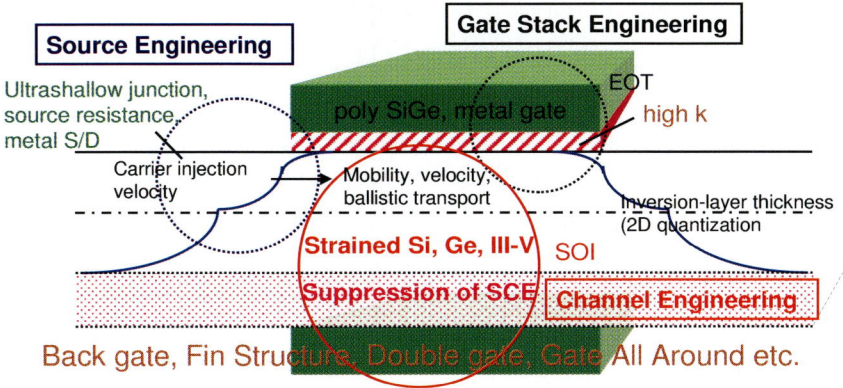

Fig. 2. Schematic diagram of three types of device engineering beyond 32 nm Node.

suppressing direct tunneling current through ultrathin gate oxides and further increasing the gate capacitance. The second one is the source/drain (S/D) engineering including the optimal design of source impurity profiles and Schotky metal source structures. The last one is the channel engineering, which includes a variety of new technologies such as carrier transport enhancement for providing high current drive and multi-gate structures for suppressing short channel effects.

Particularly, the mobility enhancement channels are recently becoming more important, because the saturation trend of the on-current in conventional Si channels, attributed partly to the rapid increase in substrate impurity concentration near the source region by halo implantation, strongly demands other paths to increase the on-current. Note here that the increase in the velocity near the source region, which could correspond to the injection velocity under ballistic transport, is essential to the increase in the on-current. Thus, we call MOS channels to provide higher on-current the carrier-transport-enhanced channels in this study. In most cases, on the other hand, higher mobility can lead to the higher velocity even in short channel devices through the velocity overshoot, less scattering probability and smaller effective mass.

Considering on the future trend of this channel engineering, there are two important issues. The first issue is that the continuous enhancement or improvement of carrier transport properties is needed for successive growth of future CMOS LSIs, because the restless increase in the on-current will be strongly demanded with a progression of technology nodes.

Table 1. Ways to enhance carrier transport properties in MOS channels.

|  | nMOSFET | pMOSFET |
|---|---|---|
| Channel Direction | — | $\langle 100 \rangle$ on (100) surface |
|  |  | $\langle 110 \rangle$ on (110) surface |
| Surface Orientation | — | (110) |
| Strain in Si/Ge | bi-axial tensile | bi-axial tensile |
|  |  | uni-axial compressive |
| Materials | (III–V) | SiGe/Ge |

The second issue is that the future CMOS structures need to combine the carrier-transport-enhanced channels with multi-gate structures, because of the stringent requirements of both the current drive and the short channel effect immunity. Therefore, device platforms allowing us to easily implement the carrier-transport-enhanced channels into the multi-gate MOSFETs are necessary for the 32 nm technology node and beyond.

Table 1 summarizes the existing concepts for enhancing carrier transport properties, which include the choices of surface orientations, channel directions, strain configurations and channel materials. As seen here, many options are available for hole transport enhancement. In contrast, application of tensile strain is the only technique, at present, to enhance carrier transport in n-channel MOSFETs, except for very high electron mobility III–V material channels, whose introduction to Si CMOS platform has recently stirred a strong interest.[2,3]

This paper reviews our recent results on the development of these carrier-transport-enhanced device structures based on global novel substrate technologies and the combination of local techniques with them. Here, there are two new directions for the development of the global substrate technologies including carrier-transport-enhanced materials. One is the emphasis on hole mobility enhancement, which is not enough in Si $p$-MOSFETs with bi-axial tensile strain. A key is an introduction of Ge into channels. The other direction is the combination of any local formation technologies, allowing us to separately optimize the strain configuration and the channel materials for n-channel and $p$-channel MOSFETs for maximizing the CMOS performance. Finally, we briefly touch on the introduction of III–V channel MOSFETs into Si CMOS platform and the critical issues.

## 2.    Strained-Si/SiGe/Ge CMOS Technologies Using Global Substrates

Recently, a variety of local strain techniques have widely been developed for boosting the electron and hole mobility and some of them have already been implemented in real products,[4] mainly because of the easier implementation into conventional CMOS processes. In contrast, global strain technologies, based on substrates using high mobility materials, are expected to provide more uniform and higher strain than the local ones, leading presumably to higher performance and higher robustness against the performance variations. On the other hand, main challenges of the global substrates consist in further reduction in crystal defects and imperfections as well as reduction in the wafer cost.

Thus, one of the most critical issues in carrier- transport-enhancement channels using global substrate is the fabrication of high quality and low cost substrates. Our original approach to fabricate such SiGe-based global substrates is based on the Ge condensation concept.[5−7] The schematic process flow of this fabricate method is shown in Fig. 3. The key fabrication step is to oxidize SiGe films grown on standard SOI substrates at high temperatures and in dry $O_2$. During the oxidation, Ge atoms are rejected from the oxide layer into the SiGe films. On the other hand, the buried oxide layers block the diffusion of Ge atoms into the Si substrate regions. As a result, as the oxidation proceeds, the Ge content comes to increase and the Ge distribution becomes uniform, because of the diffusion within the SiGe layers. It has also been found that the relaxation ratio of the condensed SiGe layers can be controlled by the thicknesses of initial epitaxial SiGe films and SOI layers. The thicker SiGe films lead to larger relaxation ratio.[8]

This fabrication method allows us to prepare various types of SiGe-based substrates, as shown in Fig. 3. A typical one is strained-Si-On-Insulator substrates, where relaxed SiGe-On-Insulator layers are used for introducing biaxial tensile strain in Si films grown on the layers. It has been demonstrated[9,10] that a strained-SOI n-channel MOSFET with gate length ($L_g$) of 70 nm fabricated on the substrate exhibits the drive current enhancement of around 15% against a control SOI MOSFET, though three-time-higher source/drain resistance in the strained-SOI MOSFETs, attributable to the formation of NiSi on SiGe, limit the enhancement of the current drive in shorter $L_g$ region. On the other hand, another essential problem of the carrier-transport enhancement due to bi-axial tensile strain is that large strain is needed for providing sufficiently high hole mobility

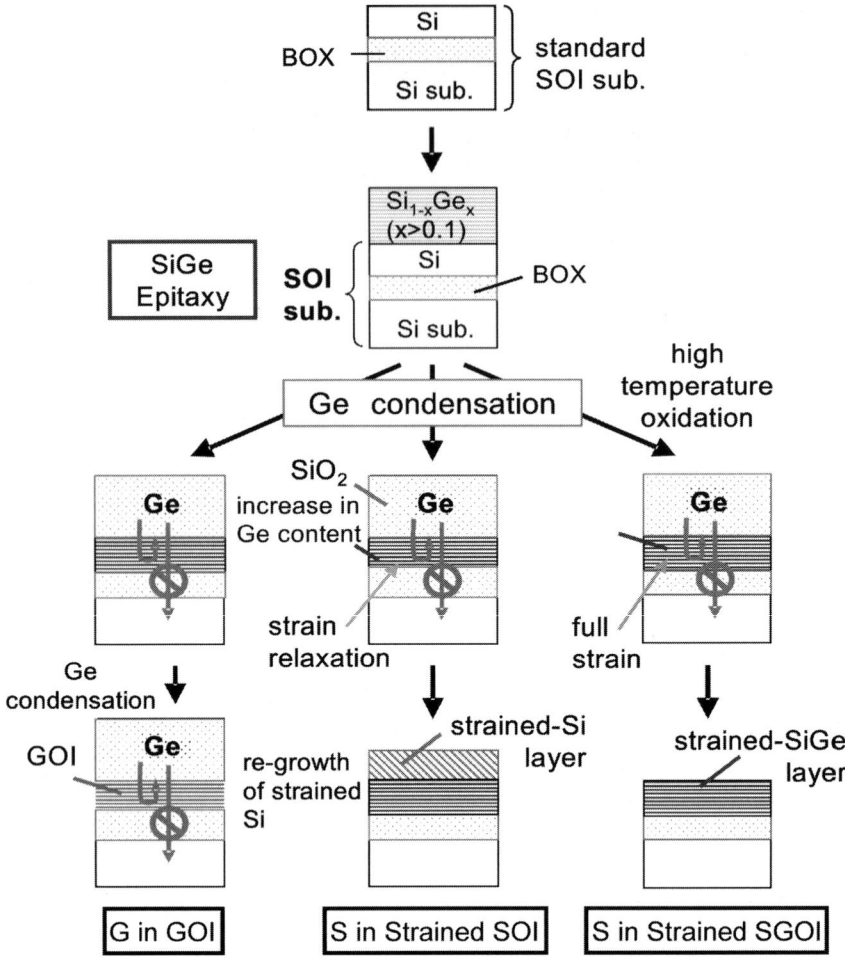

Fig. 3. Schematic diagram of Ge condensation method.

enhancement.[9,11,12] Figure 4 shows the enhancement factors of electron and hole mobility in bi-axial tensile strain Si MOSFETs fabricated on relaxed SiGe substrates, as a function of strain.[9] The symbols mean the experimental data published so far. The solid and dash curves mean the theoretical calculations.[13−16] While the electron mobility exhibits a large enhancement factor in a small amount of strain, high enhancement factor of hole mobility is obtained for biaxial tensile strain higher than 1.5%. Also, another disadvantage of *p*-MOSFETs with bi-axial tensile strain is the decrease in

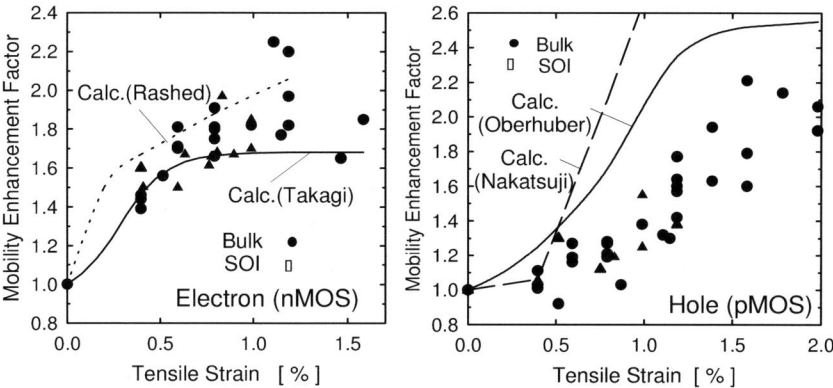

Fig. 4. Experimental and calculated enhancement factors of electron and hole mobility in bi-axial tensile strain Si MOSFETs on relaxed SiGe substrates against conventional Si MOSFETs as a function of strain in Si.

the enhancement factor in high field region, attributable to the decrease in the subband energy difference between the heavy hole and the light hole bands.[17,18] These results suggest that other ways to efficiently enhance hole mobility are desirable for maximizing the CMOS performance.

One way to improve the transport properties of holes is the introduction of high Ge content SiGe channels or pure Ge channels, because Ge is known to have the highest hole mobility in bulk among the column IV and III–V semiconductors, as shown in Table 2. Also, the application of compressive strain to Ge is effective in further increasing hole mobility.[19] There are two strategies for introducing SiGe/Ge channels into Si CMOS platform. One is to use SiGe/Ge global substrates and the other is to locally form SiGe/Ge channel regions. In this section, the global substrate and device technologies are described, while local SiGe/Ge channels will be presented in the next

Table 2. Lists of electron and hole mobilities, electron effective mass, band gap and permittivity for typical III–V compound semiconductors, Si and Ge.

|  | Si | Ge | GaAs | InP | InAs | InSb |
|---|---|---|---|---|---|---|
| Electron mob. (cm2/Vs) | 1600 | 3900 | 9200 | 5400 | 40000 | 77000 |
| Electron Mass mt/m0 | 0.19 | 0.082 | 0.067 | 0.082 | 0.023 | 0.014 |
| Hole mob. (cm2/Vs) | 430 | 1900 | 400 | 200 | 500 | 850 |
| Band Gap (eV) | 1.12 | 0.66 | 1.42 | 1.34 | 0.36 | 0.17 |
| Permittivity | 11.8 | 16 | 12 | 12.6 | 14.8 | 17 |

section. In both cases, thin body structures such SiGe-On-Insulator (SGOI) or Ge-On-Insulator (GOI) are strongly preferred, because of high short-channel effect immunity and reduction in leakage current associated with the narrow band gap materials. Also, ultrathin SGOI/GOI structures are suitable for multi-gate application.

The fabrication of SiGe global substrates and the strain control can be achieved by the Ge condensation technique, shown in Fig. 3. Figure 5 shows the hole mobility of SiGe channel MOSFETs with the Ge content of 28, 35 and 42% as a function of the effective normal field, $E_{eff}$.[20] Here, the SiGe channels after the Ge condensation are fully strained, because the SiGe films initially grown on SOI substrates are sufficiently thin. The thickness of the SGOI channels with the Ge content of 28, 35 and 42% is as thin as 33, 23 and 19 nm, respectively. After growing 5-nm Si layers on these SGOI channels, the Si layers are completely oxidized and also the SGOI channels are slightly oxidized, indicating that the MOS interface is composed of $SiO_2$/SiGe. It is found from Fig. 5 that the mobility enhancement in high $E_{eff}$ region increases with an increase in the Ge content and that the hole mobility enhancement of as high as 2 is obtained from the single-layer ultrathin SiGe p-MOSFETs with the Ge content of 42%, attributable to the higher Ge content and the higher compressive strain. On the other hand, the decrease in mobility in lower $E_{eff}$ with increasing the Ge content can be caused by the increase in the interface state density and the resulting increase in Coulomb scattering.

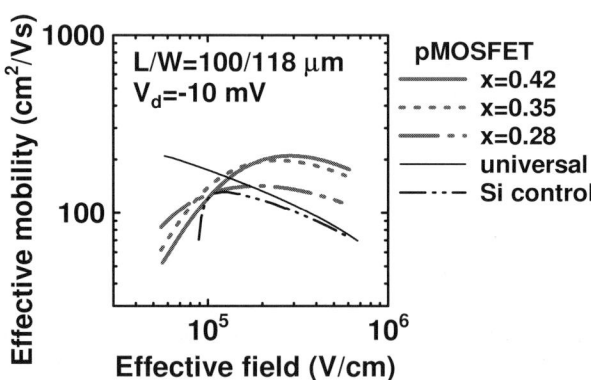

Fig. 5. $E_{eff}$ dependence of the hole mobility in SGOI p-MOSFETs (Ge content of 28%, 35% and 42%) with bi-axial compressive strain.

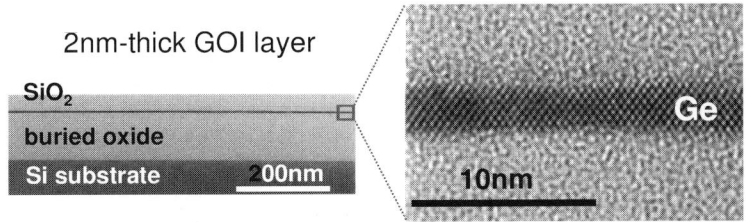

Fig. 6. TEM photograph of GOI substrates with the thickness of 2 nm.

Furthermore, pure-GOI channels are expected to provide higher current drive of *p*-MOSFETs. We have already reported that almost pure GOI substrates can also be fabricated through the Ge condensation technique by just continuing to oxidize SGOI substrates, as shown in Fig. 3.[21] Figure 6 shows a TEM photograph of an ultrathin GOI structure with GOI thickness of 2 nm. It is confirmed that the flat and uniform GOI layer can be fabricated. The residual Si concentration is estimated to be less than 0.01% by SIMS analyses, meaning the high purity of the fabricated structures. It is also confirmed that the GOI thickness can be precisely controlled by changing the amount of Ge before the condensation.[22]

We have recently succeeded in fabricating *p*-MOSFETs on the 150 mm GOI substrates by conventional CMOS processes.[23] Here, $SiO_2$ formed during the Ge condensation process and poly-Si films are used as the gate insulator and the gate electrode, respectively. Boron ion implantation is used for forming $p^+$ S/D regions. It is found that the hole mobility of the fabricated GOI *p*-MOSFETs amounts to 3 times as high as the universal one at $E_{eff}$ of $\sim 0.2$ MV/cm, which is close to the bulk hole mobility ratio of Ge to Si ($\sim 4$). This high enhancement factor is attributable to the smooth and high quality interface with the thermal $SiO_2$ gate insulators formed at temperature of as high as 900°C during the Ge-condensation process.

On the other hand, one of the most critical issues in MOSFETs on the fabricated GOI substrates is the high residual hole concentration in the GOI layers of typically order of $10^{17} \sim 10^{18}$ cm$^{-3}$, which is attributable to any defects or dislocations included in the GOI layers. It is confirmed from SIMS analyses that the residual boron concentration in the GOI layers is less than $10^{16}$ cm$^{-3}$.[24] We have actually observed the generation of a number of micro-twins generated during the Ge condensation,[25] though the relationship between the generated defects and the residual hole concentration is still not clear. Further studies for identifying the origin of residual holes and reducing the crystal defects in the GOI films are strongly needed.

## 3. Strained-Si/SiGe/Ge CMOS Technologies Combined with Local Mobility Enhancement Techniques

### 3.1. *Local formation of SiGe/Ge channel regions on SOI substrates*

In the previous section, the Ge condensation technique was utilized to fabricate global mobility-enhanced material substrates. This technique can also be applied to the local formation of SiGe/Ge channel *p*-MOSFETs by selectively oxidizing the active area of *p*-channel MOSFETs in SiGe films on SOI substrates, which we call the local condensation technique. This technique allows us to optimize the device structure of *p*-channel MOSFETs, separately from that of *n*-channel MOSFETs, which is similar with the several local strain technologies. We have successfully fabricated high Ge content surface-channel SGOI *p*-MOSFETs by using this local Ge condensation technique.[26,27] The fabrication processes are shown in Fig. 7. In this example, the S/D is formed in thick SGOI regions, leading to the reduction in the S/D resistance. In order to form S/D and gate electrodes of Fig. 7 in a self-aligned manner, the damascene gate process can be used. Also, more simply, the S/D regions can be formed in thin SGOI active area regions with high Ge content fabricated by the local condensation.

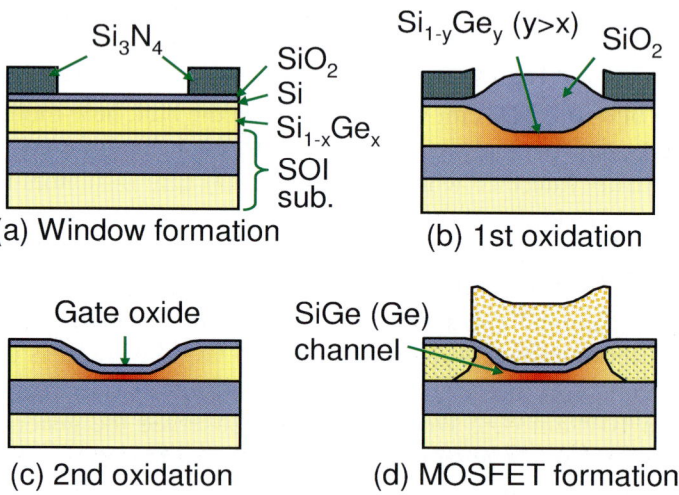

Fig. 7. Fabrication process of SGOI/GOI MOSFETs on selective areas by the local Ge condensation technique.

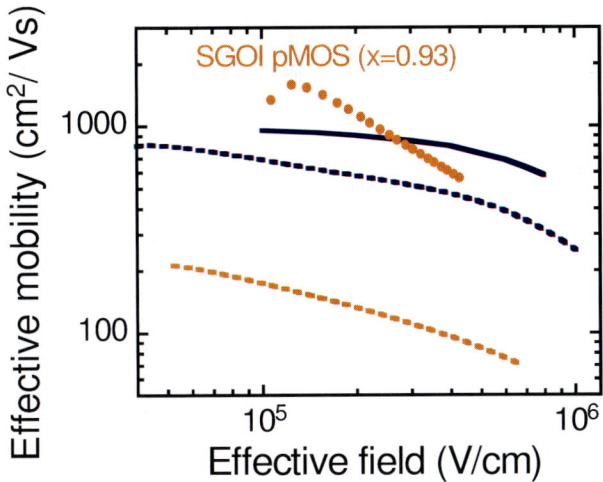

Fig. 8. Comparison of hole mobility in 93% Ge SGOI *p*-MOSFETs with electron mobility in strained Si *n*-MOSFETs and the universal electron and hole mobilities.

Figure 8 shows the hole mobility obtained in SGOI MOSFETs with the Ge content of 93% and the SGOI thickness of 25 nm, fabricated by the present processes, as a function of $E_{eff}$. The gate insulator of the SGOI MOSFETs is formed by oxidizing Si epitaxial layers grown on SGOI. Here, in addition to the Si layers, SGOI films are also oxidized to some extent, resulting in further condensation during this oxidation and the formation of $SiO_2$/SiGe MOS interfaces. The electron mobility of bi-axial tensile strain *n*-channel MOSFETs and the universal electron and hole mobilities are also plotted in this figure. It is found that the hole mobility of the SGOI MOSFETs is almost 10 times as high as the universal hole mobility and is comparable to the electron mobility of strained-Si *n*-channel MOSFETs. This high hole mobility of the SGOI MOSFETs can originate in several factors; (1) high Ge content (2) existence of bi-axial compressive strain of 1.4% due to the effect of the device geometry associated with oxidation of the restricted regions (3) better MOS interface quality associated with high temperature oxidation at 900°C.

By using this local Ge condensation technique, we have fabricated ultrathin-body CMOS structure, where GOI channel *p*-MOSFETs have been integrated with SOI channel *n*-MOSFETs.[28,29] The fabrication process is schematically shown in Fig. 9(a). Here, SiGe layers are selectively grown on the active area of *p*-MOSFETs on SOI substrates and oxidized

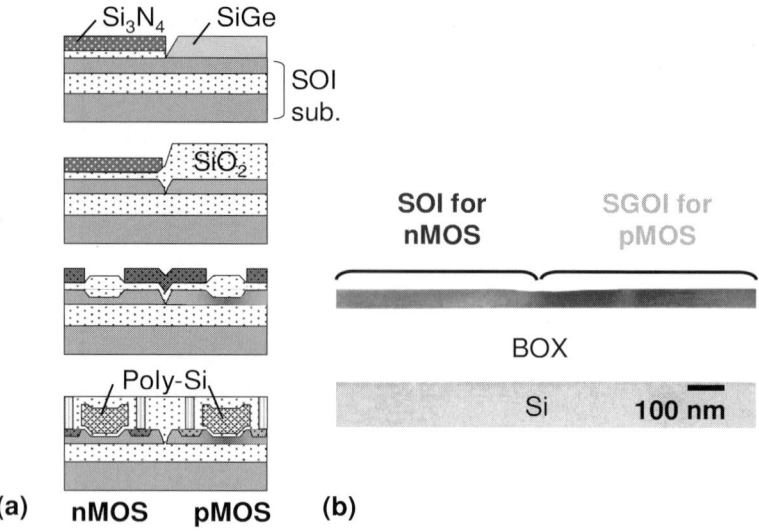

Fig. 9. (a) Fabrication procedure of a ultrathin SOI/GOI dual channel CMOS devices (b) a TEM image of the cross section of SOI *n*-MOS active region and SGOI *p*-MOS active region after the first condensation.

into the SGOI regions. The SOI regions for the active area of *n*-MOSFETs are also thinned by oxidation. A TEM photograph of this hybrid CMOS active area after the first condensation is shown in Fig. 9(b). It is confirmed that the CMOS active area of SOI *n*-MOSFETs and GOI *p*-MOSFETs has been successfully realized with the same thickness and the surface flatness. After growing thin Si layers on the active region of *p*-MOSFETs as the formation of the gate insulators, the *n*- and *p*-MOS channel regions are further recessed down to as thin as 10 nm by local oxidation, in order to form CMOS having ultra-thin body and uniform thickness channels. As a result, it is found that, in spite of ultrathin SOI/GOI channel thickness, the fabricated GOI almost pure Ge channels with the bi-axial compressive strain of 0.3% exhibit the mobility enhancement of 4 against the universal hole mobility, while there is no mobility degradation in SOI *n*-channel MOSFETs.

Furthermore, it is expected from the results of Fig. 8 that CMOS devices comprised of strained SOI nMOSFETs and strained SGOI/GOI pMOS-FETs can provide the best performance with the symmetric layout design, because of the highest and identical current drive in *n*- and *p*-MOSFETs. The schematic cross section of such a CMOS structure is shown in Fig. 10. This dual-channel CMOS structure can be fabricated by applying the local

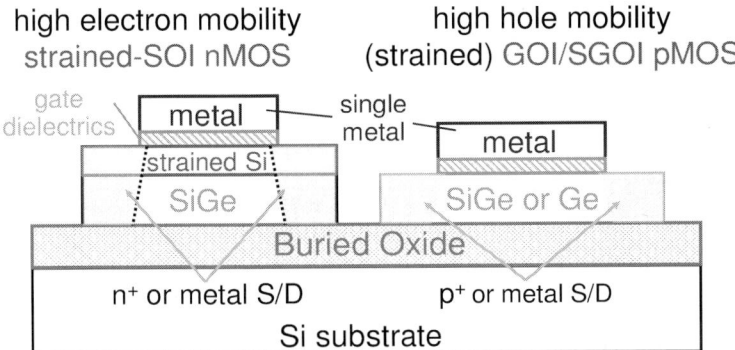

Fig. 10. Cross section of typical dual channel CMOS structures composed of strained-SOI *n*-MOSFETs and strained-GOI *p*-MOSFETs.

Ge condensation technique described above to relaxed SGOI substrates and also combining with selective growth of strained-Si layers on SGOI regions. We have recently succeeded in this integration of strained-SOI *n*-MOSFETs and strained-SGOI *p*-MOSFETs on same SGOI substrates by using the process flow shown in Fig. 11(a).[28,29] First, fully-relaxed SGOI substrates with the Ge content of 14% and the thickness of 160 nm are formed by the global Ge condensation method. Subsequently, recessed SGOI channels with the Ge content of 66% and compressive strain of 1.3% are formed on the *p*-MOS regions by the second condensation process. Strained Si layers with the thickness of 21 nm are selectively grown as channels in *n*-MOS regions on the relaxed SGOI substrates, followed by blanket growth of 5-nm-thick Si cap layers. Gate insulator layers with thicknesses of 20 nm and 30 nm are formed on the *n*-MOS and the *p*-MOS channels, respectively, by oxidizing the whole wafers. Here, whole the Si cap layers and a part of the SGOI layers on the *p*-MOS region are also oxidized, resulting in a surface-channel configuration for both *n*- and *p*-MOSFETs. The oxide thickness difference is due to higher oxidation rate in SiGe than in Si.

Figure 11(b) shows $I_d$–$V_d$ characteristics of fabricated strained-SOI *n*-MOSFETs and SGOI *p*-MOSFETs. The characteristics of control SOI CMOS are also plotted for comparison. It is found that the performances of the hybrid CMOS overcome those of control SOI CMOS. The estimated enhancement factors of the electron mobility of strained-SOI *n*-MOSFETs and the hole mobility of SGOI *p*-MOSFETs against the universal mobility amount to 1.65 and 2.1, respectively, at $E_{eff}$ of 0.5 MV/cm. These results

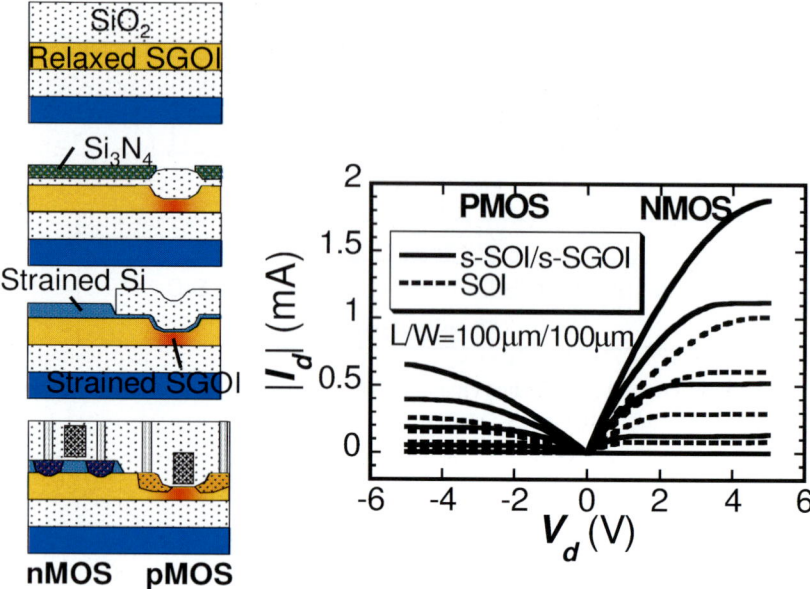

Fig. 11. (a) Fabrication procedure of a strained SOI/strained SGOI dual channel CMOS devices (b) $I_d$–$V_d$ characteristics for the long channel ($L_g = 100~\mu$m) dual-channel CMOS and the SOI-CMOS. |$V_g$–$V_{th}$| values were set to 0–5 V for SOI CMOS and strained SOI *n*-MOS, whereas to 0–7.85 V for strained SGOI *p*-MOS in order to compensate thicker gate oxide due to higher oxidation rate.

indicate that the present dual-channel concept is quite effective for boosting the CMOS performance.

## 3.2. *Formation of global and uni-axial compressive strain SiGe channels*

While the reasons of the recent success in the local strain techniques is attributable to the easier implementation into conventional CMOS processes, lower cost and fewer defects/dislocations than in the global strain techniques, another important advantage of the local strain techniques, particularly in *p*-MOSFETs, is the introduction of uni-axial strain, which is known to be more effective in enhancing hole transport properties than bi-axial tensile strain.[3,18,30] On the other hand, possible drawbacks of the local strain techniques are as follows; (1) the amount of strain is strongly dependent on $L_g$ and other device geometries, leading presumably to complexity of the circuit design and to the increase in the variation of the device characteristics (2) the amount of strain induced by stress linear tends to be smaller with a decrease in the pitch of active regions, associated with device scaling. In order to overcome these possible problems of the local strain technologies, we have proposed a novel technique to introduce un-iaxial compressive strain by using global strain SGOI substrates.[31,32] A unique feature of this technique is the co-existence of global strain and uniaxial compressive strain, which can be advantageous from the viewpoints of both the performance and the robustness against the performance variation.

The key concept for fabricating the present global uni-axial MOSFET is the uni-axial strain formation by using lateral relaxation of bi-axially-strained SGOI, as schematically shown in Fig. 12. Fully-strained SGOI substrates with bi-axial compressive stress[20] are used as starting materials. The SGOI layers are patterned into mesa structures as the active areas of *p*-MOSFETs. Here, the elastic strain relaxation is introduced from the edge of the mesa islands.[33−35] Thus, when the shape of the mesa is sufficiently narrow along the channel width direction and long along the current flow

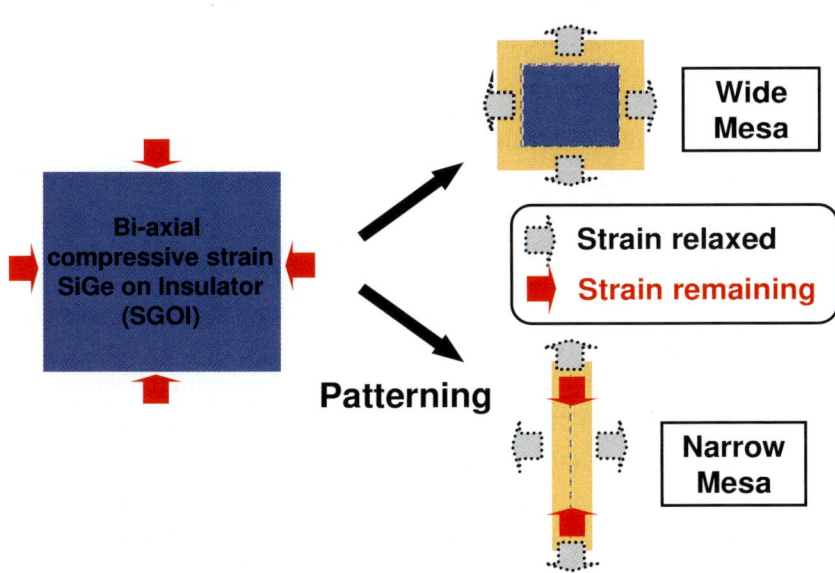

Fig. 12. Schematic illustration showing the concept of uni-axial strain relaxation. Since the elastic strain relaxation occurs from the edge of island, uni-axial compressive stress can be realized in narrow channel devices.

direction, uni-axial relaxation configuration that the compressive stress is applied only along the current flow direction can be realized in the channel.

It is found from the gate width ($W_g$) dependence of the current drive in the linear region ($I_{dlin}$) that $I_{dlin}$ of the SGOI $p$-MOSFETs increases with a decrease in $W_g$ and that the enhancement factor amounts to around 1.8 at $W_g$ of 0.4 $\mu$m. On the other hand, $I_{dlin}$ of the control SOI $p$-MOSFETs slightly decreases with a decrease in $W_g$, attributable to the compressive stress from the shallow trench isolation. Actually, the expected uni-axial strain configuration has been experimentally confirmed for SGOI $p$-MOSFETs with $W_g$ of 0.3 $\mu$m by the nano-electron diffraction method.[31,36] Therefore, the observed $I_{dlin}$ enhancement in SGOI $p$-MOSFETs is attributed to the change in the stress configuration from bi-axial to uni-axial compressive strain, associated with the lateral strain relaxation.

The unique feature of the present uni-axial strain MOSFETs is that the channels can hold the uni-axial strain even for long channel devices, allowing us to directly measure the inversion-layer mobility by using the conventional split C-V technique. It is confirmed from the results of the measured mobility that the mobility enhancement higher than 2 is obtained in high $E_{eff}$ regions, which is a common character in uniaxilly-strained hole mobility.[4,18] Also, the fact that the $I_{dlin}$ enhancement of 1.8 is still kept in $L_g$ of 50 nm and has almost no $L_g$ dependence confirms us that the compressive strain in the channels is global strain. It is found, furthermore, that the saturation current in $p$-MOSFETs with $L_g$ of 40 nm also exhibits 80% enhancement, meaning that this technique is scalable down to 40 nm or less.

Also, this MOS structure is applicable to multi-gate MOSFETs such as FinFETs and Tri-gate FETs. Recently, we have successfully fabricated multi-gate $p$-MOSFETs using uniaxially-strained SGOI channels by just applying the lateral relaxation technique to the Fin and Tri-gate channels.[37] The $G_m$ increases of 200 and 40% in the uni-axial SGOI FinFETs have been obtained against (100) SOI control $p$-MOSFET and SOI FinFETs, respectively, owing to the successful combination of the three factors of the hole mobility enhancement, SiGe channels, (110) surfaces and uni-axial compressive strain.

## 4.    Merging III–V Semiconductor MISFET Technologies into Si Platform

While, as shown in Table 1, there are a variety of ways to enhance the hole transport properties, application of tensile strain is the only available technique for enhancing the electron transport. Also, the amount of the electron mobility enhancement, obtained by this technique, can be regarded as the factor of 2 at maximum. Thus, MOS channels using III–V materials with high bulk electron mobility have recently stirred a strong interest, in order to realize higher electron mobility and resulting higher current of *n*-MOSFETs. Table 2 lists the electron and hole mobilities and the other physical parameters of typical III–V compound semiconductors, Si and Ge, suggesting that the mobility enhancement of the III–V materials can amount to 3–50, at least, in bulk.

In contrast, Ge has the highest mobility among Si and III–V materials. Therefore, the best CMOS structure in terms of the current drive, can be the combination of III–V semiconductor *n*-channel MOSFETs and Ge *p*-channel MOSFETs. On the other hand, CMOS devices aiming at the application under future technology nodes on the Si platform must meet the following requirements; (1) supporting substrates are Si (2) high immunity for short channel effects is maintained. A possible device structures to satisfy these requirements is shown in Fig.13.[2] Here, III–V materials and Ge are formed as Semiconductor-On-Insulator structures on Si substrates, allowing us to minimize the influence of materials that can be impurities from the viewpoints of the Si standard processing and apparatus. In addition, the ultra-thin body structures or multi-gate structures using the thin bodies are effective in suppressing short channel effects. Here, there can be many variations regarding III–V channel formation such as III–V MOS channels expitaxially grown GOI substrates.[38] The technologies can be regarded as local or global/local-merged channel formation techniques. As a consequence, the formation of III–V materials on Si, $SiO_2$ or Ge is a key technology to realize III–V MOSFETs on the Si platform.

One of possible processes to form III–V materials on $SiO_2$ is the microchannel epitaxy and successive lateral over-growth,[39] where windows of $SiO_2$ are opened on Si substrates and III–V materials are epitaxially grown on the limited Si areas. Since the penetration of dislocations generated at III–V/Si interfaces can be blocked by the $SiO_2$ wall and the III–V film surfaces, it is expected to grow III–V-On-Insulator (III–V-O-I) structures without any dislocations on $SiO_2$. Figure 14 shows a TEM photograph of one example of GaAs-On-$SiO_2$ structures grown by this method using molecular beam expitaxy.[40] A two-step growth technique, where GaAs directly touched on (110) Si substrates and upper GaAs layers are grown at lower and higher temperatures, respectively, is employed to pursue for both the selective growth on Si substrates and the suppression of the generation

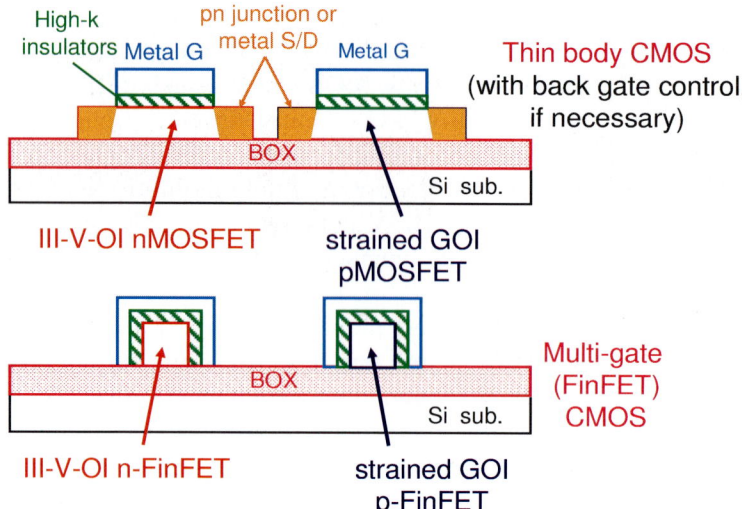

Fig. 13. Ultimate CMOS structure composed of the combination of III–V semiconductor *n*-channel MOSFETs and Ge *p*-channel MOSFETs (a) ultrathin-body CMOS (b) multi-gate CMOS.

of anti-phase domains. However, there are still many issued to be solved for forming high quality III–V-O-I layers, such as (1) further suppression of the generation of dislocations, point defects and anti-phase domains (2) III–V-O-I thickness control under ultra-thin regime (3) controls of surface flatness and edge shapes of III–V-O-I films.

On the other hand, one of the most critical and challenging issues on III–V MISFETs is known to be the realization of high quality MIS interfaces. While the successful operation of inversion-mode GaAs MISFETs has recently been reported,[41] high- mobility and stable III–V MISFETs have not been realized yet. Thus, the establishment of III–V MIS interface control technologies, particularly by using high-k materials,[42,43] is strongly needed. In addition, another essential problem in III–V MIS gate stack structures is the smaller inversion-layer capacitance ($C_{inv}$) and the resulting smaller gate capacitance than those in Si and Ge.[44,45] This is because the light effective masses of III–V semiconductors provide lower density of states and the thick inversion layer thickness, both of which lead to the reduction in $C_{inv}$.[45,46] Since the gate capacitance is the series capacitance of $C_{inv}$ and the gate insulator capacitance, equivalent oxide thickness ($T_{eq}$), where the total gate capacitance is converted into the equivalent $SiO_2$ thickness, becomes thicker in III–V MIS structures under a given thickness of

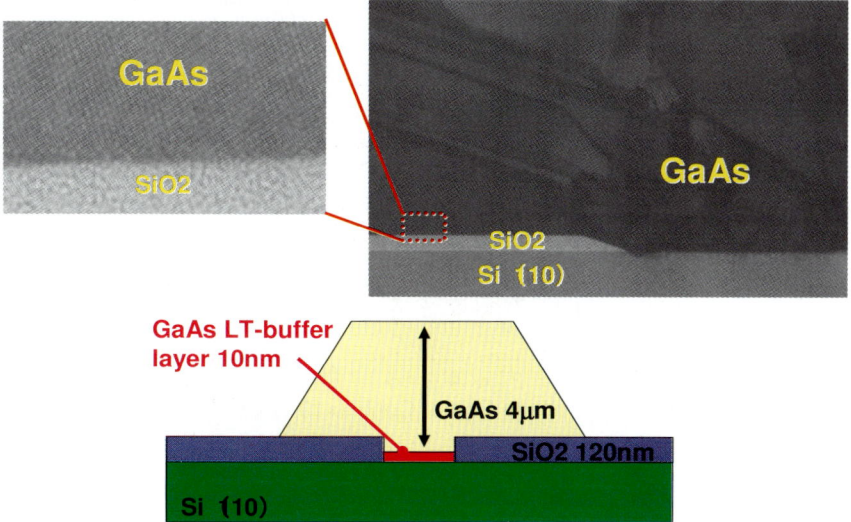

Fig. 14. Cross-sectional TEM microphotograph of GaAs on a Si substrate with SiO$_2$ ll mask. There are no dislocations in GaAs on SiO$_2$. The lattice image of GaAs on SiO$_2$ is clearly observed.

gate insulators than that in Si and Ge. Figure 15 shows the calculated results of the increase in T$_{eq}$ due to C$_{inv}$ as a function of surface carrier concentration (N$_s$). It is found that the increase in T$_{eq}$ is much thicker in the III–V materials. This thicker T$_{eq}$ can seriously affect the current drive in thin gate insulators, because the increase in T$_{eq}$ reduces N$_s$ at a given gate voltage and the resulting current drive. This fact suggests that comparatively-thick gate insulators are more suitable for III–V MIS channels. Also, ultrathin III-V-O-I channels, as shown in Fig. 14, or ultrathin quantum well structures can be expected to mitigate this degradation of the current drive to some extent.

## 5.    Conclusions

Continuous and successive enhancement of carrier transport properties is needed for boosting the CMOS performance beyond sub 100 nm technology nodes, because of physical limitations on CMOS scaling. Thus, channel strain/material engineering keep being mandatory for realizing high performance advanced CMOS. For this purpose, optimization of strain, surface orientation and channel materials including Ge and III–V materials will

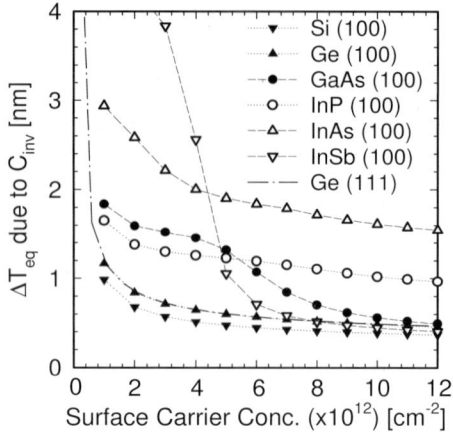

Fig. 15. Calculated results of increase in equivalent gate oxide thickness due to $C_{inv}$ as a function of surface carrier concentration, $N_S$.

be pursued through local and global process/device engineering and its combination.

Particularly, a device family including strained SOI, SGOI and GOI structures enables us to optimally design a wide variety of new CMOS structures so as to maximize the performance. Uni-axial compressive strain and Ge channels are quite effective in pMOS performance enhancement. The combination of global and local channel formation engineering, allowing us to separately optimize the strain configurations and channel materials for *n*-MOSFETs and *p*-MOSFETs, is one of useful channel design concepts.

On the other hand, III–V material channels can be one possible choice for nMOS performance enhancement. While ultrathin body III-V-O-I MOS-FETs or FinFETs are promising as electron-transport-enhanced MOS channels, an attention has to be paid to the thicker inversion layers of III–V materials associated with the light effective masses for the device design.

## Acknowledgments

The authors would like to thank T. Maeda, N. Hirashita, Y. Moriyama, A. Tanabe, Y. Miyamura, E. Toyoda, K. Ikeda, N. Taoka, Y. Yamashita, M. Harada, T. Yamamoto, Y. Shuto, S. Ohya and M. Tanaka for their cooperation and M. Hirose, T. Kanayama and T. Masuhara, for their continuous supports. This work was partly supported by the New Energy and Industrial Technology Development Organization (NEDO) and a Grant-in- Aid for Scientific Research from the Ministry of Education, Culture, Sports, Science and Technology.

## References

1. International Technology Roadmap in Semiconductors (ITRS), http://public.itrs.net/

2. S. Takagi, *Nikkei Micro Devices*, **22**(8) 54 (2005) (in Japanese); S. Takagi, T. Tezuka, T. Irisawa, S. Nakaharai, T. Numata, K. Usuda, N. Sugiyama, M. Shichijo, R. Nakane and S. Sugahara, *Solid-State Electron* **51**, 526 (2007).

3. D. Sadana, Sematech Meeting, "New Channel Materials for Future MOSFET Technology", 04 December (2005).

4. T. Ghani, M. Armstrong, C. Auth, M. Bost, P. Charvat, G. Glass, T. Hoffmann, K. Johnson, C. Kenyon, J. Klaus, B. McIntyre, K. Mistry, A. Murthy, J. Sandford, M. Silberstein, S. Sivakumar, P. Smith, K. Zawadzki, S. Thompson and M. Bohr, *Tech. Dig. IEDM*, 978 (2003).

5. T. Tezuka, N. Sugiyama, T. Mizuno and S. Takagi, *Proc. VLSI Symposium*, 96 (2002).

6. T. Mizuno, N. Sugiyama, T. Tezuka and S. Takagi, *Appl. Phys. Lett.* **80**, 601 (2002).

7. T. Tezuka, N. Sugiyama, T. Mizuno, M. Suzuki and S. Takagi, *Jpn. J. Appl. Phys.* **40**, 2866 (2001).

8. S. Nakaharai, T. Tezuka, N. Sugiyama, Y. Moriyama and S. Takagi, *Proc. Int. Conf. of Si Epitaxy and Heterostructures (ICSI3)*, 129 (2003).

9. T. Numata, T. Irisawa, T. Tezuka, J. Koga, N. Hirashita, K. Usuda, E. Toyoda, Y. Miyamura, A. Tanabe, N. Sugiyama and S. Takagi, *Tech. Dig. IEDM*, 177 (2004).

10. T. Numata, T. Irisawa, T. Tezuka, J. Koga, N. Hirashita, K. Usuda, E. Toyoda, Y. Miyamura, A. Tanabe, N. Sugiyama and S. Takagi, *IEEE Trans. Electron Devices* **53**, 1030 (2006).

11. C. W. Leitz, M. T. Currie, M. L. Lee, Z.-Y. Cheng, D. A. Antoniadis and E. A. Fitzgerald, *J. Appl. Phys.* **92**, 3745 (2002).

12. M. L. Lee, E. A. Fitzgerald, M. T. Bulsara, M. T. Currie and A. Lochtefeld, *J. Appl. Phys.* **97**, 011101 (2005).
13. S. Takagi, J. L. Hoyt, J. J. Welser and J. F. Gibbons, *J. Appl. Phys.* **80**, 1567 (1996).
14. M. Rashed, W. K. Shih, S. Jallepalli, T. J. T. Kwan and C. M. Maziar, *IEDM Tech. Dig.*, 765 (1995).
15. R. Oberhuber, G. Zandler and P. Vogl, *Phys. Rev. B* **58**, 9941 (1998).
16. H. Nakatsuji, Y. Kamakura and K. Taniguchi, *IEDM Tech. Dig.* 727 (2002).
17. T. Mizuno, N. Sugiyama, T. Tezuka, T. Numata and S. Takagi, *IEEE Trans. Electron Devices* **51**, 1114 (2004).
18. S. E. Thompson, G. Sun, K. Wu, J. Lim and T. Nishida, *IEDM Tech. Dig.*, 221 (2004).
19. M. V. Fischetti and S. E. Laux, *J. Appl. Phys.* **80**, 2234 (1996).
20. T. Tezuka, N. Sugiyama, T. Mizuno and S. Takagi, *IEEE Trans. Electron Devices* **50**(5), 1328 (2003).
21. S. Nakaharai, T. Tezuka, N. Sugiyama, Y. Moriyama and S. Takagi, *Appl. Phys. Lett.*, **83**, 3516 (2003).
22. S. Nakaharai, T. Tezuka, N. Sugiyama and S. Takagi, *ECS Symposium on SiGe: Materials, Processing, and Devices, Proceeding Volume 2004-7*, 741 (2004).
23. S. Nakaharai, T. Tezuka, E. Toyoda, N. Hirashita, Y. Moriyama, T. Maeda, T. Numata, N. Sugiyama and S. Takagi, *Ext. Abs. SSDM*, 868 (2005).
24. T. Maeda, K. Ikeda, S. Nakaharai, T. Tezuka, N. Sugiyama, Y. Moriyama and S. Takagi, *Thin Solid Films* **508**, 346 (2006).
25. S. Nakaharai, T. Tezuka, N. Hirashita, E. Toyoda, Y. Moriyama, N. Sugiyama and S. Takagi, *Abs. 3rd International SiGe Technology and Device Meeting*, 208 (2006).
26. T. Tezuka, S. Nakaharai, Y. Moriyama, N. Sugiyama and S. Takagi, *Proc. VLSI Symp.*, 198 (2004).
27. T. Tezuka, S. Nakaharai, Y. Moriyama, N. Sugiyama and S. Takagi, *IEEE Electron Device Letters* **26**, 243 (2005).
28. T. Tezuka, S. Nakaharai, Y. Moriyama, N. Hirashita, E. Toyoda, N. Sugiyama, T. Mizuno and S. Takagi, *VLSI Symp.*, 80 (2005).
29. T. Tezuka, S. Nakaharai, Y. Moriyama, N. Hirashita, E. Toyoda, T. Numata, T. Irisawa, K. Usuda, N. Sugiyama, T. Mizuno and S. Takagi, pusblished in *Semi. Sci. & Technol.*

30. M. Uchida, Y. Kamakura and K. Taniguchi, *Proc. SISPAD*, 315 (2005).
31. T. Irisawa, T. Numata, T. Tezuka, K. Usuda, N. Hirashita, N. Sugiyama, E. Toyoda and S. Takagi, *VLSI symp.*, 178 (2005).
32. T. Irisawa, T. Numata, T. Tezuka, K. Usuda, N. Hirashita, N. Sugiyama, E. Toyoda and S. Takagi, published in *IEEE Trans. Electron Devices*.
33. T. Tezuka, N. Sugiyama, T. Kawakubo and S. Takagi, *Appl. Phys. Lett.* **80**, 3560 (2002).
34. T. Tezuka, N. Sugiyama and S. Takagi, *J. Appl. Phys.* **94**, 7553 (2003).
35. J. C. Sturm, H. Yin, R. L. Peterson, K. D. Hobart and F. J. Kub, *Ext. Abs. SSDM*, 220 (2004).
36. K. Usuda, T. Numata and S. Takagi, *Materials Science in Semiconductor Processing* **8**, 155 (2005).
37. T. Irisawa, T. Numata, T. Tezuka, K. Usuda, S. Nakaharai, N. Hirashita, N. Sugiyama, E. Toyoda and S. Takagi, *Tech. Dig. IEDM*, 727 (2005).
38. M. Heyns, M. Meuris and M. Caymax, *European MRS, Symposium B* (2006).
39. T. Nishinaga, *Journal of Crystal Growth* 237–239, 1410 (2002).
40. M. Shichijo, R. Nakane, S. Sugahara and S. Takagi, *Ext. Abs. SSDM*, **1088** (2006); M. Shichijo, R. Nakane, S. Sugahara and S. Takagi, *Jpn. J. Appl. Phys.* **46**, 5930 (2007).
41. F. Ren, M. Hong, W. S. Hobson, J. M. Kuo, J. R. Lothian, J. P. Mannaerts, J. Kwo, S. N. G. Chu, Y. K. Chen and A. Y. Cho, *Solid-State Electron.* **41**, 1751 (1997).
42. P. D. Ye, G. D. Wilk, J. Kwo, B. Yang, H.-J. L. Gossmann, M. Frei, S. N. G. Chu, J. P. Mannaerts, M. Sergent, M. Hong, K. Ng and J. Bude, *IEEE Electron Device Lett.* **24**, 209 (2003).
43. M. Passlack, R. Droopad, K. Rajagopalan, J. Abrokwah, R. Gregory and D. Nguyen, *IEEE Electron Device Lett.* **26**, 713 (2005).
44. M. Fischetti, *IEEE Trans. Electron Devices* **38**, 634 (1991).
45. S. Takagi and S. Sugahara, *Ext. Abs. SSDM* **1056** (2006).
46. S. Takagi and A. Toriumi, *IEEE Trans. Electron Devices* **42**, 2125 (1995).

# 4

# High-K Gate Dielectrics

Hei Wong*, Kenji Shiraishi[†], Kuniyuki Kakushima[‡],
and Hiroshi Iwai[§]

*Department of Electronic Engineering, City University of Hong Kong,
Tat Chee Avenue, Kowloon, Hong Kong.

[†]Graduate School of Pure and Applied Physics, University of Tsukuba,
Tennodai, Tsukuba, Ibaraki, 305-8571, Japan.

[‡]Interdisciplinary School of Science and Engineering,
Tokyo Institute of Technology, Nagatsuta,
Midori-ku, Yokohama, Kanagawa, 226-8502, Japan.

[§]Frontier Collaborative Research Center, Tokyo Institute of Technology,
Nagatsuta, Midori-ku, Yokohama, Kanagawa, 226-8502, Japan.

*heiwong@ieee.org
[†]shiraishi@comas.frsc.tsukuba.ac.jp
[‡]kakushima@ep.titech.ac.jp
[§]iwai.h.aa@m.titech.ac.jp

··································

To maintain a proper control of the drain current flow in
nanoscale CMOS devices, the thickness of silicon dioxide which
has been used as the gate dielectric material for over four decades
is now pushed into its technological limit of about 1 nm and theoretical limit of 0.7 nm. Further device downsizing would require
even thinner gate dielectric films. This stringent requirement can
only be achieved by using a high-dielectric constant (high-k)
material. High-k gate dielectric together with metal gate electrode has been recognized as an effective technological option
to boost the performance of present integrated circuit technology. However, there are still a lot of issues need to be solved in

order to incorporate this new material into the existing CMOS technology. This chapter reviews the development of high-k gate dielectric materials for nanoscale CMOS device applications. We shall focus on the issues related to the electrical properties and the reliability of high-k materials used as the MOS gate dielectrics.

## 1.    Gate Dielectric Scaling

To maintain proper switching characteristics and to suppress the short-channel effects of MOS transistors, the downsizing on the gate length requires an equal factor of decrease in the gate oxide thickness. In the 65 nm technology node with gate length of 32 nm, the thickness of silicon oxide ($SiO_2$) should be about 0.9 nm. This thickness is already below the technological manageable thickness for mass production (about 1 nm) and is very close the theoretical limit (0.7 nm thick) for bulk silicon dioxide.[1−3] Further device downsizing would require even thinner gate dielectric films which can only be achieved by introducing high-dielectric constant (high-k) materials. By using dielectric with higher k value, a larger value of gate capacitance can be achieved with a thicker film. With reference to the same capacitance value as implemented using silicon dioxide, the effective thickness of high-k dielectric film is reduced by a factor of $\kappa_{ox}/\kappa_{high-\kappa}$ (where $\kappa_{ox}$ and $\kappa_{high-\kappa}$ are the dielectric constant of silicon oxide and high-k material, respectively.) That is the idea of equivalent oxide thickness (EOT). The EOT is defined as

$$\text{EOT} = \frac{\kappa_{ox}}{\kappa_{high-\kappa}} t_{high-\kappa} \qquad (1)$$

where $t_{high-\kappa}$ is the physical thickness of high-$\kappa$ dielectric film.

The introduction of high-k material does not resolve the physical constraint of the oxide thickness for further downsizing, it also help to suppress the large gate leakage current in MOS devices using tunneling gate oxide. Figure 1 illustrates the theoretical leakage current levels of some ultrathin silicon oxide films.[4] According to the theoretical study,[5] the leakage current exceeds $100 \, A/cm^2$ at 1 V gate voltage for 1.2 nm thick silicon oxide. Such large current would produce unacceptable power dissipation in large scale integrated circuits. By using physically thicker high-k materials, this issue has been overcome quite successfully. Figure 2 illustrates the leakage current of some gate dielectric films reported in the literatures. The leakage

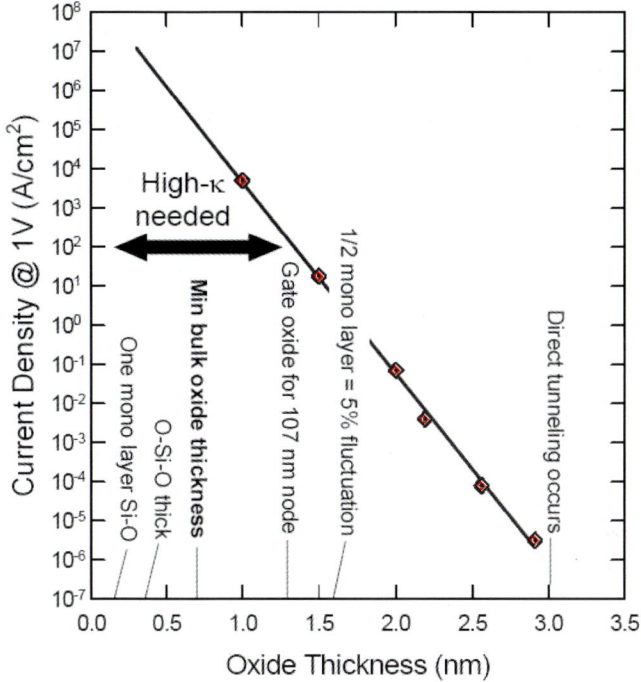

Fig. 1. Direct tunneling current in thin silicon dioxide. Oxide thinner than 1.2 nm would result in too large a gate leakage current and difficulties in process control; high-k material must be used. Reproduced from Ref. 4. Markers are theoretical data.[5]

current level could be reduced by several orders of magnitude by replacing the conventional silicon oxide with high-k materials at the same EOT values.

A good example for the benefit of using high-k material is the Intel Core 2 family of processors fabricated using 45 nm CMOS technology. By adopting hafnium-based high-k dielectric film, the power dissipation of the Core 2 microprocessor has been significantly reduced as compared to the Pentium 4 duo and with significant improvement in speed and some other performances.

## 2.    High-k Candidates

There are many high-k candidates being studied. Ionic metal oxides, having highly polarized metal-oxygen bonds, would have much larger k values

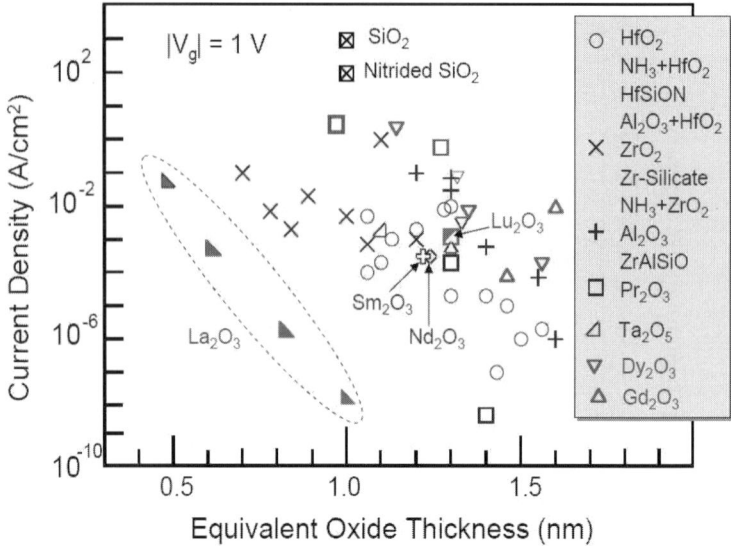

Fig. 2. Example of leakage current reduction by using some high-k materials. Data are taken from various sources.

than that of the covalent dielectric materials. Table 1 lists the major properties and problems associated with some dielectrics proposed to be used for future CMOS technology.[2] Amongst those materials, Hf-based materials, such Hf silicates, Hf aluminates, have been considered as the most promising materials and have already been used in the state-of-the-art CMOS technology.

The high-k materials listed in Table 1 still suffer from several serve problems such as the thermal instability, poor interface properties with silicon, forming interface silicate layers, low mobility, high interface trap density, high oxide trap density and large leakage current. These problems are mainly due to the fundamental properties of the transition or rare earth metal oxides.[2] In transition metals, rare earth metals also have the similar properties, the chemical and material properties are determined by the (n)d-state and (n+1)s-state valance electrons. They can be readily oxidized by transferring these electrons to oxygen 3s or 3p empty orbits and ionic metal-oxygen bonds are formed. The low d-state energy limits the bandgap size of the metal oxides. Bonding with the d-state electrons of the metal, the metal-oxygen bond will be more ionic and require less energy for oxidation. As a result, the metal oxides generally have large amount of oxygen vacancies,

Table 1. Major characteristics and problems associated with the major high-k candidates.[2]

| Dielectric | Dielectric constant (bulk) | Bandgap (eV) | Conduction band offset (eV) | Merits | Drawbacks |
|---|---|---|---|---|---|
| Silicon dioxide ($SiO_2$) | 3.9 | 8.9 | 3.15 | Excellent Si interface, Low $Q_{ox}$ and $D_{it}$ | Low $\kappa$, EOT > 0.8 nm |
| Aluminum oxide ($Al_2O_3$) | 9–10 | 8.8 | | $E_g$ comparable to $SiO_2$, Amorphous Good thermal stability | Medium $Q_{ox}$ and $D_{it}$ medium $\kappa$ |
| Tantulum pentoxide ($Ta_2O_5$) | 25 | 4.4 | 0.36 | High $\kappa$ | Unacceptable $\Delta E_C$, Not stable on Si, |
| Lanthana ($La_2O_3$) | ~27 | 5.8 | 2.3 | High $\kappa$, better thermal stability Low $D_{it}$ | Moisture absorption, instable with Si High $Q_{ox}$ |
| Gadolinium oxide ($Gd_2O_3$) | ~12 | ~5 | # | # | Crystallization |
| Yttrium oxide ($Y_2O_3$) | ~15 | 6 | 2.3 | Large $E_g$ | Low crystallization temperature, hight $D_{it}$, silicide formation |
| Hafnium oxide ($HfO_2$) | ~20 | 5.6–5.7 | 1.3–1.5 | Most suitable compared to other candidates | Crystallization, silicate and silicide formation, |
| Zirconium oxide ($ZrO_2$) | ~23 | 4.7–5.7 | 0.8–1.4 | Similar to hafnia | High $Q_{ox}$ and $D_{it}$ marginal stable with Si, crystallization, silicide formation |
| Strontium titanate ($SrTiO_3$) | ~300 | 3.3 | −0.1 | High $\kappa$ | Unacceptable $E_g$ and $\Delta E_C$, field fringing effect |

*Data from Robertson,[6] Gusev *et al.*,[7] Hubbard and Schlom,[8] and other sources. Slightly different values of those parameters were report time to time.

#Data are not available.

easy to crystallize and higher oxide trap density in the bulk. As the metal elements can also react with the substrate Si atoms at low energy, they produce silicate and silicide bonds. The interfacial metallic silicide bonds, working as interface trap precursors, can also lower the conduction band offset energy. The interface silicate has a lower k value and increases the resultant EOT. The highly-polarized metal-oxygen bonds lead to the high k values and the existence of soft optical phonons, which further induce a large leakage current and channel mobility degradation. The higher degree of ionicity of the metal-oxygen bonds also cause the conduction band to move lower with respect to the silicon conduction band.[2] Those fundamental limitations are difficult to overcome. This chapter aims to review the recent progress on the high-k dielectric research. We shall focus on the electrical characteristics the performance degradations of devices using high-k materials as the gate dielectrics.

## 3.    Nature of Defect Formation

High-k dielectrics are mandatory for further scaling as mentioned before. However, they have several intrinsic problems because of the ionic nature of the chemical bonding.[2] It has been reported that $HfO_2$ contains much higher content of O vacancies than $SiO_2$. In this section, we shall have a close look at the physics of defect formation in $HfO_2$.

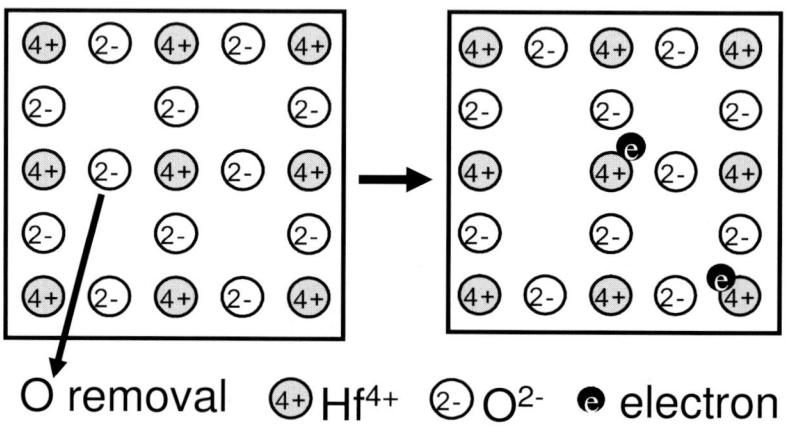

Fig. 3.   Illustrations of O vacancy formation in ionic $HfO_2$.

In the ionic $HfO_2$ crystal, Hf and O atoms are in the form of $Hf^{4+}$ and $O^{2-}$ ions, respectively (see Fig. 3). If an O vacancy is formed by removing an O atom from the $HfO_2$ network, two surplus electrons are generated as illustrated in Fig. 3. First, we investigate the case when the O atoms are in an equilibrium condition. We assumed that the formation and the elimination of O vacancies are balanced by capturing and releasing of O atoms in the network or in the gas phase. That is, the behavior of the electrons in $HfO_2$ is quite important in determining the behavior of the O vacancy. The two electrons generated after the O vacancy formation occupies the empty states at energy levels below the bottom of the $HfO_2$ conduction band.[9] This originates from the increase in the electron entropy as a result of the occupation of the empty states of the $HfO_2$ conduction band. This interesting phenomenon can be understood by comparing the following two reactions involving $O_2$. A neutral O vacancy $(V_O^0)$ can be formed when two electrons are trapped (see the reaction depicted in (2)) or a doubly positive O vacancy $(V_0^{2+})$ may be formed with the contribution of two conduction electrons (see (3)).

$$HfO_2 \leftrightarrow V_O^0 + \tfrac{1}{2}O_2 - \Delta G_1 \tag{2}$$

$$HfO_2 \leftrightarrow V_O^{2+} + 2e + \tfrac{1}{2}O_2 - \Delta G_2 \tag{3}$$

where $\Delta G_1$ and $\Delta G_2$ are the free energies required for forming the oxygen vacancies. Since the energy level of $V_O^0$, $E(V_O^0)$, is located inside the forbidden gap, $\Delta G_2( \approx \Delta G_1 + 2(E_C - E(V_O^0)))$ is larger than $\Delta G_1$ by a value of about $2(E_C - E(V_O^0))$. Here $E_C$ is the bottom of conduction band for $HfO_2$.

According to the mass action law, the $V_O^0$ concentration, governing by reaction (2), can be described as

$$N(V_O^0) \propto \exp\left(-\frac{\Delta G_1}{kT}\right) \tag{4}$$

On the other hand, by considering the facts that $N(e) = 2N(V_O^{2+})$ and $\Delta G_2 \approx \Delta G_1 + 2(E_C - E(V_O^0))$, reaction (3) can be expressed as follows:

$$N(V_O^{2+}) \propto \exp\left[-\frac{\Delta G_1 + 2\{E_C - E(V_O^0)\}}{3kT}\right] \tag{5}$$

As a result, the effective energies for forming $V_O$ according to reactions (2) and (3) are given by $\Delta G_1$ and $\{\Delta G_1 + 2(E_C - E(V_O^0))\}/3$, respectively. Hence, if $E_C - E(V_O^0) < \Delta G_1$, reaction (3) becomes dominant. Otherwise

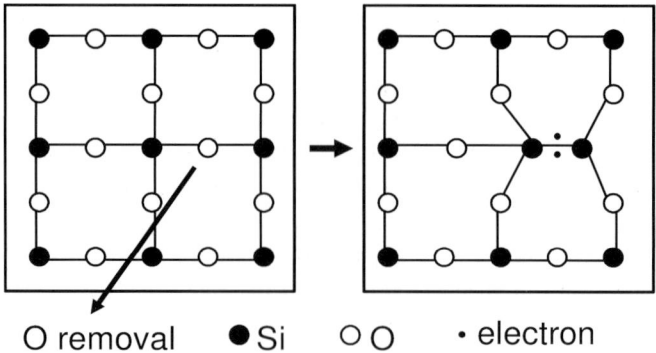

O removal    ● Si    ○ O    · electron

Fig. 4. Schematic illustrations of O vacancy formation in covalent $SiO_2$.

reaction (2) will dominate if $E_C$-$E(\mathbf{V}_O^0) > \Delta G_1$. In $HfO_2$, the calculated $\mathbf{V}_O^0$ formation energy is about 6.4 eV,[10] and the experimentally observed energy level of $\mathbf{V}_O^0$ is about 1.2 eV below the $HfO_2$ conduction band edge.[11] That is, reaction (3) is the dominant reaction in the O vacancy formation in $HfO_2$.

The estimated effective formation energy required to form an O vacancy in an $O_2$ ambient is $\sim$2.9 eV. The situation is quite different in covalent $SiO_2$. In $SiO_2$, the calculated $\mathbf{V}_O^0$ formation energy is about 5.2 eV,[10] and the $\mathbf{V}_O^0$ energy level is about 7 eV below the bottom of the $SiO_2$ conduction band.[12] This relatively lower energy level in $SiO_2$ originates from the fact that the formation of O vacancy induces a great lattice relaxation which enables the generation of a new Si–Si bond as illustrated in Fig. 4.[12] As a result, reaction (2) takes over the vacancy generation and the estimated effective energy for the formation of an O vacancy is about 5.2 eV in $SiO_2$. In summary, the effective forming energy of an $HfO_2$ O vacancy is much lower than that of $SiO_2$. Thus the O vacancy concentration in $HfO_2$ is much higher than $SiO_2$ counterpart regardless that the actual forming energy of an O vacancy in $HfO_2$ is much higher than that in $SiO_2$. The higher concentration of O vacancies originates from the ionic nature of $HfO_2$. From a microscopic view point, this is due to the fact that the relatively higher energy level of O vacancies in $HfO_2$ which lowers the effective $V_O$ forming energy.

## 4.    Dielectric and Interface Trap

The reliability issues of a gate dielectric film, such as threshold voltage shift due to charge trapping and trap generation, leakage current, and dielectric breakdown, are governed by the neutral and charged electronic defects in the dielectric film and at the dielectric/silicon interface. These defects or localized states which can trap electrons or holes and are often termed as trapping centers or simply "traps". In silicon oxide, although it is considered as the best insulator for MOS devices, there are still many kinds of oxide traps and give rise to many reliability problems.[13]

The defect structures in high-k materials are much complicated. In high-k dielectric films, the trap densities are much higher because the high-k oxides are more ionic and less stable. The large amount of oxygen vacancies ($V_O$) is primary source of oxide traps. In addition, the incorporation of Si atoms into the metal oxide networks (at the oxide/Si substrate and oxide/polysilicon interfaces) makes the bonding configuration even more complicated. Because of the different bond lengths, different numbers of bonding coordination, and different strains, significant amount of interface/bulk traps and some trap precursors were found in the high-k metal oxides. For example, on $\langle 100 \rangle$ Si surface, there are several possible bonding structures for hafnium.[14,15] The Si dangling bonds can be either terminated with excess oxygen or excess metal atoms. The oxygen terminated interface poses fourfold coordinated oxygen atoms and would contribute to the insulating property of the oxide films. For metal terminated Si dangling bonds, silicide (Hf–Si) bonds are formed and more interface traps were found. The Hf–Si bonds are amphoteric centers and have an energy level lie in the Si bandgap. The transition layer will be much thicker in the metal terminated high-k/Si interface.

One of the major reasons for having high oxide trap density in high-k materials is that the processing temperature for metal oxide is low ($<700°C$); this makes a large chance for incomplete oxidation and leads to a higher amount oxygen vacancies which produce donor levels in the bandgap and become charge traps in the dielectrics.[16] Forming gas annealing can reduce the measured defect density effectively as a hydrogen atom can substitute the O vacancy and forming more stable $V_O$-H complex and producing a positive fixed charge. This is one of the reasons for high positive fixed charge in $HfO_2$ and other high-k materials. However, hydrogen atoms can also be incorporated into the dielectric films as interstitials and bonded to threefold-coordinated O atoms. When hydrogen is bonded to a fourfold-coordinated O of the oxide network, one of the four metal-O bonds is nearly broken. This would reduce the reliability of high-k materials. It was found that the as-deposited samples have poor Hf/O stoichiometric (oxygen deficiencies).[17] Annealing in oxygen ambient can significantly improve the

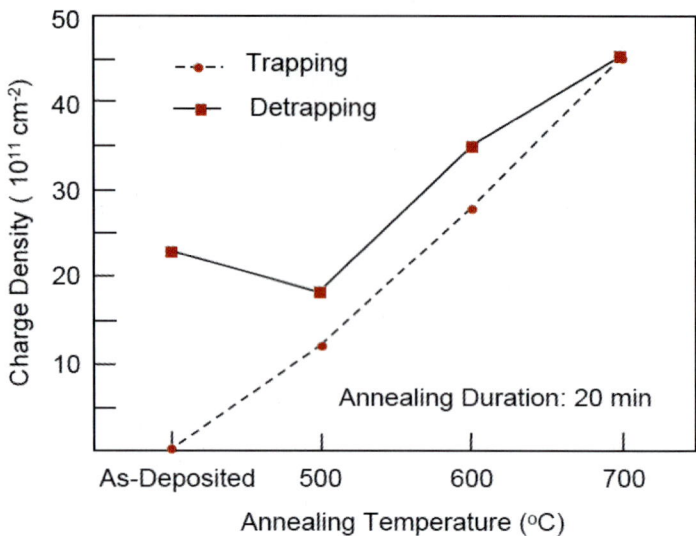

Fig. 5. Charge trapping and detrapping properties hafnium oxide filmes annealed at different temperatures. Redrawn based on Ref. 17.

stoichiometry of the samples, but the amount of the shallow traps could be increased due to the crystallization effects.

Figure 5 illustrates the effects of thermal annealing on the charge trapping and detrapping on hafnium oxide films.[17] The trapping experiments were conducted by constant voltage stressing. For as-deposited samples, most of the trapped charges cannot be discharged in the detrapping experiment. At 700°C, almost all trapped charges were de-charged indicating that the energy levels of these traps are much shallower. However, the amount of charge trapping is much larger than in the case of the as-deposited or low temperature annealed samples.

Significant improvements on both materials and electrical properties were reported by introducing some nitrogen (N) atoms into the hafnium oxide.[18–24] It was found that the nitrogen incorporation can increase the crystallization temperature[23] and the stability against thermal treatment,[22] remarkably. In addition, both the interface and bulk properties can also be improved with the nitrogen incorporation. It was also reported that leakage current can be reduced remarkably with the nitrogen incorporation.[22] The reduction in the leakage current was attributed to the suppression of $V_O$ centers which is considered as the major conduction pathway in $HfO_2$.

Theoretical calculations have shown that the incorporation of N atoms next to the O vacancy can push the vacancy level up out of the gap.[26,27] However, experiments demonstrated that it is hard to incorporate nitrogen atoms into the HfO$_2$ films. Nitrogen incorporate in HfO$_2$ is very low ($\sim$4%) but it distributes quite evenly in the film. This observation was attributed to the uniform distribution of O vacancies in the samples.[28] The incorporated N atoms fill some of the $V_O$ centers in the HfO$_2$ network and replace some of the nearest neighbor O sites to $V_O$. Fortunately the trace amount of N incorporation still gives rise significant reduction on both interface and bulk trap densities. Figure 6 plots the high-frequency (1 MHz) capacitance-voltage (C–V) characteristics for both samples with nitrogen incorporation and without nitrogen incorporation.[29] The nitrogen incorporation was done by plasma immersion ion implantation and the samples were annealed at 800°C in nitrogen ambient.[29] The large shift of the C–V curves and the smooth transition between the depletion and accumulation regions of the samples without nitrogen incorporation indicate that the bulk trap and interface trap densities are very high. The large bulk trap density was attributed to the $V_O$ centers and grain boundary states.[2,17] The $V_O$ centers in HfO$_2$ are electron traps and the energy level is about 0.3 eV in the Si bandgap. In HfO$_x$N$_y$ the $V_O$ level is reduced to about 0.2 eV as results of nitrogen induced bandgap

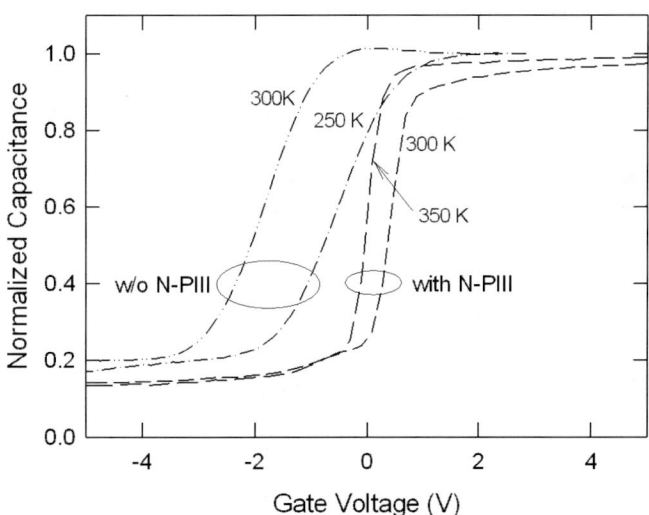

Fig. 6. Effects of ambient temperature on the capacitance-voltage characteristics for hafnium oxide and hafnium oxide with nitrogen implantation.[29]

narrowing and valence band lowering.[30,31] Thus the N-incorporation may help to suppress the leakage current only when the nitrogen atoms involves in either filling the $V_O$ centers or the replacement $V_O$ neighbor O atoms such that neutral $V_O^0$ is converted into positively charged $V_O^{2+}$. The two electrons trapped at the $V_O$ level are transferred to N 2p orbitals at the top of the valence band and the $V_O$ related gap state disappears.[31]

It is further noted that hafnium oxide is also a poor glass former and can be easily crystallized at temperature as low as 325°C.[2] The grain boundary states at micro-crystallites surface have quite shallow energy levels and may not be able to trap any electrons at room temperature. It can be filled with electron and participates in the current conduction at lower temperatures.[32] The large positive shift of the C–V curve measured at 250 K for sample without N implantation can be explained with the shallow trap effect. With nitrogen implantation, the sample has pronounced reduction in the flatband shift of the temperature-dependent C–V characteristics; namely, the amount of bulk traps has been significantly suppressed. In addition, the N-implanted sample has much steeper slope in the transition region of the C–V curves. It indicates that the interface trap density has been reduced to a very low level. This improvement is due to the combined effect of several improvements occurred at the interface.[29] Firstly, the Hf–Si bonds can be converted into Hf–N bond after nitrogen implantation. Unlike the O atom in $HfO_2$ which is six-fold coordinated, the N atom in $HfO_2$ is fourfold coordinated, this will help to reduced the average coordination number at the interface and a better interface is expected.[2] Secondly, trace amount of oxygen released from the nitrogen substitution can react with the substrate Si during the 800°C post-implant annealing and forms an interfacial $SiO_2$ layer.[29] Thirdly, although the separated Si–N phases, which can deteriorate the $SiO_2$/Si interface, were also formed, they still contributed to the interface improvement as Si–N bonding is still better than the Hf–O and Hf–Si bonding at the interface.

Large amount of defect states were also found in $La_2O_3$ films. $La_2O_3$ film poses even larger $k$ value and well suits for the half nanometer EOT applications. Figure 7 compares the 1 MHz C–V characteristics of $La_2O_3$ films at different measurement temperatures.[33] The films were deposited using an MBE system[34] and then annealing in nitrogen at 400 or 600°C. For sample with 400°C post-deposition annealing (PDA), about 1 V negative flatband shift was recorded when lowering the measurement temperature from 350 K to 200 K. For the sample with 600°C PDA, a slight positive shift of the C–V curve was found at the same measurement temperature range.

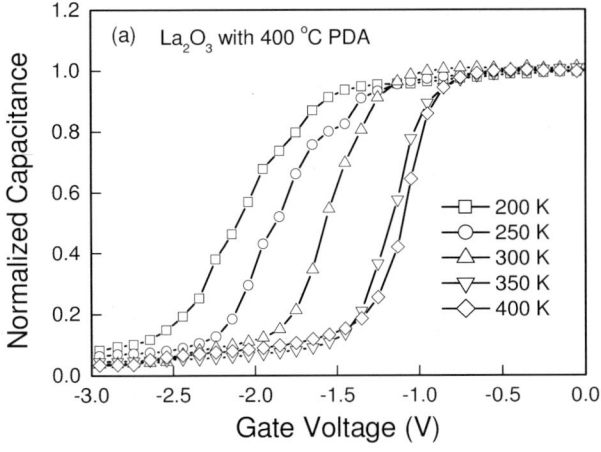

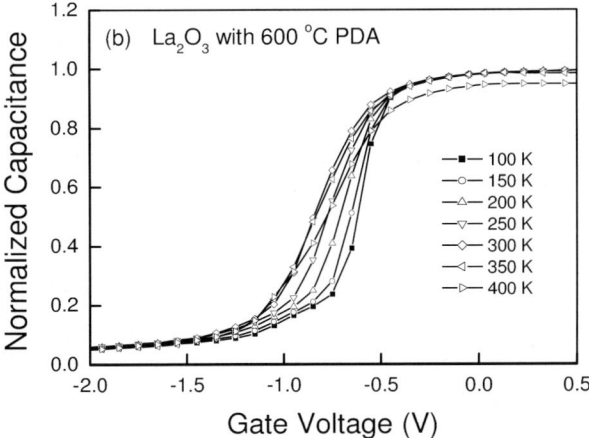

Fig. 7.  Effects of measurement temperature on the capacitance-voltage characteristics. The sample annealing temperatures was (a) 400°C and (b) 600°C. Redrawn based on Ref. 33.

The C–V shift is due to the bulk trap charging with electrons injected from the substrate when the gate voltage is swept from the accumulation to the depletion. The temperature dependence of the C–V characteristics is governed by the energy levels of the traps. For sample with 400°C PDA there exist a large amount of shallow traps[33] or deep traps allowing status transition at small energy. The negative flatband shift, representing generation of large amount of positive fixed oxide charge, are attributed to oxygen vacancies or to the presence of hydroxyl groups in the vacancy sites.[34] O vacancies ($V^0$, $V^+$ or $V^{++}$) are considered as the major electron traps in these materials. The smearing-out effect low temperatures indicates the existence of shallow interface traps.[33]

The behavior of O-vacancies in $La_2O_3$ is quite different in hafnium oxide. The O vacancy levels in $La_2O_3$ (including the positively charged ones) lie above the Si conduction band edge because of the larger conduction band offset of $La_2O_3$ with silicon. The large negative shift of the C–V curve corresponding to the sample treated with 400°C PDA can be explained with the existence of positively-charged vacancies. The origin of positive charges trapping was also attributed to protons captured by $O^{2-}$ or $OH^-$ ions in the $La_2O_3$ films. After higher temperature PDA, e.g. at 600°C, the fixed charge density reduced greatly. The removal of OH groups and O-vacancies may involve the following reaction:[33]

$$La - OH \cdots HO - La \rightarrow La - O - La + H_2O \qquad (6)$$

On the other hand, the growth of interfacial silicon oxide and silicate layers would also help to reduce the oxide charge in the traps.

## 5. Threshold Voltage Control and Fermi Level Pinning

Fermi level pinning has become an important issue for threshold voltage control in actual application of high-k material in CMOS devices. As shown in Fig. 8, it was found that the threshold voltage varies as the hafnium becomes thicker.[35] The threshold voltage values are different between $n^+$ doped poly-Si and $p^+$ doped polysilicon. The difference narrows as the deposition goes on and remains fairly constant as the sub-monolayer region is completely covered. This observation could not be due to the charging effect in $HfO_2/SiO_2/Si$ structure as the bottom $SiO_2$ is quite thick and no reaction was found between the $HfO_2$ and $SiO_2$ during the deposition.[35] The asymmetry threshold voltage shift of $p^+$ and $n^+$ polysilicon gate was explained by the existence of Hf-Si at the polysilicon/hafnium interface. The silicide bonds reduce the degree of the depletion of the gate electrode. *Ab initio* calculations showed that the interaction between Hf and Si atoms could produce surface dipoles at the polysilicon/$HfO_2$ interface which in effect modify the interface barrier height and then the flatband voltage.[36] Another explanation to the large difference of flatband voltages between $n^+$ doped polysilicon and $p^+$ doped polysilicon was proposed recently by

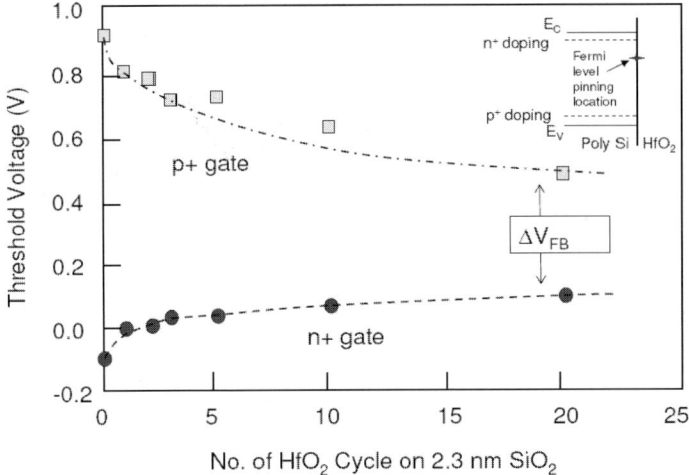

Fig. 8. Asymmetry of the threshold voltage shift during the hafnium oxide growth in n$^+$ (or p$^+$) polysilicon/HfO$_2$/SiO$_2$/Si structure. Inset illustrates the Fermi-level pinning location. Redrawn based on data published by Hobbs *et al.*[35]

Shiraishi *et al.*[37] The forming of an O vacancy would result in the generation of two electrons. If the O vacancy is near the interface, the generated electrons can transfer across the interface to the polysilcion gate electrode and an interface dipole produced in the oxide. This dipole gives rise to the large flatband shift.[37] The Fermi level pinning or interface dipole effect would result in a high threshold voltage (particularly for p-channel devices) and causes some difficulties in logic designs. Similar flatband voltage shift was also found in undoped poly-Si layer with fully silicided gates and it is suggested that the flatband shift should be owing to the presence of fixed charges in the high-k layer.[38] Diffusion of poly-Si dopants into the high-k layer could be the source of the fixed charges and the asymmetry of the flatband shifts can be explained with the different types of atoms used for the polysilicon doping.[39] Nevertheless, issue related to asymmetry flatband shift has annoying the alternative gate oxide researchers as it is difficult to achieve a low workfunction (e.g. <0.2 eV below the conduction band of polysilicon), which is a common figure used in the conventional process. This problem has received significant attentions recently and several methods have been proposed to solve it.[40−42] Replacing the polysilicon electrode by certain metal silicide can solve this problem.

Fermi level pinning also occurs in metal gate electrodes after a high temperature treatment when a thin $SiO_2$ interface layer exists between the high-k and the Si substrate.[43] This is called $V_{fb}$ roll-off. The Fermi level pinning in high work function metal (p-metal in short) is similar to that of $p^+$-doped polysilicon gate.[44] Fermi level pinning of $p^+$ gate and p-metal gate can be systematically explained with the O vacancy model.[44] Figure 9 illustrates the difference of Fermi level pinning for $p^+$ gate and p-metal gate. As shown in Fig. 9, both O and electron transport from the high-k to the gate electrode. Although detailed interface reaction is different each other, main interface reaction can be described by the same reaction equation as follows,

$$HfO_2 + \tfrac{1}{2}Si \leftrightarrow \tfrac{1}{2}SiO_2 + HfO_2 + V_O^{2+} + 2e \tag{7}$$

This naturally leads to the conclusion that pinning positions of a $p^+$ gate and p-metal gate stacks are almost the same. In the pinning situation, the interface reaction is under thermal equilibrium as shown in Fig. 10. This thermal equilibrium condition can be determined by the intrinsic nature of $HfO_2$ and Si. As a result, the position of Fermi level pinning almost neither depends on the film quality nor the processing conditions. It has been

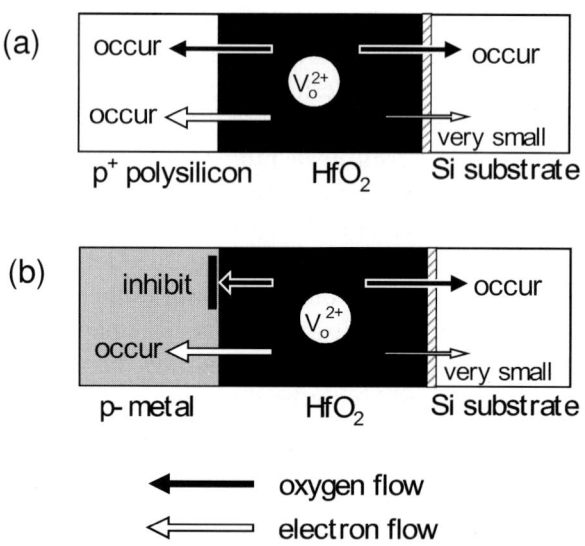

Fig. 9. Illustration of the different interface reactions between $p^+$-doped polysilicon (a) and p-metal gates with thin interface layer (b).

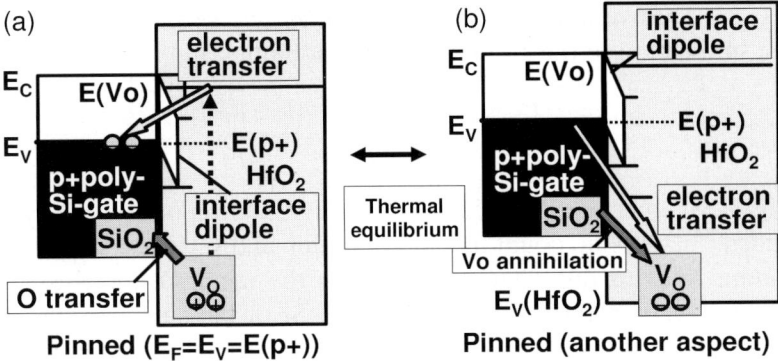

Fig. 10. Illustration of Fermi level pinning: (a) $V_O$ generation, and (b) $V_O$ annihilation are balanced, and the system reaches thermal equilibrium.

reported experimentally that injection of oxygen into the p-metal gate stacks can recover the Fermi level pinning.[45,46] Hence, if (7) governs the Fermi level pinning, following recipes should be effective in suppressing the Fermi level pinning. One is low temperature process which inhibits the system to reach thermal equilibrium. The use of FUSI can be categorized into this recipe.[47] In fact, it has been reported that high temperature treatment really causes Fermi level pinning in NiSi metal gates.[48] Another possibility is to change the thermodynamics of interface reactions. For example, the use of other high-k dielectrics with different interface thermodynamics is also hopeful.[49] Further, the interface dipole modulation between high-k dielectrics and Si substrates are also effective, since this modulation does not change the thermodynamics of interface reaction that governs the Fermi level pinning. F incorporation into Si substrate[50] or counter doping effects are categorized in this recipe which modulates the dipole at interfacial layer/Si interfaces. It has been also proposed that Al and La incorporation into Hf-related oxides can modulate the dipole at high-k/interfacial layer interfaces. Those measures also fall in this category.[51]

## 6.    Channel Mobility

The surface mobility is governed by various scattering mechanisms at the bulk silicon and at the dielectric/Si interface. The major scattering mechanisms affecting the channel mobility at the $SiO_2$/Si interface are the Coulomb ($\mu_{Coul}$), surface roughness ($\mu_{SR}$) and phonon scattering

$(\mu_{Ph})$.[52,53] The overall effective channel mobility is described by the Mathiessen summation of the aforementioned mobilities and is given by:

$$\frac{1}{\mu_{eff}} = \frac{1}{\mu_{Coul}} + \frac{1}{\mu_{SR}} + \frac{1}{\mu_{Ph}} \tag{8}$$

At high-k/Si interface, the channel mobility was reported to be greatly degraded.[54,55] It is no doubt that the Coulomb and surface roughness play important roles to this degradation. Since the metal-O, metal-Si generally have longer bond lengths than the Si–Si of the substrate, the metal oxide/Si interface would have higher degree of roughness. On the other hand, as mentioned in Sec. 4, high-k oxides have much higher oxide trap and interface trap densities than $SiO_2$, the Coulomb scattering would be more pronounced when compared to the $SiO_2$ case. The density of soft optical phonons should also be high in the high-k metal oxide because of the ionic bonds.[2] These phonons will interact with the channel electrons and produce mobility degradation. Other factors such as remote-charge scattering and top-surface roughness scattering may also induce mobility degradation.[56] The scattering mechanisms behave differently with temperature variations. Surface roughness scattering has a weak temperature dependence.[57] The Coulomb scattering has a positive temperature coefficient and the phonon scattering has a negative coefficient. Figure 11 shows

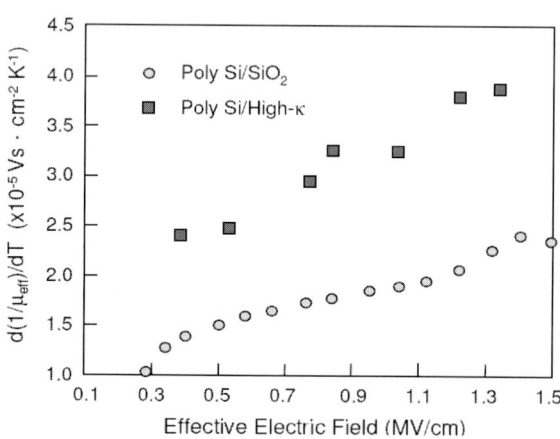

Fig. 11. In high-k samples, the temperature coefficient of the channel mobility is significantly larger than in the silicon oxide case. This is the evidence of existence of phonon scattering at the high-k/Si surface. Redraw based on Ref. 54.

the temperature coefficient $d(1/\mu_{eff})/dT$ as a function of the channel electric field.[54] When compared with the $SiO_2$ case, the temperature coefficient is significantly larger which implies that the phonon scattering contributes significantly to the mobility degradation in high-k/Si surface.

As depicted in Fig. 12, the channel mobility can be enhanced by using mid-gap metal gate electrode to screen the surface phonon scattering.[54] However, it was found that a lot of gap sates could be induced by the metal electrode. The mobility was also found to be thickness dependent. The mobility reduces as the hafnium oxide become thicker (see Fig. 13).[16] This observation can be explained with the effect of reduced metal gate screening and the increased charge trapping in thicker $HfO_2$ film. The thickness dependence charge trapping effect is illustrated in Fig. 14.

To enhance mobility and to reduce the interaction between $HfO_2$ and polysilicon, the conventional polysilicon gate electrode can be replaced with a metallic gate electrode. Figure 15 depicts the mobility curves for various $HfO_2$ gate dielectrics and gate electrode materials. The mobility of TaN-gated is significantly higher as results of increased screening of the remote-charge scattering effect by the metal gate.[54] The channel mobility can also be improved by introducing silicon, nitrogen or aluminum at the cost of a lower dielectric constant. The mobility of device with HfSiON is

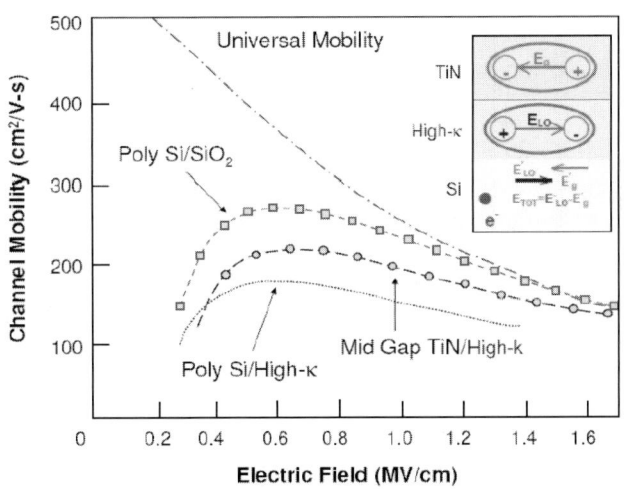

Fig. 12. Significant mobility reduction is reported for the device with high-*k*/poly-Si. The mobility can be improved with mid-gap metal gate electrode such as TiN by screening the phonons. Redraw based on Ref. 54.

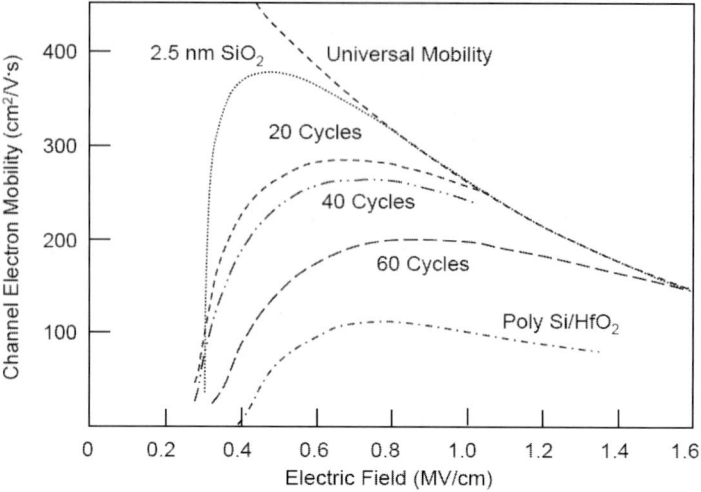

Fig. 13. Plot of effective mobility as a function of effective field for MOS transistor with different gate dielectric films. Redrawn based on Ref. 16.

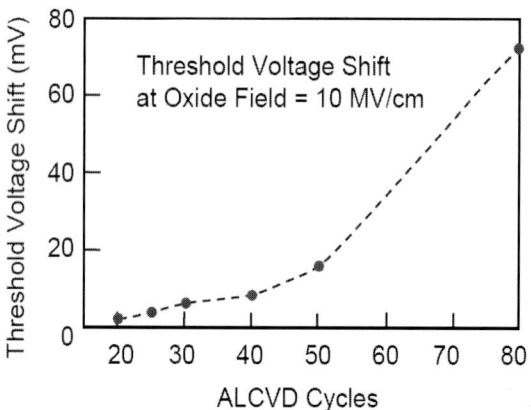

Fig. 14. The amount of oxide charge increases as the $HfO_2$ film becomes thicker. Redrawn based on Ref. 16.

much larger than the $HfO_2$ case because of lower interface trap density.[12] An excellent way to boost the channel mobility is the use of strained Si. As depicted in Fig. 15, the peak channel mobility can be maintained at a value over $300\,cm^2/V$-s by using $TiN/HfO_2/Strained$ Si structure. This

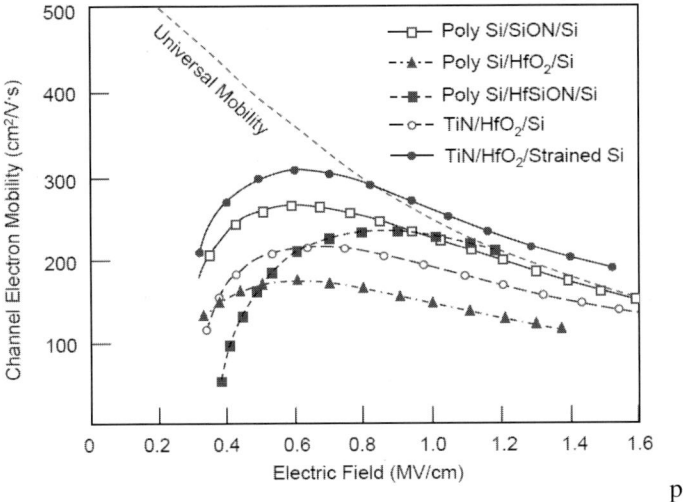

p

Fig. 15. Significant mobility reduction is reported for the device with high-k/poly-Si. The mobility can be improved with mid-gap metal gate electrode such as TiN by screening the phonons, hafnium silicate, and strained Si substrate. Data are taken from Refs. 16 and 54.

combination has been considered as the promising technological option for a CMOS device beyond the 45 nm technology node.

Figure 16 shows the effective mobility of MOSFET using $La_2O_3$ as the gate dielectrics.[34] With proper post-depostion annealing, the mobility can be improved significantly. The peak channel mobility of $La_2O_3$-gated transistor can be improved up to 261 cm$^2$/Vs at with 300°C PDA. This improvement should be related to the forming of interface silicate layer and the removal of oxide trap. The significant optical phonon scattering is another major cause for the mobility degradation. As depicted in Fig. 16, the mobility can be further improved by conducting post-metallization annealing (PMA) instead of PDA. This improvement can be explained with the reaction of Al gate metal and $La_2O_3$. The participation of Al atoms would reduce the ionicities of the bonds and then the optical phonon generation and finally leads to higher channel mobility for the PMA sample.

## 7.    Leakage Current

From the EOT point of view, high-k materials have much smaller leakage current in ultrathin EOT range. Yet this is not a good comparison because

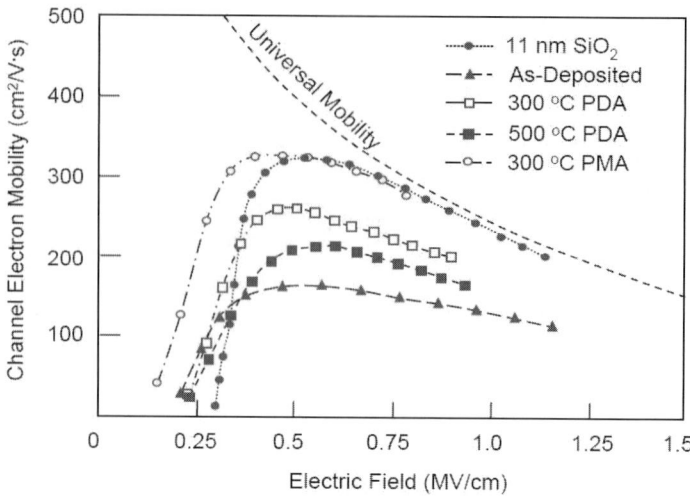

Fig. 16. Effective mobility of MOSFET with conventional $SiO_2$ and $La_2O_3$ (with or without PDA) as the gate dielectric. Higher effective mobility was achieved on MOSFET with PMA. Redrawn based on Ref. 34.

the silicon oxide in this thickness range is well below the direct tunneling limit and the high-k oxide is not.[2] The conduction band offsets ($<2\,eV$) of high-k metal oxides are generally much smaller than that of the silicon dioxide. As the direct tunneling current is exponentially governed by the tunneling barrier, when the thickness of those high-k materials are reduced close to the direct tunneling limit and the voltage across the oxide gets beyond the barrier energy, the leakage current will greatly increase.

In thin high-k metal oxides, it is often found that the measured leakage current is several orders of magnitude larger than the theoretical Fowler-Nordheim (FN) curve.[33,58–59] There are several physical mechanisms were proposed to explain the excess current in high-k metal oxides. It was found that the leakage current and the carrier emission rate in high-k oxides are strongly governed by the temperature and by the electric field and it is suggested that phonon-assisted tunneling should participate in the current conduction. The channel carriers can be polarized by the ionic metal-oxygen bonds and thus optical phonons are induced. The phonons interact with the electrons injected into the localized states of the dielectric and assist electrons in the tunneling process. In the phonon-assisted tunneling mode, the electrons do not enter the conduction band of the dielectric. Grain boundary conduction due to the polycrystalline dielectric film is another possible

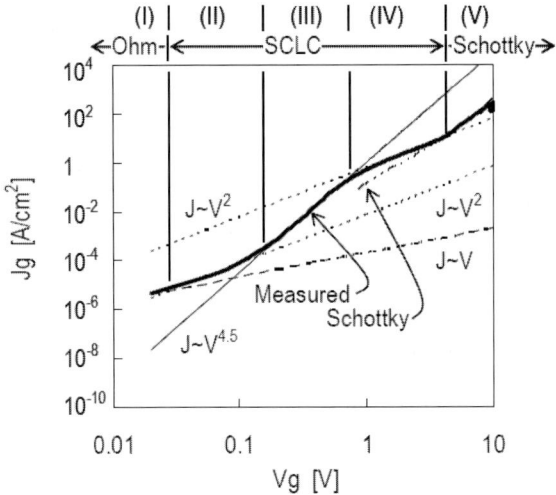

Fig. 17. An example showing the ohmic and space charge limited current conduction characteristics in a high leakage $La_2O_3$ film. Reproduced from Ref. 61.

leakage mechanism.[60] In some high trap density samples with positive flatband voltage, ohmic and space charge limited current (SCLC) conductions were also found (see Fig. 17).[61]

Probing the current-voltage characteristics of high-k oxide at different ambient temperatures is a better way to differentiate the current conduction mechanisms. Figure 18 plots the temperature-dependent current density of a $La_2O_3$ film as a function of applied voltage.[33] Although the current-voltage characteristics still follow the exponential behavior of the Fowler-Nordheim (FN) conduction mechanism, the value of the barrier extracted from the FN plot is too small and the temperature dependence is too strong to be explained by the FN mechanism.[58] Poole-Frenkel (PF) conduction mechanism cannot explain the present results either. These observations resemble the ones obtained for hafnium oxide.[2] A possible mechanism that leading to the strong temperature dependence is the thermal-assisted PF tunneling which mainly occurs at temperatures higher than 300 K. Even at low temperatures, the PF tunneling can be activated by the multi-phonon trap ionization.[2] Since the leakage current and hence the emission rate are strongly governed by the temperature and the electric field variations, it is reasonable to assume that phonon-assisted tunneling participates in the current conduction. Lanthanum oxide, having a k value of about 27 and is

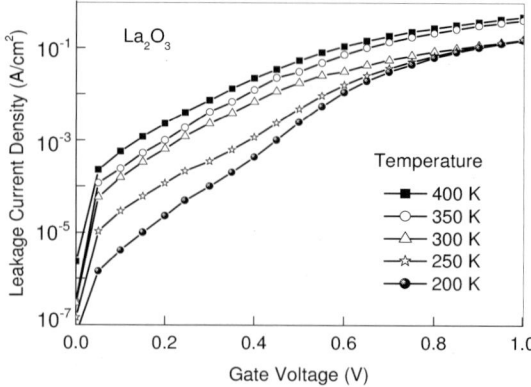

Fig. 18. Plot of leakage current characteristics as a function of the applied voltage at different sample temperatures. The sample was annealed at 400°C in nitrogen ambient. Redrawn based on Ref. 33.

more ionic, should be able to generate high amounts of optical phonons in the Si substrate.[2] The phonons interact with the electrons injected into the localized states of the dielectric and assist the inelastic tunneling of electrons.[56]

It is found that the dielectric constant and the barrier height extracted from the current-voltage characteristics often depart from the nominal values collected from other studies.[45,46] These outcomes can be explained with the two layer model of current conduction.[62] Because of the presence of an interfacial layer, with much smaller $k$ value, between the high-k/Si interface, the dominating portion of the applied electric field will be soaked up in the low-k interfacial layer. Consequently, the interface barrier is reduced and results in significant different tunneling characteristics. Figure 19 plots the effective barrier and the effective thickness of this structure with different combinations of the physical thickness of the interfacial oxide layer and the $HfO_2$ layer.[62] The effective-barrier decreases quickly as the $HfO_2$ layer grows thicker. For a thin $HfO_2$ layer, the effective thickness predicts a much smaller value than that of EOT. For the thicker $HfO_2$ film, the effective thickness exceeds the EOT, indicating that the tunneling current can be effectively suppressed if the operation voltage is low enough. According to this study, a 2.4-nm-thick stack (1.4 nm $HfO_2$ +1.0 nm $SiO_2$) may have a tunneling current larger than a 1.2-nm-thick single-layer $SiO_2$ for a large applied voltage.[62] It should also be noted that in high-k material, the values of the effective mass of the carrier may also vary greatly for different

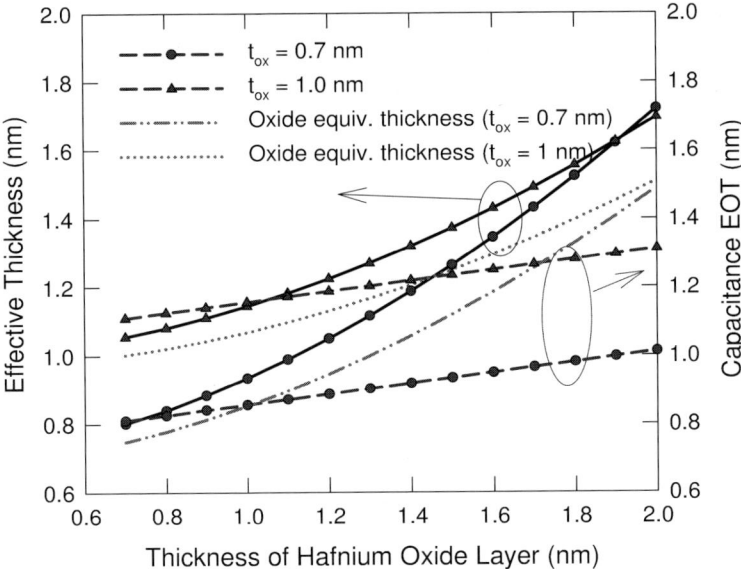

Fig. 19. Theoretical plot of the effective barrier and effective thickness of dual-layer $SiO_2/HfO_2$ film as functions of oxide and $HfO_2$ layer thicknesses. Adopted from Ref. 62.

materials and for same materials but in different modifications.[63] Density function calculations indicate the effective electron mass in different crystalline forms of $HfO_2$ may vary from $0.7\,m_0$ to $2.0\,m_0$ ($m_0$ is the mass of free electron in the vacuum). The effective mass of hole is in the range of $0.3$ to $8.3\,m_0$ depending on the crystalline structure of $HfO_2$ film.[63]

## 8.    Breakdown

High-k metal oxides are often found to have low breakdown field when compared to silicon oxide.[64−66] In high-k metal oxides, the local electric field is substantially larger than the applied electric field because of the polarization effect. This polarization effect is directly proportional to the dielectric constant.[2] The large local field distorts the molecular bonds and makes them more susceptible to breakage. According to McPherson *et al.*,[67] the intrinsic dielectric breakdown is given by

$$E_{BD} = \frac{\Delta H_0}{p_0(2 + \kappa)/3} \tag{9}$$

where $H_0$ is the activation energy required for metal ion displacement and $p_0$ is the molecular dipole-moment component opposite to the local field which is governed by the valence state, number of active dipole and bonding component.

As the dielectric constant is also a function of bandgap energy, the breakdown voltage also correlates very well with the bandgap energy if we separate the homopolar and heteropolar materials (see Fig. 20).[2] That is, there exists intrinsic breakdown field for each of the high-k materials. For $HfO_2$ film, the intrinsic breakdown field is around 4 MV/cm. However, the actual breakdown mechanism in high-k/Si structure is quite complicated. It was found in hafnium film prepared by direct sputtering method that the oxide exhibited a number of soft breakdowns before a hard breakdown to occur (see Fig. 21).[64]

Based on the time-dependent-dielectric breakdown (TDDB) results, it was found that the Weibull shape factors for soft and hard breakdown are 1.43 and 1.95, respectively.[64] These values are slightly smaller than other reports.[65] The different values of Weibull shape factor for soft and hard breakdown suggest a different scenario from the breakdown mechanism of silicon oxide. This observation is explained with the two-layer model of dielectric breakdown.[64] Since a low-k silicate interfacial layer was found between the hafnium oxide and silicon substrate, the applied electric field across this high-k/low-k stack will be largely distributed in the low-k region

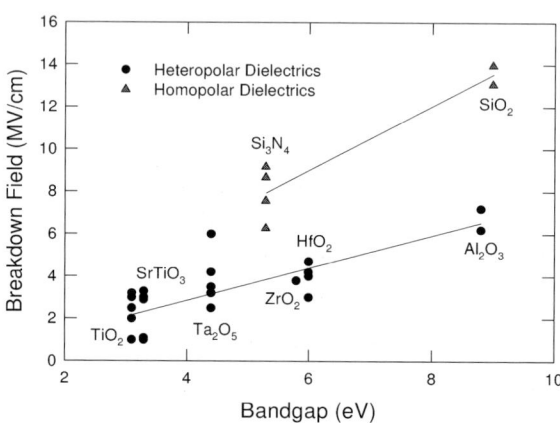

Fig. 20. Good correlations between the breakdown voltage and bandgap are obtained by separating the homopolar and heteropolar dielectric films. Reproduced from Ref. 2 and the experimental data are taken from various source.

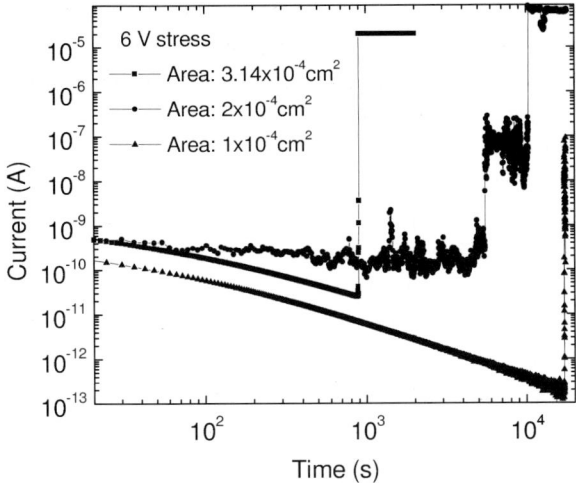

Fig. 21. Time-dependent dielectric breakdown characteristics of the capacitor with area of $3.14 \times 10^{-4}$ cm$^2$ at different stressing voltage; and area dependent TDDB of three different size capacitors stressed at 6 V. From Ref. 64.

according to the Gauss's law (i.e. $E_{low-k}/E_{high-k} = \kappa_{high-k}/\kappa_{low-k}$). In addition, the critical defect density for causing the low-k layer to break down is much lower because it is much thinner than the bulk high-k layer. As a result, the soft breakdown takes place in the low-k layer before the hard breakdown of the bulk HfO$_2$ layer.

## 9. Hot Carrier Effect and Negative Bias Temperature Instabilities

In small-sized MOS devices, hot carrier reliability was recognized as a serious issue as the channel electric field near the drain region is significantly larger than the critical field for impact ionization.[68–73] Hot carrier reliability is also a serious concern for high-k dielectric materials, not only because the new dielectric materials will be used in the ultimate nanoscale devices but also due to the weaker bond strengths of the metal oxides as compared with the conventional SiO$_2$. In addition, as the band offsets of high-k metal oxides are much smaller than SiO$_2$, greater hot-carrier induced degradations are expected. The hot carrier induced degradation in high-k materials are much complicated than the SiO$_2$ or oxynitride films.[74] It was reported that the threshold energy for hot-electron damage in HfO$_2$ is about 3.8 eV which is much smaller than that of silicon oxide. However, the capture cross-section of the hot carrier induced traps is found to be in the range of $10^{-16}\,cm^{-2}$ which is in the same order of magnitude as the neutral traps in $SiO_2$.

Negative bias temperature instability (NBTI) in p-channel transistors is another major concern.[75] Pronounced threshold voltage shift was found in p-channel MOS when the transistors are subjected to negative gate bias stressing at elevated temperature in the range of 100 to 150°C. This degradation leads to instabilities and failures in both analog and digital circuits. This instability was attributed to Si dangling bonds or $P_{b0}$ center near the interface region in $SiO_2$ or in silicon oxynitride gate dielectric. These defects can be effectively passivated by hydrogen atoms introduced during the forming gas annealing. However, at high field stressing hole capturing at the interface would result in the decomposition of hydrogen bonds. The hydrogen species will then diffuse away the interface at the elevated temperature and results in the threshold voltage shift.[70,75] Similar phenomenon is also observed in high-k gate dielectric[76–78] but the physical origins of the NBTI are much complicated. Figure 22 shows an example of threshold voltage shift of a PMOS with $HfO_2/SiO_2$ stack stressed at 125°C for different durations.[79] Significant negative threshold shift are recorded indicating the positive charges buildup at the NBT stress. The positive charges buildup can be due to trapping of positive species at pre-existing defects. As mentioned

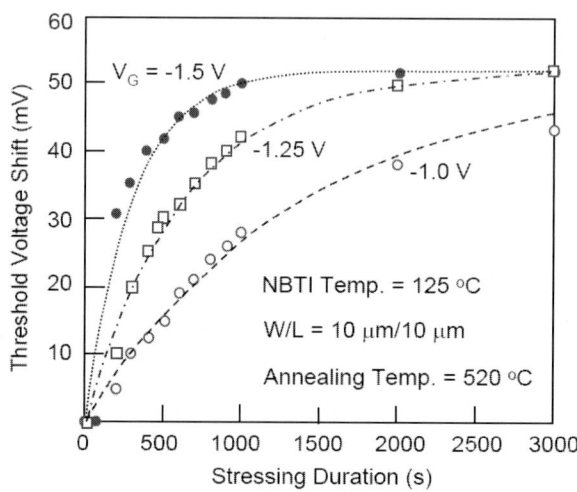

Fig. 22. Negative bias temperature instability of a PMOS transistor with $HfO_2/SiO_2$ stack. Redrawn based on Ref. 79.

in Sec. 4, high-k materials are often found to have higher bulk oxide trap and interface trap density. Oxygen vacancies which are hole trapping centers should be responsible to the threshold voltage shift during negative bias stressing. In high-k/$SiO_2$ stack the hydrogen atoms diffused from the $SiO_2$ layer during the negative gate bias stressing can fill up the oxygen vacancies in high-k layer and result in the formation of positive fixed charges. Because of the lower conduction band and valance band offsets, a significant portion of trapped charges can be readily depopulated. The stress-induced threshold voltage shift can be due to the accumulation of reversible charges instead of defect generation mechanisms.[77] It was found that the oxide trap and interface trap recoveries are about 40% and 25% respectively for HfSiON film.[78] It was also found that the higher initial trap density would lead to larger threshold voltage shift during the stressing.[76] This is an indirect evident for this conjecture. Another explanation of the NBT induced positive charge buildup observed in $SiO_2$/$HfO_2$ stacks is attributed to forming of over-coordinated oxygen centers as a result of proton trapping at strained bonds (e.g. Hf–O–Hf or Hf–Hf).

Using metal gate electrode also has significant improvements on the NBTI characteristics. Figure 23 compares the NBT stressing effects on the polysilicon and TaN-gated devices as a function of oxide field.[16] At low

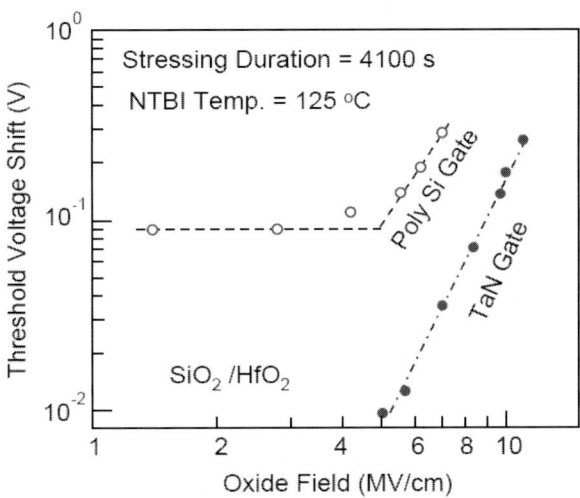

Fig. 23. Effects of gate electrode on NBT stress on the threshold voltage shift in MOS device with $SiO_2$/$HfO_2$ gate stack Ref. 16.

electric field ($<4$ MV/cm) stressing, the threshold voltage shift of polysilicon gated device remains at a constant level of about 90 mV. The defects responsible to this degradation were ascribed to the hydrogen-related bulk traps. At larger electric field, the threshold voltage shift increases exponentially due to the generation of $P_{b0}$ centers at the interface during the NBT stress. Whereas in TaN-gated device, the low-field NBT induced threshold voltage shift is much smaller. It suggests that the precursors of hydrogen-related defects are arising from the polysilicon/$HfO_2$ interface or during the polysilicon deposition process. Oxygen vacancies and Hf–Hf strained bonds may be produced in the high-temperature ($>600°C$) and reduced-pressure ambient for the polysilicon deposition. Meanwhile, the polysilicon/$HfO_2$ interface defects may also involve in the excessive NBTI degradation in the polysilicon-gated devices.

Figure 24 shows the flatband shift before and after constant-voltage stressing with positive gate bias of 1.0 V.[33] PDA temperature has profound effects on the stressing-induced flatband voltage shift. The sample with 400°C PDA in $N_2$ ambient has a much larger flatband shift during stressing than that of the sample with 600°C PDA. This result indicates that the 400°C PDA is not enough to remove the oxide traps or weak bonds in the dielectric film. For the sample with 600°C PDA, a small negative flatband

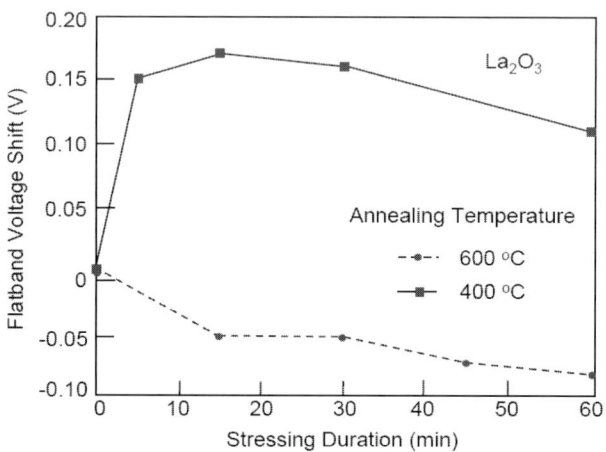

Fig. 24. Plot of the flatband voltage shift as a function of the stressing time for samples annealed at 400°C and 600°C after deposition. The stressing voltage is 1.0 V. Redrawn based on Ref. 33.

shift was found. This improvement can be attributed to the removal of hydroxyl groups from the oxygen vacancies.

## 10. Concluding Remarks

High-k gate dielectric has been recognized as a promising technology option to sustain further CMOS device downsizing to the nanoscale range and to boost the device and circuit performances for the present technological node. Particularly, the requirement for sub-nanometer EOT gate dielectric films in the nanoscale CMOS devices can only be achieved with the high-k materials. High-k dielectrics are also good for MOS transistors with gate oxide EOT in the range of 1 to 3 nm thick; the gate leakage current can be reduced by several orders of magnitude as the physical thickness of the high-k gate dielectric will be much larger than the direct tunneling limit. However, to incorporate the high-k materials into the present CMOS technology would require some major changes in the fabrication technique and the process sequence as the high-k materials must be deposited at much lower temperatures and the high-k materials themselves can react with the silicon substrate and have much lower crystallization temperature than the conventional silicon oxide or silicon oxynitride. From the device operation point of view, high-k materials often result in the performance degradations such as the Fermi level pinning and the channel mobility degradation. These performance degradations can be alleviated by using proper metal gate electrode. Reliability issues, such as high interface and oxide trap densities, low breakdown voltage, significant hot carrier-induced trap generation and negative bias temperature instabilities (NBTI), are also crucial for devices with high-k dielectrics. Significant improvements in these issues have been found by incorporating nitrogen and aluminum atoms into the metal oxide networks. However, the characteristics of high-k materials are still much poor than the conventional silicon oxide or silicon oxynitride in many aspects. There is still plenty of room for further improvement in both the material and the electrical properties of high-k dielectric films.

## References

1.  H. Wong and H. Iwai, *Phys. World*, **18**(9), pp 40–44, (2005).
2.  H. Wong and H. Iwai, *Microelectron. Engineer.* **83**, pp 1867–1904 (2006).

3.   G. D. Wilk, R. M. Wallace and J. M. Anthony, *J. Appl. Phys.* **89**, pp 5243–7275 (2001).

4.   D. Misra, H. Iwai and H. Wong, *Interface* (The Electrochemical Society, USA), **14**(2), pp 30–32 (2005).

5.   S.-H. Lo, D. Buchanan, Y. Taur and W. Wang, *IEEE Electron Device Lett.* **18**, pp 209–211 (1997).

6.   J. Robertson, *J. Vac. Sci. B.* **18**, pp 1785–1791 (2000).

7.   E. P. Gusev, E. Cartier, D. A. Buchanan, M. Gribelyuk, M. Copel, H. Okorn-Schmidt and C. D'Emic, *Microelectron. Eng.* **59**, pp 341–349 (2001).

8.   K. Hubbard and D. Schlom, *J. Mater. Res.* **11**, pp 2757–2776 (1996).

9.   K. Yamada, *Extended Abstracts of International Conference on Solid State Device and Materials, Tokyo, Japan*, pp 257–260 (1986).

10.  W. L. Scopel, Antonio J. R. da Silva, W. Orellana, A. Fazzio, *Appl. Phys. Lett.* **84**, pp 1492–1494 (2004).

11.  H. Takeuchi, D. Ha and T.-J. King, *J. Vac. Sci. Technol. A* **22**, pp 1337–1341 (2004).

12.  A. Oshiyama, *Jpn. J. Appl. Phys. Part 2*, **37**, pp L232–L234 (1998).

13.  H. Wong and V. A. Gritsenko, *Microelectron. Reliab.* **42**, pp 597–605 (2002).

14.  P. W. Peacock and J. Robertson, *Phys. Rev. Lett.* **92**, pp 576011–576014 (2004).

15.  J. Robertson and P. W. Peacock, *Phys. Stat. Sol. B* **241**, pp 2236–2245 (2004).

16.  M. Houssa, L. Pantisano, L.-A. Ragnarsson, R. Degraeve, T. Schram, G. Pourtois, S. De Gendt, G. Groesenekenb and M. M. Heyns, *Mater. Sci. Engineer. R* **51**, pp 37–85 (2006).

17.  H. Wong, K. L. Ng, N. Zhan, M. C. Poon and C. W. Kok, *J. Vac. Sci. Technol. B*, **22**, pp 1094–1100 (2004).

18.  A. Uedono, K. Ikeuchi, K. Yamabe, T. Ohdaira, M. Muramatsu, R. Suzuki, A. S. Hamid, T. Chikyow, K. Torii and K. Yamada, *J. Appl. Phys.* **98**, 023506 (2005).

19.  M. Lee, Z.-H. Lu, W. T. Ng, D. Landheer, X. Wu and S. Moisa, *Appl. Phys. Lett.* **83**, pp 2638–2640 (2003).

20.  A. P. Huang, R. K. Y. Fu, P. K. Chu, L. Wang, W. Y. Cheung, J. B. Xu and S. P. Wong, *J. Cryst. Growth* **277**, pp 422–427 (2005).

21.  J. Choi, S. Kim, J. Kim, H. Kang, H. Jeon and C. Bae, *J. Vac. Sci. Technol. A* **24**, pp 900–907 (2006).

22.  C. S. Kang, H.-J. Cho, K. Onishi, R. Nieh, R. Choi, S. Gopalan, S. Krishnan, J. H. Han and J. C. Lee, *Appl. Phys. Lett.* **81**, pp 2593–2595 (2002).
23.  K.-J. Choi, J.-H. Kim, S.-G. Yoon and W. C. Shin, *J. Vac. Sci. Technol. B* **22**, pp 1755–1758 (2004).
24.  R. J. Carter, E. Cartier, A. Kerber, L. Pantisano, T. Schram, S. DeGendt and M. Heyns, *Appl. Phys. Lett.* **83**, pp 533–535 (2003).
25.  X. Wang, J. Liu, F. Zhum, M. Yamada and D. L. Kwong, *IEEE Trans. Electron Devices*, **51**, pp 1798–1801 (2004).
26.  N. Umezawa, K. Shiraishi, T. Ohno, H. Watanabe, T. Chikyow, K. Torii, K. Yamabe, H. Kitajima and T. Arikado, *Appl. Phys. Lett.* **86**, 143507 (2005).
27.  K. Xiong, J. Robertson, S. J. Clark, *Appl. Phys. Lett.* **89**, 22907–22909 (2006).
28.  B. Sen, H. Wong, B. L. Yang, A. P. Huang, P. K. Chu, V. Filip, C. K. Sarkar, *Jpn. J. Appl. Phys.* pp 3234–3238 (2007).
29.  H. Wong, B. Sen, B. L. Yang, A. P. Huang and P. K. Chu, *J. Vac. Sci. Technol. B*, **25** pp 1853–1858 (2007).
30.  J. Choi, R. Puthenkovilakam and J. P. Chang, *J. Appl. Phys.* **99**, 053705 (2006).
31.  G. Shang, P. W. Peacock and J. Robertson, *Appl. Phys. Lett.* **84**, pp 106–109 (2004).
32.  H. Wong and Y. C. Cheng, *J. Appl. Phys.* **70**, pp 1078–1080 (1991).
33.  B. Sen, H. Wong, J. Molina, H. Iwai, J. A. Ng, K. Kakushima, C. K. Sarkar, *Solid State Electron.* **51**, pp 475–480 (2007).
34.  J. A. Ng, Y. Kuroki, N. Sugii, K. Kakushima, S.-I. Ohmi, K. Tsutsui, T. Hattori, H. Iwai and H. Wong, *Microelectron Eng.* **80**, pp 206–209 (2005).
35.  C. Hobbs, L. Fonseca, V. Dhandapani, S. Samavedam, B. Taylor, J. Grant, L. Dip, D. Triyoso, R. Hegde, D. Gilmer, R. Garcla, D. Roan, L. Lovejoy, R. Rai, L. Hebert, H. Tseng, B. White and P. Tobin, *VLSl Technol. Symp. Dig. Technical Papers*, pp 9–10 (2003).
36.  J. M. Soler, E. Artacho, J. D. Gale, A. Garcia, J. Junquera, P. Ordejo'n and D. Sa'nchez-Portal, *J. Phys. Condens. Matter* **14**, pp 2745–2779 (2002).
37.  K. Shiraishi, K. Yamada, K. Torii, Y. Akasaka, K. Nakajima, M. Kohno, T. Chikyo, H. Kitajima and T. Arikado, *VLSl Technol. Symp. Dig. Technical Papers*, pp 108–109 (2004).

38.   E. Cartier, V. Narayanan, E. P. Gusev, P. Jamison, B. Linder, M. Steen, K. K. Chan, M. Frank, N. Bojarczuk, M. Copel, S. A. Cohen, S. Zafar, A. Callegari, M. Gribelyuk, M. P. Chudzik, C. Cabral Jr., R. Carruthers, C. D'Emic, J. Newbury, D. Lacey, S. Guha and R. Jammy, *VLSl Technol. Symp. Dig. Technical Papers*, pp 44–45 (2004).

39.   A. Kaneko, S. Inumiya, K. Sekine, M. Sato, Y. Kamimuta, K. Eguchi and Y. Tsunashima, *Extended Abstracts of International Conference on Solid State Devices Materials*, p 56 (2003).

40.   J. Kedzierski, D. Boyd, P. Ronsheim, S. Zafar, J. Newbury, J. Ott, C. Cabral Jr., M. Ieong and W. Haensch, *IEDM Tech. Dig.*, pp 315–318, December (2003).

41.   C. S. Park, B. J. Cho, L. J. Tang and D. Kwong, *IEDM Tech. Dig.* pp 299–302, December (2004).

42.   M. Koyama, Y. Kamimuta, T. Ino, A. Kaneko, S. Inumiya, K. Eguchi, M. Takayanagi, A. Nishiyama, *IEDM Tech. Dig.*, pp 499–502 (2004).

43.   B. H. Lee, J. Oh, H. H. Tseng, R. Jammy and H. Huff, *Materials Today*, **9**, pp 32–40 (2006).

44.   Y. Akasaka, G. Nakamura, K. Shiraishi, N. Umezawa, K. Yamabe, O. Ogawa, M. Lee, T. Amiaka, T. Kasuya1, H. Watanabe, T. Chikyow, F. Ootsuka, Y. Nara and K. Nakamura, *Jpn. J. Appl. Phys.*, **45**, pp L1289–L1292 (2006).

45.   E. Cartier, F. R. McFeely, V. Narayanan, P. Jamison, B. P. Linder, M. Copel, V. K. Paruchuri, V. S. Basker, R. Haight, D. Lim, R. Carruthers, T. Shaw, M. Steen, J. Sleight, J. Rubino, H. Deligianni, S. Guha, R. Jammy and G. Shahidi, *VLSl Technol. Symp. Dig. Technical Papers*, pp 230–231 (2005).

46.   A. Ohta, S. Miyazaki, Y. Akasaka, H. Watanabe, K. Shiraishi, K. Yamada, S. Inumiya and Y. Nara, *Extended Abstracts of 2006 International Workshop on Dielectric Thin Films for Future ULSI Devices — Science and Technology*, Kawasaki, Japan, pp 61–62 (2006).

47.   M. Terai, K. Takahashi, K. Manabe, T. Hase, T. Ogura, M. Saitoh, T. Iwamoto, T. Tatsumi and H. Watanabe, *VLSI Technol. Symp. Dig. Technical Papers*, pp 68–69 (2005).

48.   K. Shiraishi, Y. Akasaka, S. Miyazaki, T. Nakayama, T. Nakaoka, G. Nakamura, K. Torii, H. Furutou, A. Ohta, P. Ahmet, K. Ohmori, H. Watanabe, T. Chikyow, M. L. Green, Y. Nara, K. Yamada, *IEDM Tech. Dig.*, pp 43–46 (2005).

49.  M. Kadoshima, Y. Sugita, K. Shiraishi, H. Watanabe, A. Ohta, S. Miyazaki, K. Nakajima, T. Chikyow, K. Yamada, T. Aminaka, E. Kurosawa, T. Matsuki, T. Aoyama, Y. Nara and Y. Ohji, *VLSl Technol. Symp. Dig. Technical Papers* pp 66–67 (2007).

50.  M. Inoue, S. Tsujikawa, M. Mizutani, K. Nomura, T. Hayashi, K. Shiga, J. Yugami, J. Tsuchimoto, Y. Ohno and M. Yoneda, *IEDM Tech. Dig.* pp 425–428 (2005).

51.  K. Iwamoto, A. Ogawa, Y. Kamimuta, Y. Watanabe, W. Mizubayashi, S. Migita, Y. Morita, M. Takahashi, H. Ito, H. Ota, T. Nabatame and A. Toriumi, *VLSl Technol. Symp. Dig. Technical Papers* pp 68–69 (2007).

52.  Y. C. Cheng and E. A. Sullivan, *J. Appl. Phys.*, **45**, pp 187–192 (1974).

53.  S. C. Sun and J. D. Plummer, *IEEE Trans. Electron Devices* **27**, pp 1497–1508 (1980).

54.  S. Datta, G. Dewey, M. Doczy, B. S. Doyle, B. Jin, J. Kavalieros, R. Kotlyar, M. Metz, N. Zelick and R. Chau, *IEDM Tech. Dig.*, pp 653–656 (2003).

55.  J. Zhu, J. P. Han and T. P. Ma, *IEEE Trans. Electron Devices*, **51**, pp 98–105 (2004).

56.  M. V. Fischetti, D. A. Neumayer and E. A. Cartier, *J. Appl. Phys.*, **90**, pp 4587–4608 (2001).

57.  Y. C. Cheng and E. A. Sullivan, *J. Appl. Phys.*, **44**, pp 3619–3625 (1973).

58.  K. L. Ng, N. Zhan, C. W. Kok, M. C. Poon and H. Wong, *Microelectron. Reliab.*, **43**, pp 1289–1293, (2003).

59.  B. L. Yang, P. T. Lai and H. Wong, *Microelectron. Reliab.*, **44**, pp 709–718, (2004).

60.  H. Wong, P. G. Han and M. C. Poon, *J. Korean Phys. Soc.*, **35**, pp S196–S199 (1999).

61.  Y. Kim, K. Miyauchi, S. Ohmi, K. Tsutsui and H. Iwai, *Microelectron. J.* **36**, pp 41–49 (2005).

62.  H. Wong and H. Iwai, *J. Vac. Sci. Technol. B*, **24**, pp 1785–1793, (2006).

63.  T. V. Perevalov, V. A. Gritsenko, S. B. Erenburg, A. M. Badalyan, H. Wong and C. W. Kim, *J. Appl. Phys.*, **101**, 053704 (2007).

64.  N. Zhan, M. C. Poon, H. Wong, K. L. Ng and C. W. Kok, *Microelectron. J.*, **36**, pp 29–33 (2005).

65. Y. H. Kim, K. Onishi, C. S. Kang, H.-J. Cho, R. Nieh, S. Gopalan, R. Choi, J. Han, S. Krishnan and J. C. Lee, *IEEE Electron Device Lett.*, **23**, pp 594–596 (2002).

66. H. Wong, N. Zhan, K. L. Ng, M. C. Poon and C. W. Kok, *Thin Solid Films*, **462**, pp 96–100 (2004).

67. J. McPherson, J. Kim, A. Shanware, H. Mogul and J. Rodriguez, *IEDM Tech. Digest*, pp 633–636 (2002).

68. D. J. DiMaria, *Solid-State Electron*, **41**, pp 957–965 (1997).

69. E. Y. Wu, E. J. Nowak, R. P. Vollertsen and L. K. Han, *IEEE Trans. Electron Devices*, **47**, pp 2301–2309 (2000).

70. H. Wong and Y. C. Cheng, *J. Appl. Phys.*, **74**, pp 7364–7368 (1993).

71. Z. Cui, J. J. Liou, Y. Yue and H. Wong, *Solid-State Electron*, **49**, pp 505–511 (2005).

72. H. Wong and M. C. Poon, *IEEE Trans. Electron Devices*, **44**, pp 2033–2035 (1997).

73. S. Lombardo, A. La Magna, C. Spinella, C. Gerardi and F. Crupi, *J. Appl. Phys.*, **86**, pp 6382–6391 (1999).

74. A. Kumar, M. V. Fischetti, T. H. Ning and E. Gusev, *J. Appl. Phys.*, **94**, pp 1728–1737 (2003).

75. C. H. Liu, M. T. Lee, J. C. Y. L. Chen, K. Schruefer, J. Brighten, N. Rovedo, T. Hook, M. Khare, F. H. Shih, C. Wann, T. C. Chen and T. H. Ning, *IEDM Tech. Dig.*, pp 39.2.1–39.24 (2001).

76. G. Bersuker, J. H. Sim, C. D. Young, R. Choi, P. M. Zeitzoff, G. A. Brown, B. H. Lee and R. W. Murto, *Microelectron. Reliab.*, **44**, 1509–1512 (2004).

77. S. Zafar, B. H. Lee and J. Stathis, *IEEE Electron Device Lett.*, **25**, 153–155, (2004).

78. M. Aoulaiche, M. Houssa, R. Degraeve, G. Groseneken, S. De Gendt and M. M. Heyns, *Microelectron. Engineer.*, **80**, pp 134–137 (2005).

79. M. Houssa, S. D. Gendt, J. L. Autran, G. Groeseneken and M. M. Heyns, *Appl. Phys. Lett.*, **85**, pp 2101–2103 (2004).

# 5

# Fabrication of Source and Drain — Ultra Shallow Junction

Bunji Mizuno

UJT lab., Ultimate Junction Technologies Inc.
P.O. Box 570-8501, 3-1-1, Yakumonakamachi, Moriguchi Osaka, Japan

Mizuno@ujtlab.com

....................................

Semiconductor devices have been successfully produced by the miniaturization of planar transistors and their transformation into a 3D structure. This innovation will realize ideal performance in electric devices. In this article, plasma doping combined with several state-of-the-art rapid thermal processing is shown to be a technology for enabling the fabrication of miniaturized 2D devices and advanced 3D structures. Plasma Doping provides superior performance of physical and electrical characteristics.

## 1.     Introduction

In the semiconductor business, the International Technology Roadmap for Semiconductors (ITRS[1]) describes the guidelines to develop the new technologies and devices to ensure not to invest in wrong way. In such a constructive discussion, semiconductor will progress in the next ten years along the Moore's law and "more Moore" directions.

In this group effort, several most energetic companies have announced aggressive ways. Their announcement is that they will develop three dimensional devices (3D). Since the development of planar devices more than 20 years ago, we have progressed by miniaturizing the source-gate-drain structure of MOS transistors. But these planar technologies face big "red walls" in the quite near future (almost this year).

By switching to a 3D structure, transistors will get their small foot print and stability that will come from ideal control by surrounded 3D gate structure. Managements have invested fancy lithography machines to miniaturize LSI's. The Plasma Doping (PD) technology is a major option to fullfill the 3D issues requirments and will thus allow to develop ideal performance of miniaturized LSI's.

J. Kavalieros *et al.*[2] have proposed a FinFET with tri-gate structure for use in the 32 nm technology node. In order to minimize access resistance, a tall Fin height was necessary and device performance was improved. In this chapter, the author emphasizes the way to dope a large number of fins conformally and uniformly.

## 2.    Doping Technologies

We have to re-consider again how to dope 3D structures to realize industrialization of 3D devices. In other words, without a new doping technology, 3D devices will not be industrialized and the progress of LSI will stop. Because conventional doping technology — ion implantation (II) — has been developed as a key technology for 2D (planar) devices but faces strong challenges for 3D structures. For the miniaturization of planar transistors, II technology was modified with lower energy and higher beam current.[3] Meanwhile for lower energy and 3D structure, we developed and proposed Plasma Doping (PD) as an alternative technology. The technologies that can be adapted to 3D structures are gas phase doping and PD only. For industrialization in the IC fab environment, the author thinks that only PD has a capability in terms of reality and maturity of its technology.

PD technology has been developed as a new semiconductor technology.[4−7] It took 20 years to develop PD[8] from the first experiment.[4] The first work on plasma doping was done by Shockley who shared the Novel prize for the invention of the transistor. H. Strak, of Shockley laboratory, developed plasma doping in a glass tube reactor.[9] This experiment showed that energetic ions can penetrate into the semiconductor materials, such as Si and Ge, to form p-n junctions. After that, Cockcroft-Walton type accelerators were modified for ion implantation to manufacture CMOS transistors. Shockley's work was patented in 1954. Twenty years later, in 1970 ion implantation was utilized in semiconductor fabs. In a similar fashion, the first results using PD were presented in 1987[4] and 20 years later, PD is utilized as a doping technology for DRAM fabrication.

## 3. PD Experimental Conditions

In the following example, we used a PD tool "A"[10] equipped with a Hericon wave plasma source, which had the characteristics of a high plasma density. The source power was 1000–2250 W. The total gas pressure was 0.9–2.5 Pa of a $B_2H_6$/He gas mixture with concentrations varied from 5%/95% to 0%/100%. In the He Pre Amorphization (He-PA) process, He plasma (the gas concentration of 0%/100%) irradiated a Si substrate, in which bias voltage was 30–310 V, process time was 7 s, the typical plasma density and electron temperature were $5.5 \times 10^{10}$ cm$^{-3}$ and 6.5 eV. In the following PD process, bias voltage was 30–100 V, dosage of B was $8 \times 10^{14}$–$5 \times 10^{15}$ cm$^{-2}$ and process time was 7–60 s. These two processes were carried out continuously in the same process chamber. After these processes, the as-doped wafers were annealed by using Flash Lamp Annealing (FLA) or All Solid Laser Annealing (ASLA). In the FLA, the intermediate temperature was 700–725°C, front side peak temperature was 1275–1306°C and flash lamp irradiating time was 1 ms. In the ASLA, a green frequency-doubled diode pumped solid state laser ($\lambda = 0.53$ $\mu$m) irradiated for 100 ns with the energy density of 1400–1500 mJ/cm$^2$. Thickness and optical absorption parameters of the surface amorphous layers were evaluated by ellipsometry. The thickness of the amorphous layer was also measured by TEM.

## 4. PD Physical and Electrical Characterization

Figure 1 shows a cross-sectional TEM image just after the He-PA process. An amorphous layer was found to be formed on the surface of Si substrate.

Figure 2 shows the relationship between bias voltage in the He-PA process and the thickness of the amorphous layer. The thickness was controlled from 2 nm to 17 nm by changing the He-PA bias voltage. Additionally, the thickness of 22 nm was obtained when He-PA process time was 30 s.

Figure 3 shows a comparison of optical absorption spectra. The optical absorption coefficient of the He-PA layer was as large as that of the Ge preamorphization implantation (PAI) (5 kV, $1 \times 10^{15}$ cm$^{-2}$) layer and it was 5 to 45 times larger than that of c-Si at the wavelength from 400 to 800 nm. Amorphization by bombarding with light atoms or their plasma such as He

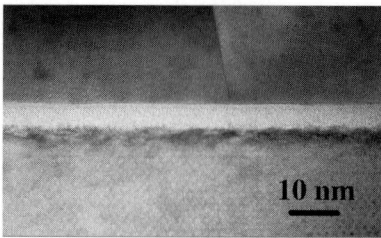

Fig. 1. Cross-sectional TEM image for after He-PA sample in which bias voltage was 60 V and process time was 7 sec.

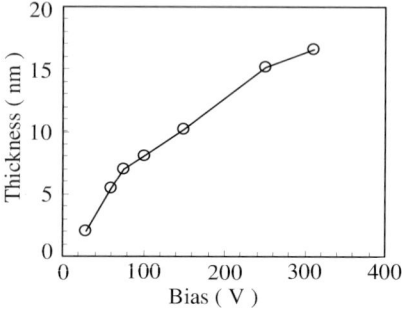

Fig. 2. Relationship between bias and thickness of the amorphous layer formed by He-PA process for process time of 7 sec.

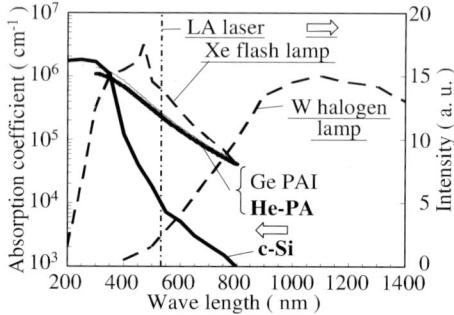

Fig. 3. Optical absorption spectra of Si surface prepared by He-PA or Ge PAI compared with that of c-Si. Spectra of some light source are also represented as references.

is not well known. Si amorphization by He bombardment was first reported by the author of this chapter in 2004.[11]

Thanks to the large optical absorption a highly activated dopant concentration after annealing can be achieved. Figure 4 shows the optical

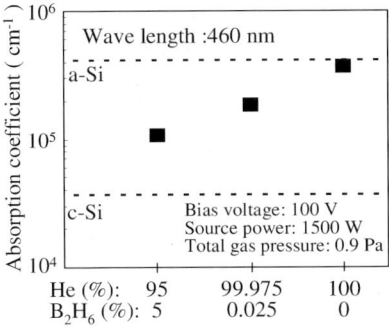

Fig. 4. Optical absorption coefficients of surface layers depending on $B_2 H_6$ /He gas concentration ratio.

absorption coefficient depending on $B_2H_6$/He gas concentration. The optical absorption coefficient was controlled by $B_2H_6$/He gas concentration ratio. It is considered to be related with plasma density and electron temperature since higher He gas concentration results in larger plasma density and electron temperature.

The as-doped profile obtained in this work was compared with those by the Ge Pre Amorphizaztion Implant (Ge PAI) $+BF_2$ Ion Implantation ($BF_2$ I/I)[12] reported so far as shown in Fig. 5. The He-PA+PD method achieved a steeper profile abruptness and higher dose with a shallow depth.

Figures 6 and 7 show SIMS profiles before and after the FLA and the ASLA, respectively.[10] *Rs* of 1000 ohm/sq and 588 ohm/sq were obtained, while the diffusion length of B during the annealing processes was only 2–2.5 nm. Figure 8 shows of the dopant profiles differences after the ASLA

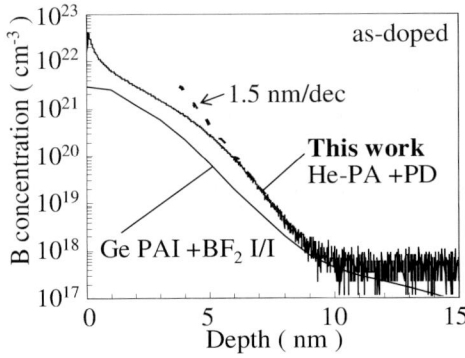

Fig. 5. SIMS profiles for this work (He-PA +PD) and reported works of Ge PAI+BF$_2$ I/I.[12]

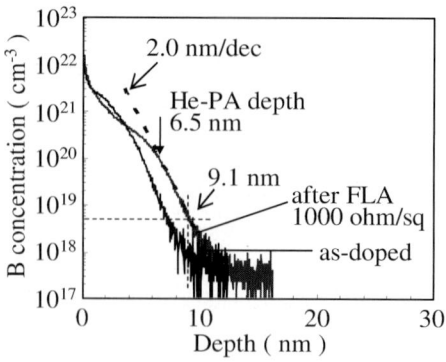

Fig. 6. SIMS profiles before and after FLA. Doping process was He-PA +PD (bias: 60 V).

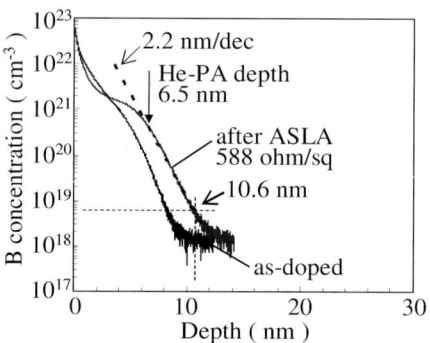

Fig. 7. SIMS profiles before and after LA. Doping process was He-PA +PD (bias: 60 V).

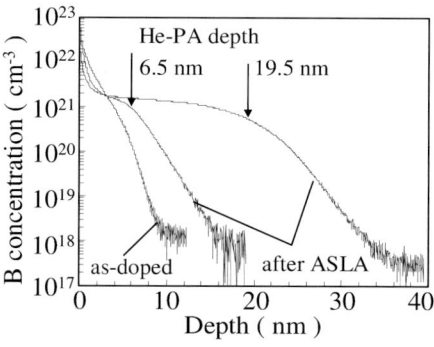

Fig. 8. Variation of junction depth depending on He-PA thickness for constant LA energy density ($1500 \, mJ/cm^2$).

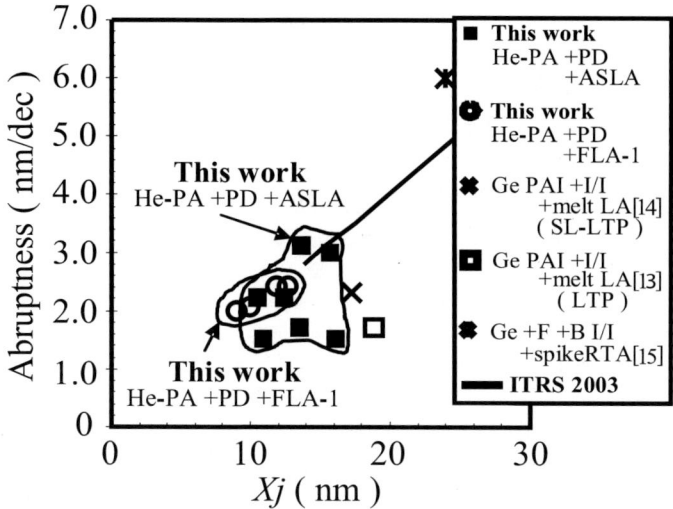

Fig. 9. Abruptness for this work (He-PA +PD +FLA-1 or ASLA), reported works[5−7] and ITRS 2003 required value.

depending on the thickness of amorphous layer formed by the He-PA for the same ASLA condition. This shows that $Xj$ is able to be controlled by changing the He-PA depth.

Figure 9 shows the comparison of junction abruptness between this work and earlier results.[13−15] Good abruptness of 1.5–2.4 nm/decade at $Xj$ of around 10 nm was obtained in this work.

Figure 10 shows the $Rs$-$Xj$ plots using FLA. The $Rs$ was reduced by 30% for the same $Xj$ with the use of PD compared to those for the Ge PAI +BF$_2$ I/I +FLA-2.[12] Figure 11 shows similar plots using LA. The $Xj$ of this work was much shallower than those for the Ge PAI +I/I +melt-LA.[13,14] The $Rs$ of this work was much lower than those for the submelt-LA.[16] These results indicate the superiority of our new method combining PD and advanced annealing.

We also investigated the integratability of our doping technique. The contained He was almost completely out-gassed after the FLA process because of the high diffusivity of He in the Si substrate, as shown in Fig.12. Hydrogen behaves in a similar fashion. Surface roughness was almost the same as that of the initial Si substrate throughout the He-PA, the PD and the final FLA processes. The sputtering rate was found to be less than 0.08 nm/s in the He-PA process.

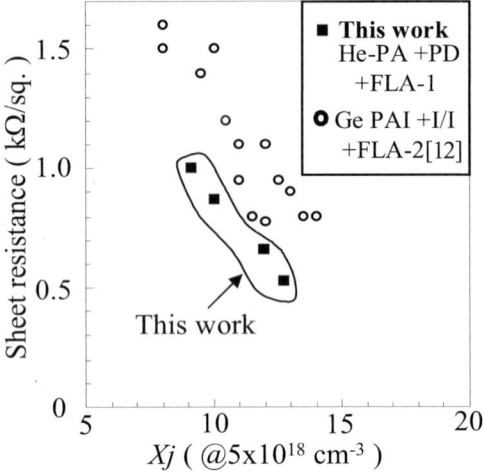

Fig. 10. Relationship between Rs and $Xj$ for this work and reported works using FLA.

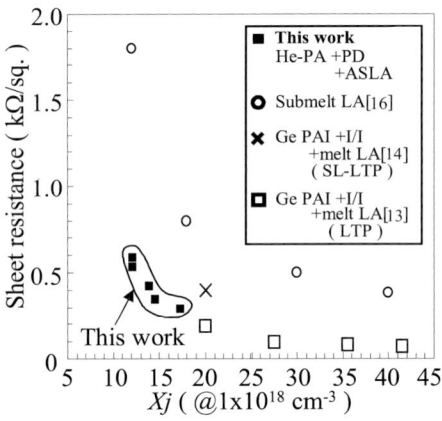

Fig. 11. Relationship between Rs and $Xj$ for this work and reported works using FLA.

## 5.    Recent Application to ULSI Devices

S.H. Lee *et al.* demonstrated the application of PD to the fabrication of NAND flash memory with 3D fin transistor structures[17]: 70% cell current improvement was attributed to fin structure and an additional 30% to PD.

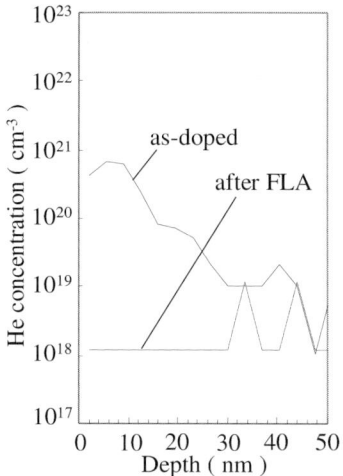

Fig. 12. SIMS profiles of He before and after FLA.

D. Lenoble *et al.* described the requirements of required junctions depths in the future.[18] MOSFET drive current is expected to increase whilst junction depth to decrease. As an example, two devices show the same drivability at 500 ohm/sq with Xj of 25 nm and 1450 ohm/sq with Xj of 15 nm. The Source and Drain Extension (SDE) profile and depth are of primary importance to control the drivability and short channel effects of MOS transistors. According to recent results,[10,19] PD will be used with conventional RTA for the 45 nm node and beyond by improving annealing technologies i.e. FLA or LA. However, they also pointed out that PD technology has to overcome the issues of uniformity, repeatability and accuracy in terms of dose control.

## 6.    Industrialization Issues

The process requirements of PD are uniformity, repeatability and accurate dose control. After clearing these tough questions, its effect, i.e., improvement of transconductance, define and show benefit, is an over-joy world citing Japanese literature edited by Prof. Nishizawa.[20]

Measuring the dosage has been one of the major difficulties associated with the use of PD. This problem was overcome by introducing the Self Regulating Plasma Doping (SRPD) process.[8] The basic behavior of the SRPD is schematically described in Fig. 13. For a given plasma condition, the boron dose increases rapidly as a function of time in the initial stage.

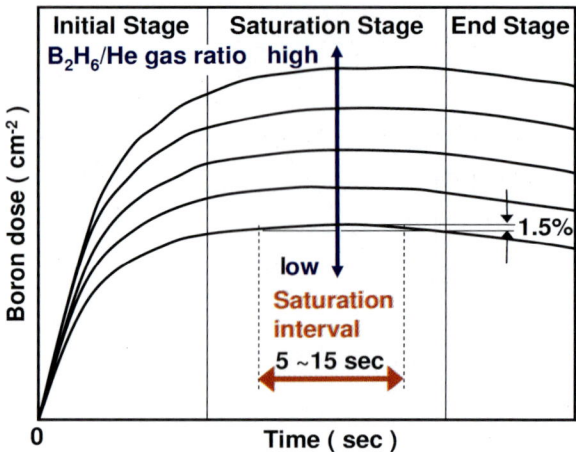

Fig. 13. The basic characteristics of the SRPD process.

The increase in dose begins to slow down and finally peaks at a unique value in the dose saturation stage, as long as the $B_2H_6$/He ratio is as low as below $1 \times 10^{-2}$. During this stage, the dose remains almost constant for typically 5–15 seconds, within 1.5%, which makes it possible to control the dose with remarkably high accuracy. The value of the saturating dose can be controlled over a wide range typically between the orders of $10^{14}$–$10^{16}$ cm$^{-2}$ by changing the gas ratio (Figs. 13 and 14). The SRPD

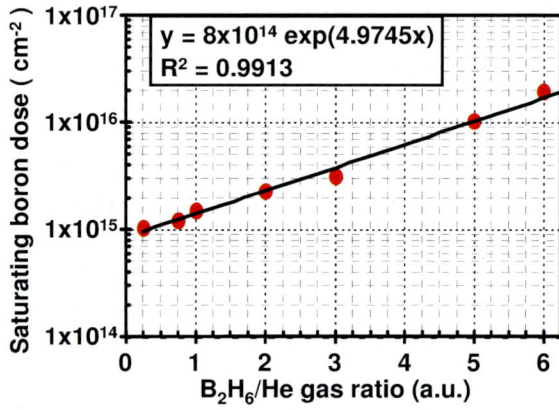

Fig. 14. Relationship between $B_2H_6$/He gas ratio and the saturating boron dose.

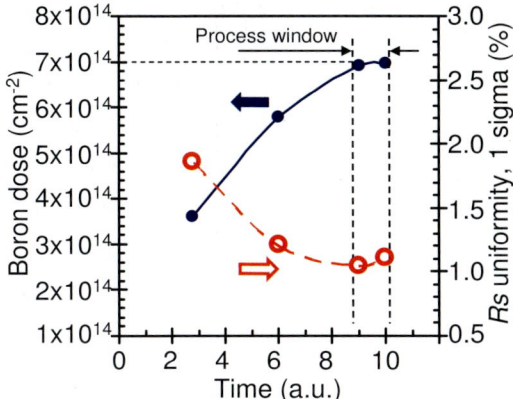

Fig. 15. The dosimetry and the within-wafer uniformity on *Rs* of SRPD process performed by Tool B. The anneal condition is 1075°C for 20 sec.

process eventually moves to the end stage where the dose starts to decrease due to self-sputtering.

The relatively long saturation time also helps to improve the dose uniformity across the wafer. Figure 15 shows the dose and the uniformity plots using Tool B. The dose saturation stage is seen at around time 10. The uniformity was about 2% in the initial stage and was improved steadily as a function of time to 1.0–1.1% towards the dose saturation stage (Figs. 15 and 16). Figure 17 shows the typical example of the plasma uniformity measured at approximately 10 mm above the wafer plane in Tool B. The plasma density dropped significantly near the edge of the wafer and the overall uniformity at one sigma was 8.8%. Rs values over a 300 mm wafer processed by Tool B using the same plasma condition are also plotted in Fig. 17 demonstrating a uniformity of 1.04%. This would mean that, even if the dose uniformity in the initial stage is poor, the entire wafer surface reaches the same dose during the saturation interval (Fig. 13).

The depth of the dopant profiles is predominantly controlled by the bias potential to the wafer as shown in Fig. 18. The abruptness of the as-doped profiles is as steep as 2.0 nm/decade at 10 nm using the SRPD process because of the simultaneous amorphization at ultra shallow depth by using very low $B_2H_6$/He gas ratio plasma, which is significantly steeper than those by II and the conventional PD methods[21,22] (Fig. 19). The formation of the well-defined amorphous layer on surface is seen after the PD process

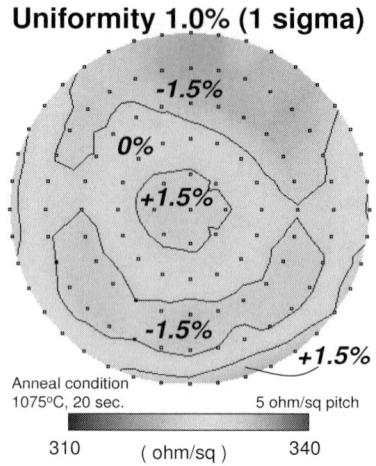

Fig. 16. The distribution map on *Rs* of SRPD process performed by Tool B.

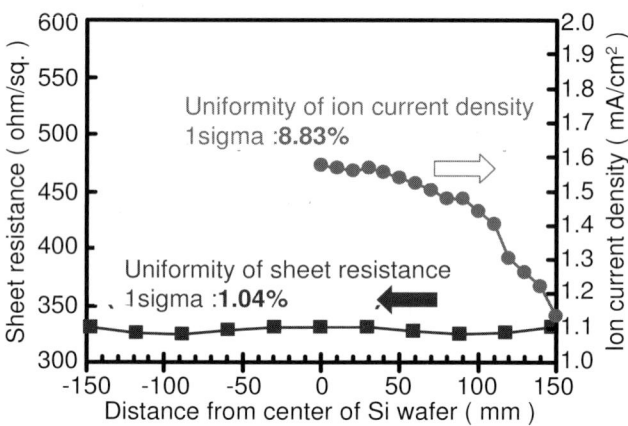

Fig. 17. The uniformity of the ion current density in the $B_2 H_6$ /He gas plasma of Equipment B and the *Rs* uniformity. The anneal condition is 1075°C, 20 sec.

(Fig. 20(a)), however, no remaining defects were observed after spike RTA (Fig. 20(b)).

Metal contamination was evaluated by ICP-mass analysis as shown in Table 1. The results of the PD show successful suppression of the contamination down to the level equivalent to II by conducting an appropriate

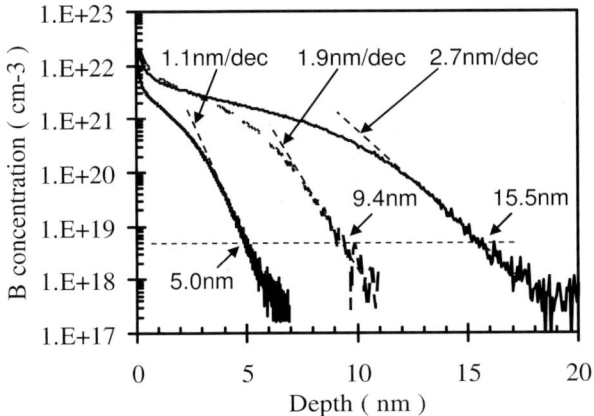

Fig. 18. SIMS profiles (SRPD) at various bias voltages.

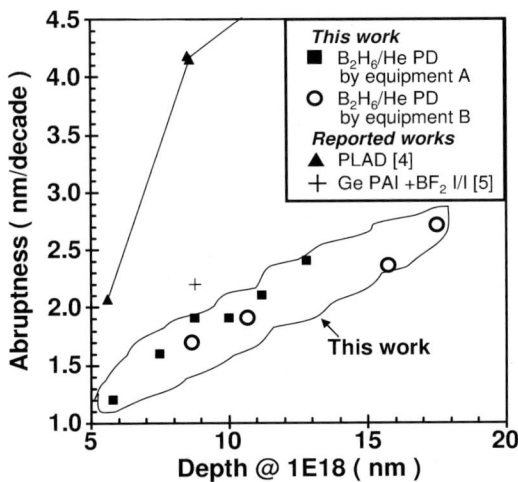

Fig. 19. Abruptness for this work and previously reported data.[4,5]

coating on the PD chamber. The leakage current on pn diode prepared by PD is in the order of $10^{-9}$ A/cm$^2$ at room temperature at the most when the dopant (ND) concentration in substrate is $4 \times 10^{14}$ cm$^{-3}$ (Figs. 21(a) and (b)). The leakage increases to the order of $10^{-5}$ A/cm$^2$ when ND exceeds $10^{18}$ cm$^{-3}$ due to band-to-band tunneling (Fig. 21(c)). The leakage current values values at both low and high ND conditions are comparable to those obtained by II.[22]

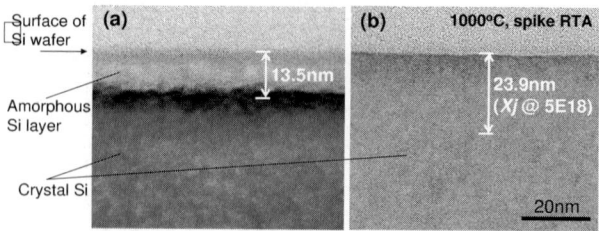

Fig. 20. Cross sectional TEM images; (a) after PD process and (b) after spike RTA process.

Table 1. Metal contamination on bare Si wafer after PD process at $1 \times 10^{15}$ cm$^{-2}$ of boron dose and 10 nm of the as-doped depth.

| Fe | Ni | Zn | Cr | Na | K | Ca | Al | Mg | Cu | Co |
|------|---------|------|---------|--------|--------|-----|-----|-----|---------|--------|
| 0.68 | < 0.081 | 0.18 | < 0.091 | < 0.21 | < 0.12 | 0.3 | 1.8 | 0.2 | < 0.075 | < 0.08 |

## 7.    Future Prospects for Plasma Doping

Plasma doping will be a major solution to the control of ultra shallow junction depth and for 3D applications. The difficulties to overcome are mainly non-uniformity and non-accuracy on the dose control. However, we proposed the SRPD method has overcome almost all difficulties. Consequently, as an important issue, FLA will be strongly needed without pattern effects and LA will be the main player with an appropriate wavelength of laser light for silicon or hybrid materials. Afterwards, atomic scale manipulation will have to be envisaged. This concept is mostly important to fabricate channel portions precisely controlled in terms of atomic placement and numbers. For well or Fin materials large area doping technologies, allowing relatively important depths and quite precise dose control will be needed. The solution will come from an in-situ monitoring by using a squid technology.[23]

Two categories of Implantation technologies will have to be considered in the future:

1) The so called co-implantation technology. Recently, S. Felch *et al.*[24] presented P implantation following Si and C implantations. This co-implantation method realized quite sharp profiles of 20 nm depth after 1050C RTA. The co-implantation cost is however 3 times larger than conventional implantation. Thus PD development is a very intersting option, co-implantation being kept as an important back-up technology.

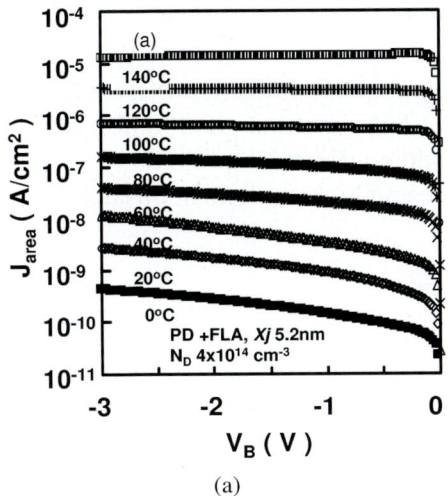

(a)

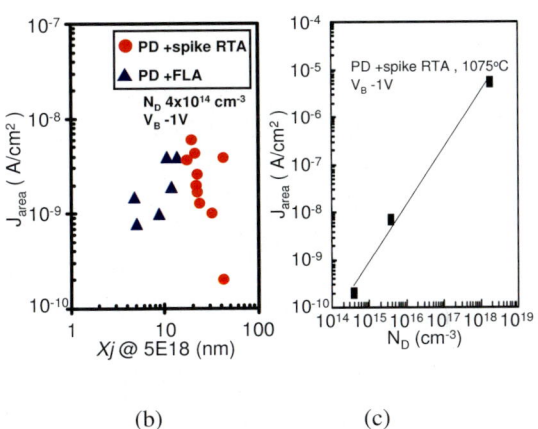

(b)                                        (c)

Fig. 21. (a) Temperature dependence of the leakage current of pn diode fabricated by plasma doping process. (b) Leakage current of pn diode fabricated by plasma doping process; (b) relationship between $Xj$ and the leakage current (PD +FLA, PD +spike RTA) and (c) relationship between $N_D$ and the leakage current (PD +spike RTA).

2) cluster or molecule ion implantation that equivalently realizes ultra low energy implantation. Several new results using new development on ion source[25–27] have already been reported. However, the doping of 3D structures could be problemlatic with this technique.

# 8.     Conclusions

The author of this chapter has developed Plasma Doping that is capable to realize efficiently, at low cost, and good uniformity ultra shallow doping for planar devices and conformal doping for 3D structures both with high devices performance and high equipment through-put.

# Acknowledgments

The author greatly thanks Prof. Iwai and Prof. Tsutsui of Tokyo Institute of Technology. Dr. Michael Current of Current Science and Dr Simon Deleonibus of LETI for their helpful discussions. He thanks Dr. Gelpey of Mattson Canada and Dr. Kudo of SHI Japan for collaboration. He also thanks Mr. Sasaki and UJT members for PD development.

# References

1.     ITRS  Home  Page:*http://www.itrs.net/Links/2005ITRS/Home2005. htm*
2.     J. Kavalieros, B. Doyle, S. Datta, G. Dewey, M. Doczy, B. Jin, D. Lionberger, M. Metz, W. Rachmady, M. Radosavljevic, U. Shah, N. Zelic and R. Chau *Symp. VLSI Tech., Digest of Technical Papers*, p 62 (2006).
3.     A. Hori and B. Mizuno, *Technical Digest of IEEE International Electron Devices Meeting (IEDM)*, p 641 (1999).
4.     B. Mizuno, I. Nakayama, N. Aoi and M. Kubota, *Extended Abstract of the 19th Solid State Devices and Materials, (SSDM)*, p 319 (1987).
5.     B. Mizuno, I. Nakayama, N. Aoi, M. Kubota and T. Komeda, *Appl. Phys. Lett.*, **53**, 2059 (1988).
6.     B. Mizuno, M. Takase, I. Nakayama and M. Ogura, *Symp. VLSI Tech., Digest of Technical Papers,* p 66 (1996).
7.     M. Takase, K. Yamashita, A. Hori and B. Mizuno, *Technical Digest of IEEE International Electron Devices Meeting (IEDM), Washington D. C., 1977,* p 475.
8.     Y. Sasaki, H. Ito, K. Okashita, H. Tamura, C. G. Jin, B. Mizuno, T. Okumura, I. Aiba, Y. Fukagawa, H. Sauddin, K. Tsutsui and H. Iwai, *Ion Implantation Technology (IIT),* p 524 (2006).

9.    H. Strack, *J. Appl. Phys.* **34**, 2405 (1963).

10.   Y. Sasaki, C. G. Jin, H. Tamura, B. Mizuno, R. Higaki, T. Satoh, K. Majima, H. Sauddin, K. Takagi, S. Ohmi, K. Tsutsui and H. Iwai, *Symp. VLSI Tech., Digest of Technical Papers,* p 180 (2004).

11.   Sasaki *et al. Ion Implantation Technology (IIT), Taiwan, 2004,* 12 (2004).

12.   T. Ito, K. Suguro, T. Itani, K. Nishinohara, K. Matsuo and T. Saito, *Symp. VLSI Tech., Digest of Technical Papers,* p 53 (2003).

13.   S. P. McCoy, *et al., Ultra Shallow Junction,* p. 104 (2003).

14.   T. Yamamoto, *et al., Symp. VLSI Tech., Digest of Technical Papers,* p 53 (2003).

15.   A. Shima, *et al., Technical Digest of IEEE International Electron Devices Meeting (IEDM),* p 493 (2003).

16.   B. J. Pawlak, *et al., Ion Implantation Technology (IIT)* p 21 (2002).

17.   S. H. Lee, J. J. Lee, J.-D. Choe, E. S. Cho, Y. J. Ahn, W. Hwang, T. Kim, W.-J. Kim, Y.-B. Yoon, D. Jang, J. Yoo, D. Kim, K. Park, D. Park and B.-I. Ryu, *Technical Digest of IEEE International Electron Devices Meeting (IEDM), San Fransisco,* p 33 (2006).

18.   D. Lenoble *Semiconductor Fabtech,* 30th *Edition,* 2006.

19.   F. Lallement, B. Duriez, A. Grouillet, F. Arnaud, B. Tavel, F. Wacquant, P. Stolk, M. Woo, Y Erokin, J. Scheuer, L. Godet, J. Weeman, D. Diataso and D. Lenoble, *Symp. VLSI Tech., Digest of Technical Papers,* p 178 (2004).

20.   M. Takase, *Study on Semiconductor* edited by Jun-ichi Nishizawa, Vol 46, issued at 10th May, p 304 (2000).

21.   D. Lenoble, *et al., Ion Implantation Technology (IIT)* p 36 (2002).

22.   S. Severi, *et al., Technical Digest of IEEE International Electron Devices Meeting (IEDM),* p 99 (2004).

23.   T. Watanabe, S. Watanabe, T. Ikeda, M. Kase, Y. Sasaki, T. Kawaguchi and T. Katayama, *Supecond. Sci.Tech,* **17**, 450 (2004).

24.   S. Felch, B. J. Pawlak, T. Hoffmann, E. Collart, S. Severi, T. Noda, V. Parihar, P. Eyben, W. Vadervorst, S. Thirupapuliyur and R. Schreutelkamp, *Proceeding of International Workshop on INSIGHT in Semiconductor Device fabrication, Metrology and modeling, Napa,* 2007.

25.   A. Renau, *International Workshop on Junction Technology (IWJT),* p 107 (2007).

26. L. M. Rubin, M. S. Ameen, M. A. Harris and C. Huynh *International Workshop on Junction Technology (IWJT), Kyoto*, p 113 (2007).
27. N. Hamamoto, S. Umisedo, Y. Koga, T. Matsumoto, T. Nagayama, M. Tanjo, N. Nagai, T. Horsky, D. Jacobson and G. Glavish, *International Workshop on Junction Technology (IWJT)*, p 125 (2007).

# 6

# New Interconnect Schemes: End of Copper, Optical Interconnects?

Suzanne Laval*, Laurent Vivien*,‡, Eric Cassan*,
Delphine Marris-Morini* and Jean-Marc Fédéli†

*Institut d'Electronique Fondamentale, CNRS,
Université Paris-Sud 11, F91405 ORSAY cedex France.

†CEA-Grenoble/LETI/DOPT/SIONA/LPS,
17 Rue des Martyrs, 38054 GRENOBLE Cedex France.

‡laurent.vivien@u-psud.fr

..............................

With the increase of integration density and complexity in CMOS circuits for microprocessors, enhancement of operating frequency becomes limited by the electrical wiring. Alternative solutions have to be found, and among these, optical interconnects can bring improvements for signal synchronization, power dissipation and noise immunity. Basic devices for optical links include light sources and modulators, optical waveguides and photodetectors. Possible ways for integration of optics with electronics are discussed.

## 1.    Introduction

As predicted by Moore's law, transistor size continuously decreases with time, leading to a strong increase of the integration density and simultaneously of the number of transistors and of the circuit size. This results in an enhancement of the complexity and poses increasingly difficult challenges in terms of physics and materials. Not only the number of interconnects increases, but their mean length scales up with the circuit size. To benefit from the increase of the switching speed with decreasing the transistor channel length, signal propagation delays in interconnects should also decrease. However, decreasing the metal cross-section increases the wiring resistance and increasing the wiring density increases capacitances between the wires. In consequence the resistance by capacity product (*RC* time constant) increases and this is responsible for an increase of the propagation delay, which becomes unacceptable for the longest interconnects on the chip. In a projected 35 nm Cu/low-$\kappa$ technology generation,[1] the transistor delay will be $\sim 1.0$ ps, and the *RC* delay of a 1 mm line will be $\sim$250 ps. Bandwidth limitations of the electrical wiring appear to be the main blocking point for increasing the clock frequency in microprocessors. Although new architectures such as multi-core processors have been introduced to improve the processor performances without increasing the operating frequency and thus to partially overcome this problem, the ITRS roadmap[1] clearly states that alternative solutions have to be developed: "For the long term, material innovation with traditional scaling will no longer satisfy performance requirements. Interconnect innovation with optical, radio frequency (RF), or vertical integration combined with accelerated efforts in design and packaging will deliver the solution".

RF technology seems closer to microelectronics one than optics, but the main drawbacks of RF interconnects are the difficulty to miniaturize the antennas and the need for complex shielding.

Optics can intrinsically handle a huge data rate, and also presents significant advantages in terms of synchronization, crosstalk and dissipated power. Optical interconnects have aroused a growing interest in the recent years. Silicon-On-Insulator (SOI) is the choice substrate to develop microphotonics. It is compatible with microelectronics technology and allows very compact integration of the main optical elements needed for optical links. A toolbox of devices is developed to progressively build optical networks on a chip. Low loss light distribution through submicron waveguides including splitters and bends has been demonstrated.[2,3] As silicon is not favourable for light emission, an off-chip optical source could be used at first. The signal will be encoded thanks to an integrated silicon-based modulator, and ultra-fast germanium photodetectors[4,5] are available to convert the optical signal back to an electrical one at the end of the link. Wavelength multiplexing/demultiplexing can also be used to increase the data rate transfer between electronics modules.

Several ways are considered to integrate the optical devices with microelectronic circuits,[6,7] either monolithically in front-end process or through an optical layer bonded on the CMOS circuit as in a 3D-integration process.

## 2. The Metallic Interconnect Limitations

The growing demand for higher performances and greater functionalities in CMOS integrated circuits (IC) has resulted in a shrinking of transistors coupled with an increase of chip size. This leads to an increasing number of wires within the chip. As the chip area is finite, interconnects have to be distributed among more than 10 metal levels (Fig. 1).

The shortest interconnects, between neighbour transistors, are provided by the first level. The intermediate levels ensure connections between modules within the chip. The upper levels distribute global signals, including clock and power, over the entire chip with the longest wires which are needed. As the wire length increases, scaling leads to an increase of the interconnect resistance and capacitance and the *RC* delay of the global interconnects becomes a performance bottleneck. Even with the introduction of copper to increase the conductivity and of low-$\kappa$ dielectrics to decrease the capacitance,[8] this will be more and more important in the future, with the process technology node decrease. This is illustrated in Fig. 2. The gate delay decreases with the transistors size shrinking. Metal 1 levels are relatively unaffected by scaling, but the *RC* delay due to global

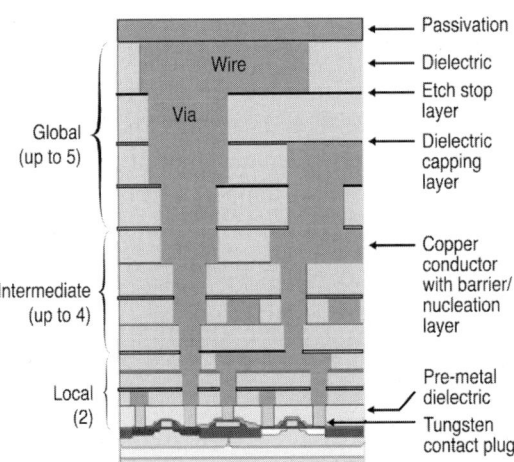

Fig. 1. Cross-section of interconnect hierarchical scaling (from ITRS[1]).

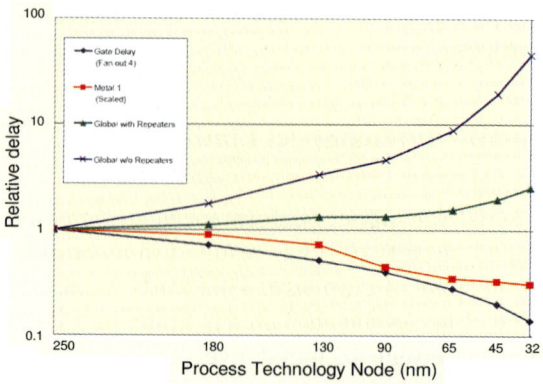

Fig. 2. Delay for Metal 1 and global wiring versus feature size (from ITRS[1]).

interconnects strongly increases and communication across the chip will require an increasing number of clock cycles. Repeaters can be incorporated but they consume power and silicon area, and increase synchronization uncertainties.[9]

In order to alleviate the problem, scaling is generally not applied to global interconnects. To ensure a minimum $RC$ value, the longest interconnects tend to be fabricated with wires wider and thicker than minimum geometry and the pitch tends to be larger, but this implies to add metal levels.[10,11] The capacitances intra- and inter-levels increase with the number of connections, leading to an enhancement of the dynamic power dissipation. Furthermore, at high frequencies the skin effect will impact the wire resistance, and inductance has also to be taken into account.[12]

Many efforts are made to improve the electrical wiring characteristics, but technological issues are still challenging. The lowering of the dielectric constant proved to be more problematic than predicted. The integration of low-$\kappa$ materials with copper processing is still difficult. Diffusion barriers have been introduced to avoid copper migration. As the wire size decreases, electron scattering on the grain boundaries and the interfaces leads to an increase of copper resistivity (Fig. 3) which is detrimental to the propagation delay. Moreover, it is also worth noting that copper resistivity strongly depends on temperature: it changes by about 40% for a temperature variation of 100 K.

Repeaters are widely used to reduce interconnect delay, transition times and crosstalk noise. Their number is planned to grow as the technology node changes. As a consequence, the needed silicon area will increase and the

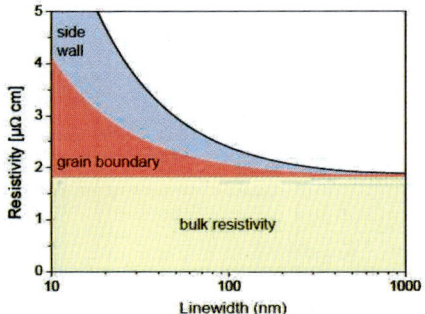

Fig. 3. Increase of copper resistivity as the feature size is reduced (from ITRS[1]).

associated power dissipation will represent a significant part of the global circuit dissipation.[13] Furthermore, via blockade will appear as an additional problem for circuit fabrication. Repeaters also introduce synchronization uncertainties, which become all the more important as the clock period decreases.[9,14]

With respect to these problems, optics presents several advantages.[14,15] The high bandwidth removes the high frequency limitation. As signal distortion along the propagation is negligible, even for high frequencies ($\gg 10\,\text{GHz}$) and long distances ($> 1\,\text{cm}$), no repeaters are needed. This allows saving silicon area and reducing design complexity and power dissipation. Optics can also ensure reduced latency and smaller skew and jitter, which are important parameters, in particular for clock signal distribution. It permits larger synchronous zones and better signal synchronization. Furthermore, optical signals are insensitive to electrical perturbations yielding noise immunity and voltage insulation properties.

## 3.    Building Blocks for Optical Interconnects

Optical interconnects could be used to replace some of the global interconnections, either for clock signal distribution or more generally for signalling. The skeleton of optical interconnects is represented in Fig. 4 for both kinds of applications. The building blocks include a light source, a modulator to encode the signal, optical waveguides with splitters and turns to distribute light on the chip and photodetectors at the end of distribution to convert the optical signal back to an electrical one.

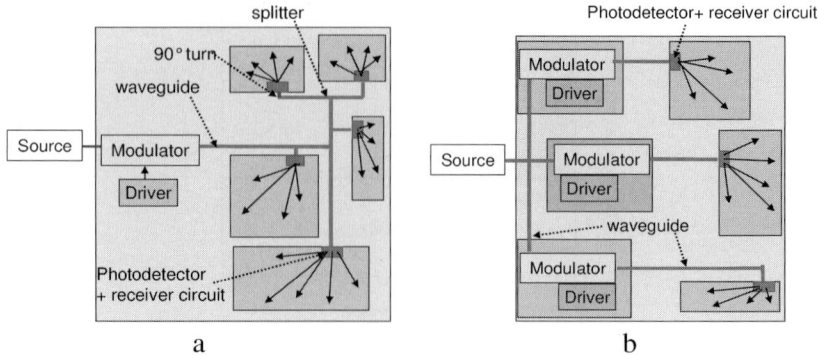

Fig. 4. Schematic representation of what optical interconnects could be for clock signal distribution (a) or signalling (b).

As discussed later, monolithic integration of a silicon emitter is the less mature point among the building blocks and an external source can be considered at first. The other active elements like modulators and photodetectors have to be integrated with the passive optical distribution. So the waveguide material and geometry is preferentially chosen to be consistent with the constraints induced by the active element requirements and by the compatibility with CMOS process. Submicron waveguides are widely used and splitter and turn areas are reduced to ensure compactness. Modulators are designed to operate at frequencies larger than 10 GHz and ultrafast and efficient photodetectors are developed. The photocurrent delivered by photodetector is generally amplified through a transimpedance amplifier (TIA).

## 3.1. Light distribution

### 3.1.1. Optical waveguides

An optical waveguide consists of a high refractive index layer surrounded by lower refractive index materials. A simplified ray theory shows that the light can be trapped in the waveguide core if total reflection occurs on the interfaces[16] (Fig. 5).

As boundary conditions imply that the longitudinal component of the wavevector is continuous at the interfaces, light is guided in the core when $\beta$ is larger than the light wavevector in any of the surrounding material, i.e. in such a way that light can propagate neither in the substrate nor in the cladding.

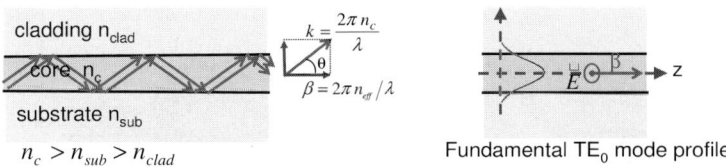

$n_c > n_{sub} > n_{clad}$          Fundamental $TE_0$ mode profile

Fig. 5. Optical waveguide.

$$\beta = \frac{2\pi n_{eff}}{\lambda} > \frac{2\pi n_{sub}}{\lambda} > \frac{2\pi n_{clad}}{\lambda}$$

Due to interferences, only discrete values of $\beta$ are allowed, defining the propagation modes in the waveguide. Each guided mode is characterized by an effective index $n_{eff}$. The thinner the waveguide, the lower the number of guided modes. Single mode waveguides are the most widely used, as they avoid mode coupling and ensure minimum propagation loss. The waveguide core is limited laterally to define a strip constituting a 2D optical waveguide.

The main required features are low propagation loss and compactness. The former needs materials which are transparent at the used wavelength to avoid absorption, and have a good optical quality to prevent scattering loss. High refractive index difference $\Delta n = n_c - n_{sub}$ between the core and the other materials enhances the electromagnetic field confinement, allowing small waveguide cross-sections, sharp bends and low crosstalk to ensure compactness.

Silicon-On-Insulator (SOI) is a choice material as it fulfils these requirements and it is compatible with CMOS technology. It consists of a silicon film separated from the silicon substrate by a buried silicon oxide layer (BOX). Silicon is transparent at the telecommunication wavelength (1.3–1.6 $\mu$m) and the crystalline character ensures a high optical quality. Its refractive index is around 3.5, compared to 1.5 for silicon oxide which is generally also used for cladding. This very large refractive index contrast leads to strong electromagnetic field confinement which is quite favourable for increasing the integration density. However, the BOX thickness must be larger than 1 $\mu$m to avoid light leakage towards the silicon substrate through the buried silicon oxide layer. The silicon thickness usually ranges from 0.2 to 0.4 $\mu$m.

Silicon wires, made by etching the silicon film down to the BOX, have been widely studied. However, these 2D waveguides suffer from propagation loss of at least a few dB/cm due to light scattering on the sidewall roughness[17] and are not favourable for integration of active devices. Rib

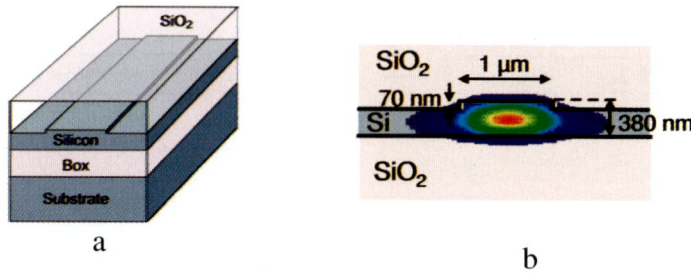

Fig. 6. Rib SOI waveguide geometry (a) and mode profile in the waveguide cross-section ($\lambda = 1.3\mu$m) (b).

waveguides made by shallow etching of the silicon film (Fig. 6a) offer potentialities for optical interconnect applications. Light is well confined under the rib. An example of calculated mode profile for a slightly etched single mode waveguide is shown in figure 6b for $\lambda = 1.3$ $\mu$m.

As the mode interacts only slightly with the etched sidewalls, the measured propagation loss for such waveguides[18] is as small as 0.1 dB/cm.

### 3.1.2. *Light injection in submicron waveguides*

Efficient light injection in the submicron waveguides is needed for developing applications. The optical mode of a single mode optical fibre has a typical diameter of 8 $\mu$m, which is much larger than the waveguide cross section. A direct coupling from the fibre to the waveguide on the chip edge requires a polished or cleaved facet and leads to high insertion loss ($> 30$ dB). Even if a lensed fiber is used to reduce the fibre mode diameter to about 3 $\mu$m, insertion loss is still as high as 12 dB. A classical way to couple a large amount of the incident light is to use a diffraction grating. It consists of grooves regularly etched on the silicon surface (Fig. 7). It allows to add a component to the tangential component to the light wavevector and then, for a given incidence angle, to adjust the wavevector of the diffracted beam to the propagation constant of the guided mode. The grating size and the groove depth are adjusted to the beam diameter.[19,20] A taper reduces the width of the guided beam to the waveguide size. Optimization[21] can lead to a taper length of the order of 15 $\mu$m. The measured coupling efficiency[22] is larger than 50% (loss $< 3$ dB).

The same device can be used in a reverse way to decouple the light from the waveguide if necessary. It is worth noting that such couplers can be inserted at any place on a chip, which allows versatile designs and can

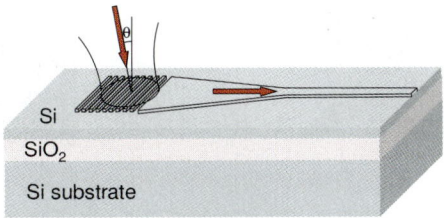

Fig. 7. Grating coupler with taper for light injection in submicron waveguide.

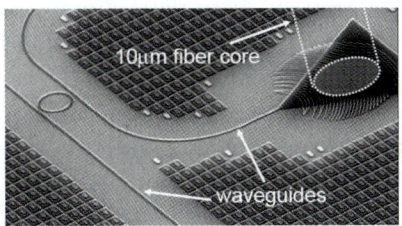

Fig. 8. Holographic lens for light coupling from an optical fiber to a submicron waveguide (from Ref. 23).

be useful for wafer testing before packaging. The set formed by the grating and the taper can also be replaced by a curved grating or holographic lens as developed by Luxtera[23]. The best published results give insertion loss of 1.5 dB in the 1530–1560 nm wavelength range.[23]

### 3.1.3. *Turns and splitters*

Compact 90° turns in slightly etched SOI rib waveguides can be made by etching silicon down to the BOX to obtain a mirror facet at the angle between two perpendicular waveguides (Fig. 9).

a                      b

Fig. 9. Etched mirror for 90° turn of rib waveguides: FDTD calculation of the field amplitude (a) scanning electron microscope (SEM) view after removal of the silicon oxide (b).

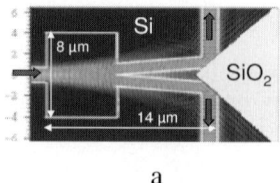

a                                    b

Fig. 10. T-splitter for rib waveguides: 3D-FDTD calculation of the field amplitude (a) ; SEM view after removal of the silicon oxide (b).

Theoretical loss determined from three dimensional Finite Difference Time Domain (3D-FDTD) numerical calculations is 0.1 dB, and the measured value[18] is under 1 dB.

Low loss and compact T-splitters can be made by collecting the light in two waveguides after it has diffracted in a wider slab region[24] (Fig. 10). Mirrors are added for convenience to deflect the beams perpendicularly to the incident waveguide. The whole splitter is only 14 $\mu$m long and 8 $\mu$m wide. The measured excess loss is 0.7 dB[2], compared to the 0.15 dB theoretical value from 3D-FDTD calculations.

### 3.1.4.   Crosstalk

One advantage of optics is that waveguides can intersect on the same level. Crosstalk between two perpendicular rib SOI waveguides has been calculated using the FDTD method (Fig. 11). The calculated transmission at $\lambda = 1.3\mu$m is larger than 98% and the measured value is 93%, which corresponds to 0.3 dB loss.[25]

Crosstalk between parallel neighbour waveguides was also estimated.[25] For the slightly etched rib waveguides as previously considered, a distance of 2 $\mu$m is enough to prevent from coupling from one waveguide to the other over propagation distances of several centimeters. This is illustrated in Fig. 12 where the modes in the two waveguides are clearly distinct.

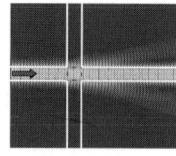

Fig. 11.  FDTD calculation of the field amplitude at perpendicular waveguide crossing.

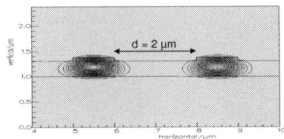

Fig. 12. EM field intensity profile in parallel rib SOI waveguides.

## 3.1.5. On-chip optical signal distribution

The various components described here above have been used to demonstrate low loss light distribution at chip scale. A H-tree distribution with 1 cm long branches, each including 4 splitters and 2 mirrors, has validated an equal power repartition between 16 output points.[3] Up to 10 successive divisions by 2 of the waveguide still allows signal detection, which demonstrates the feasibility of low loss light distribution from one input to 1024 outputs.[2] Assuming 10 mW optical input power, the measured output power is 3 $\mu$W after 10 divisions.

## 3.2. Integrated emitters

Due to its indirect band structure, silicon is not favourable for light emission. Radiative processes are very slow and most of the excited carriers recombine non-radiatively. The luminescence efficiency is very low in bulk silicon, of the order of $10^{-6}$. Many efforts have been made to reduce the non-radiative channels by improving the material,[26,27] structuring the surface,[28] using silicon nanocrystals[29] or nanostructured pn junctions.[30] On the other hand, luminescence efficiency has been improved by introducing erbium.[31] However, monolithic integration of a silicon-based emitter remains a challenge up to now.

A LED based on erbium doped nanocrystals in a MOS structure, emitting at $\lambda = 1.54 \mu$m with an efficiency of about 10%, i.e. comparable to III-V semiconductor LEDs,[32] was presented by STMicroelectronics in 2002.

In 2005, Intel claimed that laser emission has been obtained from silicon at $\lambda = 1.686 \mu$m, but this was from Raman effect, needing an external source at $\lambda = 1.55 \mu$m for optical pumping.[33] Efforts are now mainly focused on hybrid lasers made with InP diodes bonded on SOI substrate (Fig. 13).[34–37]

In 2005, the heterogeneous integration of an InP-based microdisk laser diode on a silicon substrate led to laser emission at 1544 nm.[35] A laser cavity

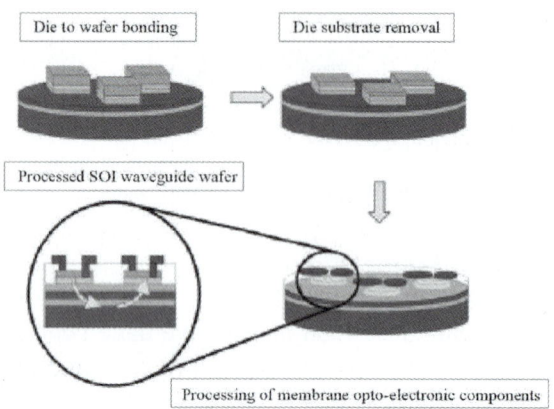

Fig. 13. Processing sequence for heterogeneous integration of III–V components and SOI waveguides (from Ref. 34).

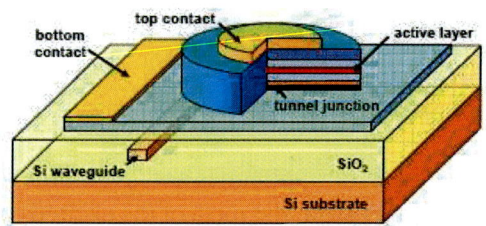

Fig. 14. Schematic cross section of the hybrid microdisk InP/silicon laser.

is formed by etching the heterostructure as a microdisk. The whispering-gallery modes excited electrically are evanescently coupled to a buried silicon waveguide just below (Fig. 14).

In 2006, Intel and the University of California, Santa Barbara (UCSB) announced the demonstration of an electrically driven hybrid silicon laser.[36] The indium phosphide-based wafer is bonded directly on pre-patterned silicon photonic chip, with no needs for precise alignment. Electrical contacts are then patterned onto the device. When a voltage is applied to these contacts, light is generated in the InP-based materials and coupled by evanescent waves into the silicon waveguide just below. The performances of the hybrid laser, in particular the emission wavelength, are determined by the cavity which is either formed by the facets of a straight waveguide[36] or by a racetrack resonator.[37]

Such bonded lasers could bring a solution for integrating emitters on silicon, before the all-silicon laser demonstration.

## 3.3. Silicon-based integrated modulators

Whatever the laser source used, the optical signal has to be encoded to ensure information transmission. To make the optical interconnects worthwhile, the frequencies which are aimed at are larger than 10 GHz. Direct modulation of an integrated laser source will not be efficient enough to meet the performance requirements. Impressive progresses have been obtained in the recent years for silicon-based modulators, although silicon is not an ideal material for modulation. The linear electrooptic effect, known as "Pockels effect" and commonly used in LiNbO$_3$ modulators, does not exist in centrosymmetric crystal like silicon. Due to the indirect bandgap, the absorption edge slope is not as steep as in III-V semiconductor structures and electroabsorption cannot be used to get large modulation depth and low insertion loss. The only viable mechanism is the free carrier dispersion properties. Both the real part $\Delta n$ and the imaginary part $\Delta \alpha$ of the refractive index change with the free carrier density $\Delta N$ for the electrons and $\Delta P$ for the holes according to the following relations for $\lambda = 1.3 \mu m$[38]:

$$\Delta n = -6.210^{-22} \Delta N - 6.010^{-18} \Delta P^{0.8}$$

$$\Delta \alpha = 6.010^{-18} \Delta N + 4.010^{-18} \Delta P.$$

For $\lambda = 1.55 \mu m$, the coefficients are slightly different:

$$\Delta n = -8.810^{-22} \Delta N - 8.510^{-18} \Delta P^{0.8}$$

$$\Delta \alpha = 8.510^{-18} \Delta N + 6.010^{-18} \Delta P$$

Several means can be used to vary the carrier density: carrier injection in a PIN diode, carrier accumulation in a MOS structure or carrier depletion in a PIN diode. Each structure is integrated in a SOI rib waveguide and the refractive index variation induces a phase shift of the guided wave. An interference device such as Mach-Zehnder interferometer, Fabry-Perot microcavity or microring resonator is used to convert the phase modulation into an intensity one. The best published results are summarized hereafter.

### 3.3.1. Injection-based modulator

The main advantage of carrier injection is that large variations of the carrier density, up to a few $10^{18}$ cm$^{-3}$, can be obtained. The induced index change is then of the order of $10^{-3}$ which allows characteristic lengths $L_\pi$ of a few hundreds of microns to get a phase variation equal to $\pi$. The main drawback arises from the operation frequency limitation to a few hundreds of MHz related to the carrier recombination time. However, high speed operation can be achieved by applying a reverse bias voltage, after the direct one, to extract the carriers from the active region.[39] The structure consists in a microring resonator with a radius of 5 $\mu$m with a PIN diode embedded in and side-coupled to a straight waveguide (Fig. 15). The waveguide rib has a width of 400 nm and a height of 200 nm. The distance between the ring and the waveguide is of the order of 200 nm. At the resonant wavelength $\lambda_0$, light is coupled into the ring resonator and undergoes loss, mainly due to scattering on the sidewall roughness, which in turn induces a dip in the transmission spectra. When a direct bias voltage is applied, injected free carriers are responsible for a decrease of the effective index in the ring. The resonance wavelength is shifted towards shorter wavelengths and the transmission at $\lambda_0$ significantly increases. The output power can then be modulated by applying a given voltage on the device.

The operation speed is enhanced by increasing the current to reach more rapidly the global charge necessary to get a high transmission, and then by applying a reverse bias to extract the carriers in the ring,[40] which however needs a rather complex driver. Modulation at 12.5 Gbit/s has been demonstrated experimentally using peak-to-peak voltage of 8 V, with an extinction ratio about 9 dB.

The microring structure allows very compact devices, insuring low capacitance and reduced access resistances. The main drawback is the sensitivity to temperature fluctuations which induce parasitic refractive index variations.

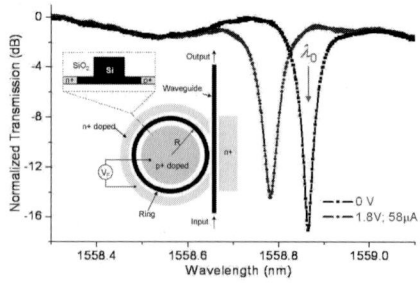

Fig. 15. Modulator structure and transmission spectra (from Ref. 23).

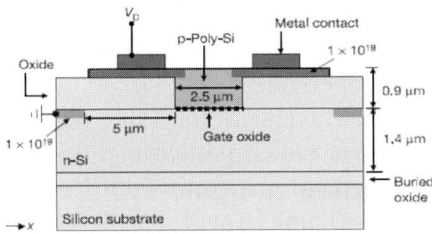

Fig. 16. MOS capacitor designed as waveguide phase shifter (from Ref. 41).

### 3.3.2. *Modulation by accumulation in a MOS device*

MOS capacitor working in the accumulation regime is a unipolar device, and this avoids the speed limitation due to carrier recombination. The phase shifter structure[41] is schematically represented in Fig. 16. The gate oxide thickness is 12 nm. The waveguide is formed by the n-Si layer and the p-doped poly-silicon rib. The latter is replaced by crystalline silicon obtained by epitaxial lateral overgrowth (ELO) to reduce optical loss.[42] When a positive voltage is applied to the p-type silicon, thin charge layers accumulate on both sides of the gate oxide. The refractive index change in the accumulated charge layers induces an effective index variation of the guided mode. As the charge layer thickness is very small, i.e. of the order of 10 nm on both sides of the gate oxide, the overlap with the optical mode is reduced and the phase change efficiency is limited.

A figure of merit for phase efficiency is commonly defined as the product $V_\pi L_\pi$, where $V_\pi$ and $L_\pi$ are the voltage swing and device length required to achieve *pi*-radian phase shift. With reduced waveguide dimensions ($1.6\,\mu\mathrm{m} \times 1.6\,\mu\mathrm{m}$) $V_\pi L_\pi = 3.3\,\mathrm{V.cm}$ has been demonstrated.[42] A Mach-Zehnder interferometer (MZI) has been used to achieve intensity modulation (Fig. 17).

The incident light is split between two arms. If both arms are identical, the two guided beams are in phase and recombine with a maximum intensity. If a $\pi$- phase shift is introduced in one of the arms, destructive interferences

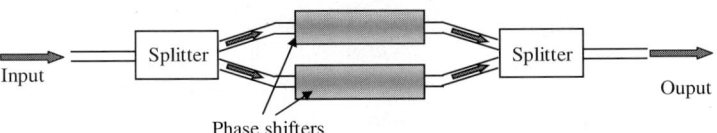

Fig. 17. Mach-Zehnder interferometer.

occur and the resulting intensity cancels. A main advantage of MZI is the insensitivity to temperature fluctuations when the two arms are close enough to be kept at the same temperature.

Owing to the rather high value of the $V_\pi L_\pi$ product, the MZI length is 15 mm, including the input and output waveguides. The measured optical loss is 10 dB. The estimated intrinsic bandwidth due to RC cutoff is 10 GHz, and data transmission at 10 Gbits/s with 3.8 dB extinction ratio (ER) was reported.[42]

### 3.3.3. *Depletion-based modulators*

Electric-field induced carrier depletion is also a unipolar process which allows high speed operation. One further main advantage is that the overlap between the depleted layer and the optical guided mode can be optimized to increase the phase efficiency.[43,44] Quite recently, high speed modulation up to 20 GHz has been demonstrated.[45] The phase shifter consists in a PN junction integrated in SOI rib waveguide (Fig. 18). It is inserted in the arms of a Mach-Zehnder interferometer. A reverse bias induces carrier depletion. RC constants due to device capacitance and access resistances are the main source of frequency limitation.[43] To overcome this problem and achieve high speed operation, a travelling-wave design has been used so that the electrical and optical signals propagate with similar speeds along the phase shifter.[45]

The measured characteristics[45] give on-chip insertion loss of about 7 dB and a $V_\pi L_\pi$ product about 4 V.cm. This relatively high value is still due to the limited thickness of the depleted space charge and its position near the PN junction. This can be improved by localizing the free carriers in a specific layer included in a PIN diode and centered on the optical guided mode, which yields a $V_\pi L_\pi$ experimental value of 3.1 V.cm before

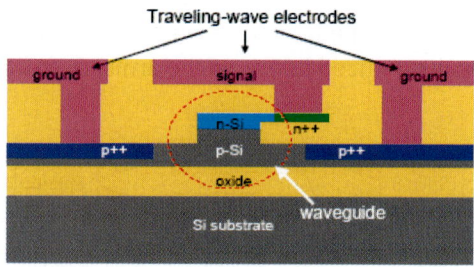

Fig. 18. Schematic cross-sectional view of the SOI waveguide phase shifter (from ref. 45).

optimization.[44] With such a device, a 10 GHz bandwidth has been obtained, with a modulation depth of 14 dB and insertion loss as low as 5 dB.[46]

However, with the careful design of the modulator reported in reference 45, a 3-dB roll-off frequency of $\approx 20$ GHz has been measured, and feasibility of data transmission of 30 Gb/s at telecommunication wavelengths was demonstrated.[45]

The tremendous progress made on integrated silicon-based modulators during the very last years gives to expect still better performances which could be quite competitive with respect to III-V semiconductor modulators. The semiconductor company Luxtera[47] already published results about a manufacturable 10 Gb/s modulator integrated with electronics in 0.13 $\mu$m SOI CMOS.[48]

## 3.4.   *Integrated germanium photodetectors*

At the end of the link, the optical signal has to be converted back into an electrical signal. A compact integrated photodetector is then needed, which must be compatible with the microelectronics technology. Strong absorption in the 1.3–1.6 $\mu$m range is required and cannot be insured by silicon itself as it is transparent and used for light guiding. The best silicon-compatible material is germanium, which is absorbent at the considered wavelengths and already introduced in microelectronics. The main issue with pure germanium is the large lattice mismatch with silicon, which is 4.2%. However, high quality crystalline germanium layers can be grown on silicon using a two-step epitaxy process.[49−51] The measured absorption spectrum of thin germanium layers is very close to the bulk germanium one (Fig. 19). A red shift of the absorption edge is observed, which allows

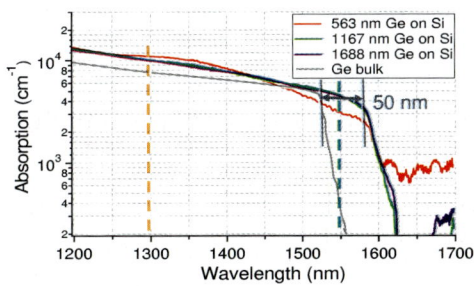

Fig. 19.  Absorption spectrum of germanium layers epitaxially grown on silicon with various thicknesses.

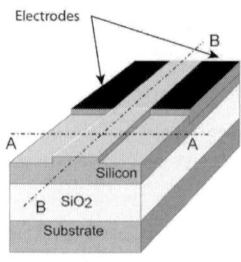
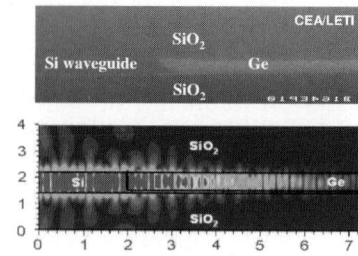

Fig. 20. Schematic view of the integration of Ge photodetector in a rib waveguide SEM view and electric field amplitude from FDTD calculations in the longitudinal cross-section ($\lambda = 1.31\,\mu$m).

detection up to $1.6\,\mu$m. It is due to a tensile strain that appears during the cooling from the epitaxy temperature (730°C) to room temperature.[52]

To integrate the photodetector in a SOI waveguide, a selective epitaxial growth of Ge is made in a recess etched in silicon (Fig. 20). Calculation of the electric field amplitude by the 3D-FDTD method shows that 95% of the light is absorbed over a distance of $4\,\mu$m. This allows very short detector length and is favourable for high speed operation.

Either metal/semiconductor/metal (MSM) photodetector or pin photo-diode can be made according to this scheme. The first one benefits by simple technological process but suffers from relatively large dark currents. On the other hand, photodiode requires more sophisticated technology but presents low dark currents.

MSM photodetectors integrated in the rib SOI waveguides have been fabricated and tested. The measured 3 dB-bandwidth reaches 25 GHz under a 6 V bias voltage. High responsivity is achieved with a measured value[4] of $1 \pm 0.2$ A/W at $1.55\,\mu$m.

Integrated PIN germanium photodetectors have also been reported with a bandwidth close to 30 GHz and an efficiency of 93%.[53]

The integration in SOI-CMOS technology of the transimpedance amplifier (TIA) used to process the detector signal with the photodetector has been recently announced.[47]

## 4.    Optical Versus Metal Interconnects

Accurate comparison between optical and metallic interconnect performances is quite difficult. It presupposes that realistic forecast of CMOS circuit performances and of optoelectronic devices are available. This is particularly unreliable for example for modulators which do not present definite characteristics. Most of the predictions are pessimistic for optics, but they assume either very large photodetector capacitance[54] (250 fF instead of a few fF[55]) or rather large photocurrents (100 $\mu$A) and an overestimation of the detector + TIA delay[56,57] (65 or 40 ps vs a few ps[55]). This leads to a critical length, above which optical interconnects are advantageous over electrical ones, of a few mm. This length is increased if non-scaled Cu interconnects are considered,[57] i.e. if the dimensions of Cu interconnects are maintained constants for every technology node. In this case, several interconnect layers are to be added to provide an equivalent bandwidth. Considering device models taking into account the recent developments of photonic devices, more optimistic results are obtained.[55,58,59]

In order to quantify some of the performances of optical interconnects, the optical link shown in Fig. 21 is considered. It consists of an off-chip laser, a transmitter composed of an optical modulator with its driver circuit, a waveguide and a receiver including the photodetector and the transimpedance amplifier.

For the transmitter, the RC limitation of the frequency response of the Mach-Zehnder interferometer modulator is minimized by using travelling wave electrodes, which also reduce power consumption. Small power consumption is also expected with the microring configuration, due to the low capacitance associated to the small size of the device.

The waveguide size determines the interconnect pitch and is driven by the optical wavelength. The high index contrast of SOI waveguides allows a rather high optical wiring density and the waveguide pitch can be smaller than 2 $\mu$m. Light propagation velocity in a SOI waveguide is of about one third of the velocity in vacuum (c/3), which leads to a reduced delay of 10 ps for a 1 mm line, compared to several hundreds ps for electrical global

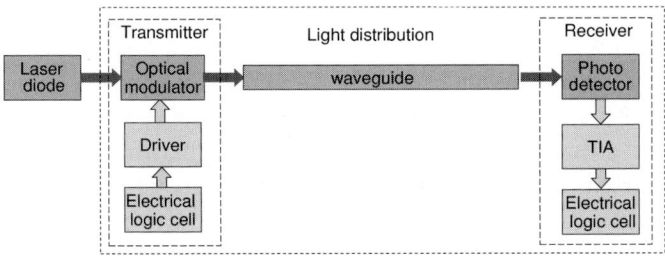

Fig. 21. Diagram of an on-chip optical link.

interconnects.[1] Furthermore, the optical signal propagates with negligible attenuation, and signal distortion is also very small as chromatic dispersion is only of the order of 10 fs/nm.cm. So the optical signal can propagate over distances of several cm without any power dissipation and any distortion whatever the operation frequency is. Furthermore, the propagation delay is almost insensitive to temperature change. A $\Delta T = 100°C$ variation makes the delay increases by 0.3 ps for a propagation length of 1 cm. It is worth noting that copper resistivity changes by 40 % for the same $\Delta T$.

High speed and high responsivity photodetectors are now available. For capacitance values smaller than 1 fF, a simple model for the TIA shows that the maximum bandwidth can be larger than 50 GHz and the dissipated power is a few tenths of mW.[55] The required optical power to ensure a BER of $10^{-15}$ is of the order of 50 $\mu$W.

The important points for clock distribution are delay, skew and jitter, and dissipated power. The delay is already smaller for optical interconnects than for electrical wiring. A potential source of skew or jitter on the optical distributed clock signal arises from possible differences between the optical intensity incident on each photodetector, due either to local process variations in the optical distribution or to random time variations of the optical source intensity. In the worst case, time uncertainty is about 3% of the clock period even for 50 GHz operation,[55] This is well under the values for conventional clock distribution, the requirements being typically of the order of 20% of the clock cycle. In the same way, a 100 $\mu$m difference of the optical path between two branches of an optical distribution leads to a 1 ps skew, which is only 2% of the period for a 50 GHz frequency. Thus optical interconnects can insure better synchronization over the entire chip.

Power consumption in microelectronic circuits like microprocessors strongly increases with clock frequency. The clock signal distribution itself consumes typically more than 50% of the total circuit power and is close to 60 W for the Pentium 4 operating at 2.53 GHz. For optical interconnects, power dissipation only occurs in the optoelectronic interfaces. Considering an optical clock distribution, it needs one modulator and a given number of photodetectors. The power consumption of the transmitter can be assumed to be smaller than 100 mW.[58] For the receivers, the static power dominates and it is of the order of 0.2 mW per receiver. Light distribution towards 64 points on the chip is quite realistic.[2,3] This yields total power consumption well under 1W for the optical clock distribution. Even if only 20% of the clock power is dissipated in the global interconnects which may be replaced by an optical distribution, this is in favour of optics. The advantage also comes from the fact that no repeaters are needed, and this saves power and silicon area and allows to get rid of a source of skew and jitter.

## 5.     Integration of Optics with CMOS Electronics

Although optical interconnects present significant advantages over global metal wiring, several issues are still to be considered for an effective integration with CMOS circuits.[6] Although the devices described here above are made using the same technology process steps than for CMOS circuits, monolithic integration implies the full compatibility of the SOI substrates and of the process flow including the thermal budget. To alleviate the constraints, some of the components can be fabricated on separate substrates and assembled in hybrid integration.

Although SOI is now commonly used for CMOS circuits, the tendency is to reduce the BOX and silicon layer thicknesses well beyond the minimum values required for optics. Monolithic integration of photonics with electronics would imply either a compromise or local increases of the layer thicknesses. The latter would make more difficult chemical Mechanical Polish (CMP) process which is often used. Preliminary studies show that a process flow taking into account the temperature constraints could be defined.[7]

One way to avoid interference with front end process is to design a photonic layer which is then bonded on the CMOS wafer in a 3D integration process.[7] The optical waveguides and the optoelectronic devices are fabricated on the photonic wafer. Oxide cladding with CMP and perfect cleaning of the two wafers allows their molecular bonding. Removing the back side of the optical wafer leaves a flat surface of oxide. Photonics can be introduced at one of the upper metal levels. Etching through the top layer is needed to connect the electro-optic components with the CMOS circuit.

In the near term, hybrid integration using silicon bench technology can be cost-effective and insure reasonable yields.[6] The SOI photonic wafer is attached by the flip-chip method on the electronic silicon wafer using metal pads.

## 6.     Conclusion

Tremendous progress has been noted in the very recent years in silicon photonics. Low loss light distribution towards at least 64 points, including rib waveguides with losses smaller than 0.1 dB, compact turns and splitters, is now available. Germanium provides high speed and high responsivity integrated photodetectors. Silicon modulators begin to reach the expected performances. Innovative research is still needed concerning silicon-based sources, but hybrid emitters have been successfully demonstrated. Appropriate testing methods have also to be developed to screen out bad devices as early as possible. Photonics technology is obviously much less mature than silicon electronics one. The present evolution can be compared to microelectronics twenty years ago. The clear advantages of silicon photonics are still tempered by manufacturing issues and need for high yields and low cost developments, but the recent advances let assume that silicon photonics will become a serious asset for future technologies.

## References

1.      International Technology Roadmap for Semiconductors, http://www.itrs.net
2.      D. Marris, L. Vivien, D. Pascal, M. Rouvière, E. Cassan, A. Lupu and S. Laval, *Applied Physics Letters* **87**, 211102 (2005).
3.      L. Vivien, S. Lardenois, D. Pascal, S. Laval, E. Cassan, J.-L. Cercus, A. Koster, J. M. Fédéli and M. Heitzmann, *Applied Physics Letters* **85**, 701–703 (2004).
4.      L. Vivien, M. Rouvière, J.-M. Fédéli, D. Marris-Morini, J.-F. Damlencourt, J. Mangeney, P. Crozat, L. El Melhaoui, E. Cassan, X. Le Roux, D. Pascal and S. Laval, *Optics Express* **15**, 9843 (2007).
5.      D. Ahn, C. Hong, J. Liu, W. Giziewicz, M. Beals, L. C. Kimerling and J. Michel, *Optics Express* **15**, 3916 (2007).
6.      N. Izhaky, M. T. Morse, S. Koehl, O. Cohen, D. Rubin, A. Barkai, R. Cohen and M. J. Paniccia, *IEEE Journal of Selected Topics in Quantum Electronics* **12**, 1688 (2006).
7.      J.-M. Fedeli, R. Orobtchouk, C. Seassal and L. Vivien, *SPIE Photonics West 2006, San Jose, USA*, 6125–15 (2006).
8.      R. H. Havemann and J. A. Hutchby, *Proceedings of the IEEE* **89**, 586 (2001).
9.      P. Kapur, *IITC 2002 Short Course, San Francisco* (2002).
10.     T. N. Theis, *IBM Journal of Research and Development* **44**, 379 (2002).
11.     A. Naeemi, R. Venkatesan and J. D. Meindl, *IEEE Transactions on Electron Device*, **50**, 980 (2003).

12. Y. I. Ismail, *Proceedings of the ACM/IEEE Design Automation Conference*, 721 (1999).
13. P. Kapur, G. Chandra and J. P. McVittie, *IEEE Trans. ED* **49**, 598 (2002).
14. A. V. Mule, E. N. Glytsis, T. K. Gaylord and J. D. Meindl, *IEEE Transactions on Very Large Scale Integration (VLSI) Systems* **10**, 582 (2002).
15. D. A. B. Miller, *Proceedings IEEE* **88**, 728 (2000).
16. See for example K. Okamoto, *Fundamental of Optical Waveguides*, (Academic Press, (2000).
17. F. Grillot, L. Vivien, S. Laval and E. Cassan, *IEEE Journal of Lightwave Technology* **24**, 891–896 (2006).
18. S. Lardenois, D. Pascal, L. Vivien, E. Cassan, S. Laval, R. Orobtchouk, M. Heitzmann, N. Bouzaida and L. Mollard, *Optics Letters* **28**, 1 (2003).
19. D. Pascal, R. Orobtchouk, A. Layadi, A. Koster and S. Laval, *Applied Optics* **36**, 2443 (1997).
20. N. Landru, D. Pascal and A. Koster, *Optics Communications* **196**, 139 (2001).
21. W. Bogaerts, D. Taillaert, B. Luyssaert, P. Dumon, J. Van Campenhout, P. Bienstman, D. Van Thourhout, R. Baets, V. Wiaux and S. Beckx, *Optics Express* **12**, 1583 (2004).
22. D. Pascal, S. Lardenois, E. Cassan, A. Koster, S. Laval, M Heitzmann., L. Mollard , B. Dal'zotto, N. Bouzaïda and R. Orobtchouk, *Integrated Photonics Research Proceedings* TOPS **78**, (2002).
23. C. Gunn, *2005 IEEE International SOI Conference Proc.*, 7 (2005).
24. A. Koster, E. Cassan, S. Laval, L. Vivien and D. Pascal, *Journal of Optical Society of America A.* **21**, 2180 (2004).
25. E. Cassan, S. Laval, S. Lardenois and A. Koster, *IEEE Journal of Selected Topics in Quantum Electronics* **9**, 460 (2003).
26. M. A. Green, J. Zhao, A. Wang and T. Trupke, *Physica E* **16**, 351 (2003).
27. M. A. Lourenço, M. Milosavljević, R. M. Gwilliam, K. P. Homewood and G. Shao, *Applied Physics Letters* **87**, 201105 (2005).
28. T. Trupke, J. Zhao, A. Wang, R. Corkish and M. A. Green, *Applied Physics Letters* **82**, 2996 (2003).
29. L. Pavesi, *Proceedings SPIE*, Photonic West San Diego (2003).
30. M. J. Chen, J. L. Yen, J. F. Chang, S. C. Tsai and C. S. Tsai, *Applied Physics Letters* **84**, 2163 (2004).

31.  S. Libertino, S. Coffa and M. Saggio, *Material Science in Semiconductor Processing***3**, 375 (2000).
32.  M. Castagna, S. Coffa, L. Caristia, A. Messina and C. Bongiorno, *ESSDERC Proc.*, 439 (2002).
33.  H. Rong, R. Jones, A. Liu, O. Cohen, D. Hak, A. Fang and M. Paniccia, *Nature,* **433**, 725 (2005).
34.  G. Roelkens, D. Van Thourhout and R. Baets, *Journal of Lightwave Technology* **23**, 3827 (2005).
35.  P. Rojo Rome, J. Van Campenhout, P. Regreny, A. Kazmierczak, C. Seassal, X. Letartre, G. Hollinger, D. Van Thourhout, R. Baets, J. M. Fedeli and L. Di Cioccio, *Optics Express* **14**, 3864 (2006).
36.  Intel Inc., http://www.intel.com/research/platform/sp/hybridlaser.htm.
37.  A. W. Fang, R. Jones, H. Park, O. Cohen, O. Raday, M. J. Paniccia and J. E. Bowers, *Optics Express* **15**, 2315 (2007).
38.  R. A. Soref and B. R. Bennett, *IEEE Journal of Quantum Electronics* **QE-23**, 123–129 (1987).
39.  Q. Xu, B. Schmidt, S. Pradhan and M. Lipson, *Nature* **435**, 325 (2005).
40.  Q. Xu, S. Manipatruni, B. Schmidt, J. Shakya and M. Lipson, *Optics Express* **15**, 430 (2007).
41.  A. Liu, R. Jones, L. Liao, D. Samara-rubio, D. Rubin, O. Cohen, R. Nicolaescu and M. Paniccia, *Nature* **427**, 616 (2004).
42.  L. Liao, D. Samara-Rubio, M. Morse, A. Liu, D. Hodge, D. Rubin, U. D. Keil and T. Franck, *Optics Express* **13**, 3129 (2005).
43.  S. Maine, D. Marris-Morini, L. Vivien, D. Pascal, E. Cassan and S. Laval, *Proc SPIE* **6183**, 618360D1-6 (2006).
44.  D. Marris-Morini, X. Le Roux, Laurent Vivien, E. Cassan, D. Pascal, M. Halbwax, S. Maine, S. Laval, J.-M Fédéli and J.-F Damlencourt, *Optics Express* **14**, 10838 (2006).
45.  A. Liu, L. Liao, D. Rubin, H. Nguyen, B. Cifcioglu, Y. Chetrit, N. Izhaky and M. Paniccia, *Optics Express* **15**, 660 (2007).
46.  D. Marris-Morini, L. Vivien, J.-M. Fédéli, E. Cassan, P. Lyan and S. Laval, *Optics Express* **16**, 334 (2008).
47.  http://www.luxtera.com
48.  A. Huang, G. Gunn, G.-L. Li, Y. Liang, S. Mirsaidi, A. Narasimha and T. Pinguet, *ISSCC Technical Digest*, Session 13, Communication 13.7 (2006).

49.  S. Fama, L. Colace, G. Masini, G. Assanto and H.-C. Luan, *Appl. Phys Lett.* **81**, 586 (2002).
50.  M. Halbwax, M. Rouviere, Y. Zheng, D. Debarre, L. H. Nguyen, J.-L. Cercus, C. Clerc, V. Yam, S. Laval, E. Cassan and D. Bouchier, *Optical Materials* **27**, 822 (2005).
51.  J.-M. Hartmann, A. Abbadie, A.M. Papon, P. Holliger, G. Rolland, T. Billon, J.-M. Fédéli, M. Rouvière, L. Vivien and S. Laval, *Journ. Appl. Phys.* **95**, 5905 (2004).
52.  Y. Ishikawa, K. Wada, D. D. Cannon, J. Liu, H.-C. Luan and L. C. Kimerling, *Appl. Phys. Lett.* **82**, 2044 (2003).
53.  T. Yin, R. Cohen, M. M. Morse, G. Sarid, Y. Chetrit, D. Rubin and M. J. Paniccia, *Optics Express* **15**, 13965 (2007).
54.  P. Kapur and K. C. Saraswat, *Physica E* **16**, 620 (2003).
55.  E. Cassan, D. Marris, M. Rouvière, S. Laval, L. Vivien and A. Koster, *Optical Engineering* **44**, 105402 (2005).
56.  K.-N. Chen, M. J. Kobrinsky, B. C. Barnett and R. Reif, *IEEE Transactions on electron devices* **51**, 233 (2004).
57.  M. J. Kobrinsky, B. A. Block, J.-F. Zheng, B. C. Barnett, E. Mohammed, M. Reshotko, F. Robertson, S. List, I. Young and K. Cadien, *Intel Technology Journal* **08**, 129 (2004).
58.  G. Chen, H. Chen, M. Haurylau, N. Nelson, P. M. Fauchet, E. G. Friedman and D. Albonesi, *SLIP'05 San Francisco*, April 2–3 (2005).
59.  M. Haurylau, G. Chen, H. Chen, J. Zhang, N. A. Nelson, D. H. Albonesi, E. G. Friedman and P. M. Fauchet, *IEEE Journal of Selected Topics in Quantum Electronics* **12**, 1699 (2006).

# Sub-section 1.2

..........................................

# Memory Devices

# 7

# Technologies and Key Design Issues for Memory Devices

Kinam Kim* and Gitae Jeong

Memory Business Division,
Samsung Electronics Company Ltd.,
Korea.

*kn_kim@samsung.co.kr

· · · · · · · · · · · · · · · · · · · · · · · · · · · · ·

For the last three decades, semiconductor memory has greatly advanced towards high density and high performance due to the tremendous progress of electronic data processing (EDP). More recently, with the advent of mobile era, power consumption has become an important aspect of memory applications. Of particular interest among the many key features for future applications are *low power consumption, high speed, and high density that* will be equally important or one/two of which will be dominantly important depending on applications. To understand the direction for future technology of semiconductor memory, the key technologies and design issues including critical technical barriers and corresponding solutions will be reviewed in terms of density (scalability), performance, and power consumption.

## 1.     Introduction

Over the last three decades, semiconductor memory industry has greatly grown due to the tremendous progress of electronic data processing (EDP) mainly led by outstanding evolution of personal-computer (PC) technology. In those days, incumbent semiconductor memories such as DRAM,

SRAM, and Flash have successfully evolved towards high density, high performance, and low cost. This trend is being accelerated and is expected to continue in the future. In addition, with the advent of mobile era from the late 1990s, mobile applications have geared up in many areas, varying from hand-held phone, digital still camera (DSC), music player (MP3) and so on. As the mobile appliances are prevailing in our daily lives, memory technologies rapidly equip with low power consumption technology, which strongly indicates that versatile low power consumption memory technologies will flourish in future.

On the other hand, in views of growing technical complexity, ever-increasing fabrication cost, and approaching ultimate limits, there have been concerns about whether this successful progress can be maintained in future nano era. The uneasiness which incumbent memories will face in future caused many research groups and companies to develop new types of alternative memories, hopefully possess longer lifetime (better scalability), less technical barriers, and more ideal memory characteristics such as non-volatility, high density, high speed, and low power consumption. Through the many dedicated efforts, even though their longevity and technical barriers are not fully known yet, ferroelectric RAM (FRAM) and phase change RAM (PRAM) have appeared to be the most promising candidates for future main memory devices. In this article, we will review important incumbent memories and new emerging memories such as PRAM and FRAM in terms of two most fundamental key aspects of scalability and switching speed. In addition, each memory will be systematically evaluated in terms of power consumption, which is considered as a critical decision making factor in mobile applications.

## 2.    Scalability (Low Cost, High Density)

Memory cell area scaling has played a crucial role in semiconductor memory progress and success, which produces commercial memories with higher density and lower cost. As the device becomes smaller and smaller, it will be much more difficult to satisfy the cell requirements as shown in Table.1 because of many limitations which are the key subjects to be discussed in this section.

For DRAM, important requirements for a DRAM cell can be categorized into two important parameters: *sensing signal margin* and *data retention time*. In order to guarantee proper device operation, sensing signal

Table 1. Cell requirements for important key memories.

| | Cell Size | Cell Requirements | Intrinsic Switching Time |
|---|---|---|---|
| DRAM | 4F2 ~ 8F2 | $C_S > 25\text{fF}$<br>Ion >~ a few uA<br>I(leakage) <~ fA<br>Charge loss <~0.1Q(stored) | ~ ns |
| NAND | ~ 4F2 | Coupling ratio ($\alpha$) ~ 0.5 ~ 0.6<br>$Q = \Delta V_{TH}^* C_{CS}$ | ~ 0.2 ms |
| FRAM | 10 ~ 15 F2 | Ion >~ a few uA<br>2Pr > 30fC/cell | < 1ns |
| PRAM | ~ 4F2 | Ireset < Ion<br>Rset* $C_{BL} < t_{READ}$ | ~ 50 ns |
| MRAM | 10 ~ 20F2 | B (nearest cell) < Hc<br>M.R. > $\Delta$ R/R | < 1ns |

should be larger than sensing noise. As well known, the sensing noise arises from many sources such as Vth imbalance of sense amplifier,[1] interference noise between bit lines,[2] unselected word line generated noise,[3] power line noise and etc.

The sum of these sensing noises is close to several tens of mV. Therefore, in order to make sure the successful sensing in mid of noise environments, the sensing signal greater than 70 ~ 100 mV is preferable. However, taking into account the fact that almost half of stored charges are lost by various leakage currents, the sensing signal should be larger than 150 mV, which requires cell capacitance larger than 25fF/cell by the following relationship;

$$\frac{C_S}{C_{BL} + C_S} \frac{V_{CC}}{2} \geq 150\,\text{mV} \rightarrow C_S \geq 25fF/CELL \qquad (1)$$

where $C_S$ is the cell capacitance, $C_{BL}$ is the bit line parasitic capacitance and $V_{CC}$ is array voltage.

Since the cell capacitor area decreases by $1/k \sim 1/k^2$ with technology scaling where k denotes the scaling factor ($k > 1$), in order to maintain almost non-scalable requirement of cell capacitance of more than $25fF/cell$ regardless of technology node, the cell capacitor structure and cell capacitor dielectric material have been continuously evolved into novel structures and high-k dielectric materials[4,5] in accordance with technology migration.

A novel capacitor structure such as mesh type cell capacitor can increase the cell capacitor height without undesired mechanical instability problem.[6] Taking into account the recent advances of DRAM cell capacitor technology, it is expected that DRAM cell capacitor technology will be available at least down to 30 nm node.[7]

As indicated in (Table 1), the requirements for cell access transistor should be satisfied in order to have the proper charging and discharging times of cell capacitor and data retention time, respectively,

$$I_{ON} \geq \text{a few} \quad \mu A, \quad I_{OFF} \leq fA \qquad (2)$$

This requires the cell access transistor to have low leakage current while maintaining the proper on-current. The on-current of the memory cell array transistor should be at least greater than a few $\mu A$ in order to achieve reasonable read and write speed. This becomes difficult to meet as technology scales down. However, this requirement can be fulfilled down to deep nano scale dimension. The only concern is the leakage current from the DRAM cell because the data retention time is mainly determined by the leakage currents arising from sub-threshold current and gate-induced drain leakage (GIDL) of cell array transistor as well as junction leakage current from storage node. As the transistor channel length is scaled down, the increased channel doping concentration to suppress short channel effects increases electric field across the storage node junction. This increases junction leakage current, leading to the eventual degradation of data retention time.[8] The degradation of data retention time becomes significant below 100 nm node due to the rapid increase of junction electric field.[9] This issue can be overcome by introducing 3-D cell transistors to DRAM, where the junction electric-field can be greatly relieved due to the lightly doped channel. One example of those newly developed structures is RCAT (Recess Channel Array Transistor) whose channel detours around some part of Si substrate so that the elongated channel can be embodied in the array transistor.[9] According to our calculation of RCATs, one can extend incumbent DRAM technology down to a 50 nm node with minor modifications. Beyond the 50 nm node, we may need another revolution in DRAM array transistors. Some studies have shown that using a body-tied FinFET[10] as a cell array transistor is very promising due to its superb transistor performances: excellent immunity against the SCE; high trans-conductance; and small sub-threshold leakage. It is believed that the body-tied FinFET could extend conventional DRAM technology down to the 30 nm node, as illustrated in Fig. 1. The strong aiming for ideal cell array transistor reflects that vertical

| | Planar Tr. | RCAT | FinFET | Vertical Tr. |
|---|---|---|---|---|
| Technology | > 100nm | 90nm ~50nm | 50~30nm | < 30nm |
| Structure | 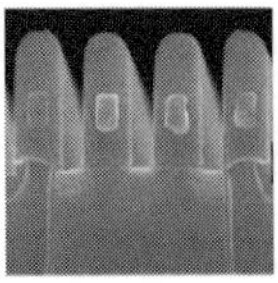 |  | 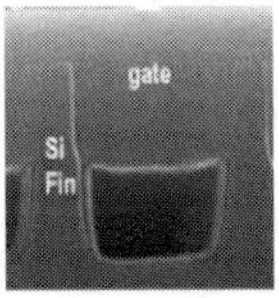 | 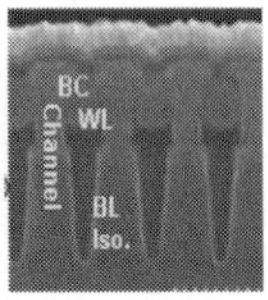 |

Fig. 1. A prospect on the cell array transistor evolution in DRAM.

transistor with a surrounding gate[11] will eventually replace the FinFET cell array transistor beyond 30 nm node because it is so far known as the best transistor structure and, unlike conventional transistors, it is not constrained with lateral dimensional scaling.

For NAND, over the last decade, NAND flash memory has been remarkably advanced in terms of cell size and density. Today, it reaches to $4 \sim 16$ Gb density which is commercially available. Its feature size (F) is expected to be smaller than 40 nm at the year of 2010. In this regime, we will confront various scaling obstacles such as difficulty to meet the optimum coupling ratio, pronounced floating gate coupling, and extremely small tolerance of charge loss. All of which would be unacceptably significant due to its intrinsic properties of current floating gate NAND Flash.

For its proper operation, NAND cell should satisfy write and read constraints. The first restriction comes from progrmaing the cell. To program a cell, the appropriate electric field strength should be applied between the floating gate and the channel of the cell [Eq. (3)] so that sufficient Fowler-Nordheim (FN) tunneling current can be injected into the floating gate.

$$\frac{1}{T_{OX}} \times \gamma \times V_{PGM} \geq \sim 10\,\text{MeV/cm, coupling ratio } \gamma = \frac{C_{ONO}}{C_{TUNNEL} + C_{ONO}} \tag{3}$$

During programming a cell, the unselected cells which share same bit line or same word line with the selected cell as illustrated in Fig. 3 should be prevented from unwanted programming. This requires the electric field on the floating gate of unselected cells on the same bit line (it is called $V_{pass}$ stress cells) and unselected cells on the same word line (it is called $V_{pgm}$ stress cells) as small as possible so that electron injection into or ejection from the unselected cells is completely prohibited.

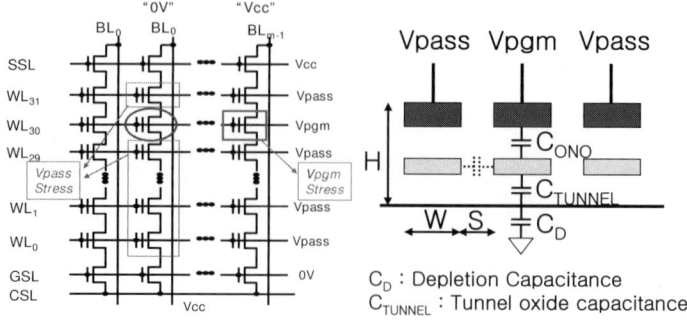

Fig. 2. NAND Cell array schematic and programming condition.

$$\frac{1}{T_{OX}} \times \gamma \times V_{pass} \leq a \ few \ MeV/cm, V_{pass \ stress} \tag{4}$$

$$\frac{1}{T_{OX}} \gamma \left( V_{PGM} - \frac{V_{PASS}/\gamma}{1 + \dfrac{C_D}{\gamma \times C_{TUNNEL}}} \right) \leq a \ few \ MeV/cm \tag{5}$$

In read operation, the voltage of the floating gate should be higher than the highest threshold voltage of cell string in order to pass read current through the string where 32 cells are serially connected.

$$\gamma \times V_{READ} \geq V_{TH} \tag{6}$$

where $V_{READ}$ is the read voltage and $V_{TH}$ is the threshold voltage of the cell transistor.

Like programming disturbance, read disturbance might occur on the unselected cells on same string and it should be completely eliminated by selecting the appropriate pass voltage and the tunnel oxide thickness.

From the constraints in programming and reading, the optimum value of coupling ratio is in the range of 0.5 ~ 0.6,

$$\gamma(coupling \ ratio) = 0.5 \sim 0.6 \tag{7}$$

and optimum tunnel oxide thickness is around 80 A. Unfortunately, as the device dimension shrinks, the optimum value of the coupling ratio becomes difficult to meet. [12] The $V_{pass}$ window determined by both $V_{pass}$ stress and $V_{pgm}$ stress indicates that unselected cells are free from program stress within $V_{pass}$ window. Again, as device dimension shrinks, $V_{pass}$ window becomes narrow due to ever-increasing depletion capacitance ($C_D$).

As the device dimension shrinks, the distance between adjacent cells becomes so small that the influence from neighboring cells can not be ruled out. Therefore, the interference between adjacent cells becomes naturally more severe as technology scaling proceeds. In order to circumvent the disturbance between cells, the floating gate width (W) tends to be more aggressively squeezed than the space (S) between floating gates, leading to the increased aspect ratio (H/W) of gate stack height. As a result, device fabrication becomes more difficult which might induce mechanical instability. Since the interference originates from the coupling between floating gates, the revolutionary structures where charge storage media do not take forms of continuum of charge like floating gate but take discrete forms have been sought. Typical examples are TANOS[13] or nano-crystal dots.[14,15] To successfully implement these structures, it is essential to meet the requirements of coupling ratio and Vpass window aforementioned. For Fowler-Nordheim tunneling based NAND type SONOS, several issues such as poor charge retention and slower erase speed should be resolved. Charge redistribution between charge traps should be properly managed. By using low leakage and high-k dielectrics as a top blocking oxide, its erase speed and charge loss can be significantly improved. The charge redistribution can also be minimized by optimizing the charge storage media.

The fundamental limitation of NAND Flash comes from the number of stored charges because the available number of storage charge rapidly decreases with technology scaling.[12] Considering the voltage difference between the nearest states in 2-level-cell is less than 1V, the threshold voltage shift due to loss of charge becomes less than 0.5V, putting a limit on the charge loss tolerance,

$$\Delta Q \leq C_{CS} * \Delta V_{TH} =\sim 0.1Q \qquad (8)$$

where $C_{CS}$ is the capacitance between control gate and storage media. In case of floating gate, $C_{cs}$ is Cono.

Thus, at most 10% charge loss is tolerable, implying the number of electron loss permitted is less than 10 over 10 years to conform with the 10 years retention requirement below the 40 nm technology node. To minimize SILC (stress induced leakage current) and suppress the trap generation within the tunnel oxide is of importance in reducing charge loss. Another way to overcome the stringent requirement of charge loss tolerance is to increase the storage charge by increasing the capacitance between the control gate and the storage media whether it is continuum or discrete. However, it may not be possible to scale down inter-poly ONO in the case of continuum media like floating gate, because it has already reached to its scaling limit of 13 nm.[16] Therefore, it is imminently required to develop new high-k dielectrics although the proper high k-dielectric suitable for NAND appears difficult to obtain. To complete discussing NAND Flash, another important figure of merit in measuring non-volatile memory, that is, endurance should be properly mentioned. Since the endurance is set by the number of cycles of programming and erasing, the key approaches are again to improve the tunnel oxide quality by minimizing trap generation at interface and bulk of the tunnel oxide together with reducing the high electric field stress during programming and erasing by tweaking design windows.

*For FRAM*, the most important parameter in FRAM is a sensing signal margin like in DRAM. The sensing signal of FRAM is proportional to the capacitor area and the remnant polarization charges (Pr) of a ferroelectric film as expressed below

$$\Delta V_{BL} = \frac{2P_r \times A}{C_{BL}} = \frac{2\varepsilon_0\varepsilon_r A}{C_{BL}d} \tag{9}$$

where $d$ is the film thickness, $A$ is the capacitor area, $\varepsilon_0$ is the vacuum dielectric constant, $\varepsilon_r$ is the dielectric constant of a ferroelectric film.

As expressed in Eq. (9), in principle, the remnant polarization can be increased with the decreasing thickness of a ferroelectric film. Thus, we may compensate the area reduction with increased polarization when the technology scales down. Consequently, we can maintain the almost same capacitance in spite of the technology scaling. However, in reality, when the thickness of PZT ferroelectric thin films decreases, the degradation of polarization tends to appear from the ferroelectric capacitor due to "dead layer" between PZT and bottom and top electrode. By eliminating "dead layer", the thickness of PZT ferroelectric thin film can be considerably reduced.

By pushing the ferroelectric film thickness, we can extend the use of planar cell capacitors down to the 130 nm node.[17,18] For further scaling down the cell area, we may need to follow the similar path which DRAM capacitor has already taken. In other words, we need three dimensional capacitor structures as shown in Fig. 3, with which we can push the FRAM technology scaling as well as FRAM cell size to $6F^2$ or $8F^2$, leading to the huge enhancement in cost competitiveness of FRAM. For this purpose, it will be essential to develop novel ferroelectric film technology securing ferroelectric film technology with nano-scale thickness and excellent conformal deposition capability along the inside walls of high aspect-ratio trench as illustrated in Fig. 3. Since FRAM is a non-volatile memory, the cell

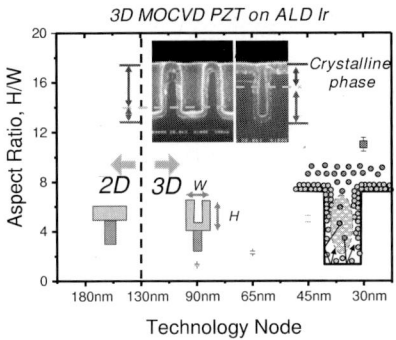

Fig. 3. FRAM cell area scaling trend with a schematic 3-D.

array transistor is not constrained from the requirement of leakage current, but constrained from the requirement of on-current which is at least greater than a few $\mu$A for reasonable read and write speed performance. Thus, this will greatly relieve the technology scaling and enable fast technology migration because the cell array transistor limits technology migration in incumbent memories such as DRAM and NAND/NOR Flash.

The retention time of FRAM is closely related to the remnant polarization decay of a ferroelectric capacitor as expressed in formula (10).

$$\frac{P(t)}{P(0)} \sim \left(\frac{t}{t_0}\right)^{-\exp\left(-\frac{AU^*}{k_BT}\right)} \tag{10}$$

where $P_0$ is initial remnant polarization, $P(t)$ is remnant polarization at time $t$ and $t_0$ is initial time, $k_B$ is Boltzmann constant, $A$ is constant and $U*$ is energy barrier. In most of interesting nano-ferroelectrics with thickness ranging from 5 to 30 nm, an energy barrier against the domain formation was evaluated to $\sim$150 $k_BT$ which is far above the energy barrier which leads to 50% of polarization decay after 10 years, $40\,k_BT$. Endurance life of the FRAM is related to the fatigue of ferroelectric capacitor which is known to occur by the generation of unwanted vacancies (acts as donors in the film) by the repeated polarization reversal. Donors in a ferroelectric film induce depolarization field which in turn diminishes remnant polarization of ferroelectric capacitor. The level of donor concentration which starts to influence the degradation of ferroelectric polarization is expressed as following Eq. (11).

$$N_D \sim \frac{2P_r}{q\varepsilon_0\varepsilon_r d} \tag{11}$$

where 2Pr is the remnant polarization and $d$ is the thickness of a film. As the thickness of a ferroelectric film becomes thinner, tolerable $N_D$ of film interior with proper ferroelectricity tends to increase. Thus, there is no issue to deliberate about pertinent endurance conundrums on ferroelectric memories as long as a ferroelectric film is encountered within several tens of nanometers. It is concluded the retention time and the endurance of FRAM can be maintained even in the deep nano-scale dimension because the dipole strength to determine data retention time and endurance is not influenced by dimensional shrinking.

*For PRAM*, phase change RAM (PRAM) has been considered as one of the promising nonvolatile memories owing to its scalability, fast read and moderately fast write time, good endurance for repetitive writing, and easiness for embedded memory.[19] Since PRAM senses the resistance difference, it has great advantage in cell scaling compared to conventional memories such as DRAM or NAND Flash memory where the charge storage is strongly influenced by leakage currents.[20]

When we consider the cell operation of PRAM, the most important requirement is that the on-current of cell access switch should be larger than the programming current to transform the crystalline state into the amorphous state.

$$I_{ON} \geq I_{PROGRAM} \qquad (12)$$

The on-current of cell access switch decreases by $1/k \sim 1/k^2$. Therefore the reduction of programming current is the most important key factor in PRAM scaling and the required programming current decreases by the scaling factor of $1/k \sim 1/k^2$.[21]

There are several different methods to reduce the programming current. The first approach is to increase the resistance of chalcogenide (GST) module.[22,23] Since the heating of GST element is caused by current flowing, heating power increases with increasing GST module resistance. There are two heating elements in the GST module — GST itself and a electrode. The resistance of GST can be increased by impurity doping. Nitrogen doping can reduce the writing current by increasing the GST resistance. At the same time, the writing current can be further reduced by employing an electrode material with higher resistivity. The writing current was observed to reduce by about 70% with nitrogen doping and a high resistive electrode.[22] The second approach is to increase the heating efficiency by changing the cell structure.[24,25] The cell structure is modified to confine GST module into a smaller volume, consequently localizing the current path. The typical examples for this approach are edge contact structure,[23] ring type contact,[26]

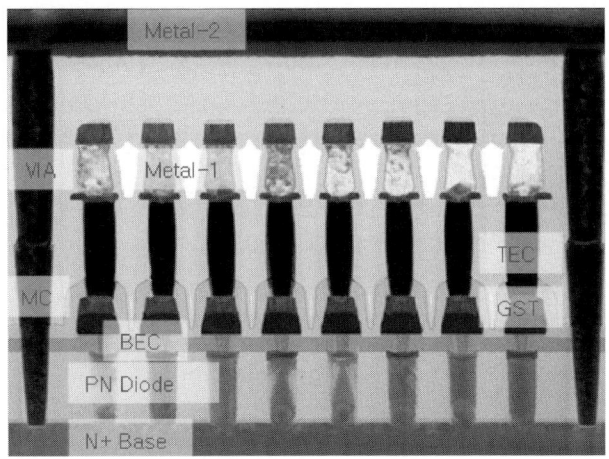

Fig. 4. A vertical structure of a PRAM cell.

micro-trench,[27] and so on. It was reported that the writing current can be reduced as much as 50% with a confined structure.[28] The third approach is to reduce the heat dissipation, which can be achieved by wrapping the GST module with isolation materials with low thermal conductivity. The reset current can be scaled down by the ratio of $1/k \sim 1/k^2$ using aforementioned approaches. There will be no known limit in the aspect of scaling the programming current.

Another concern in the scaling of PRAM is thermal disturbance. The temperature of programmed cells (crystalline state) should be raised to the melting temperature of GST before it was rapidly cooled down to transform into amorphous state. However, this may cause the temperature of nearest cells to increase to the point of disturbance. The rise in temperature of nearest cells can be expressed by the following heat flow equation. [29]

$$\nabla^2 T(x, y, z, t) = \alpha \frac{\partial}{\partial_t} T(x, y, z, t) + H(x, y, z, t), \qquad (13)$$

where $H$ is a heat source. This has no general analytic solution except for special cases. In case of a delta function type point source with spherical symmetry, the temperature impact on neighboring cells can be estimated from the following Eq. (14).

$$T(x, y, z, t) = \frac{H}{4\pi K} \frac{1}{r} \, erfc \left( \frac{r}{\sqrt{4\alpha t}} \right), \qquad (14)$$

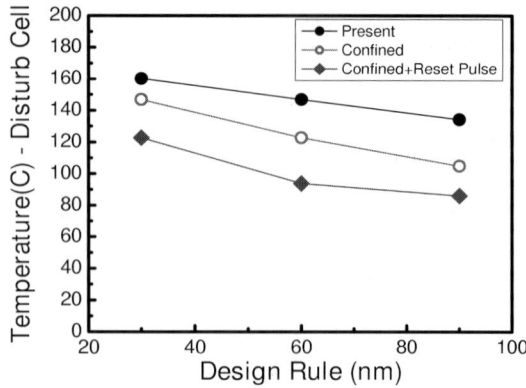

Fig. 5. The temperature in nearest cell operating at 85°C.

where $K$ is thermal conductivity, $\alpha$ is thermal diffusivity (K/C), C is heat capacity. Using this simple calculation, the estimated temperature increase of nearest cells with scaling-down of the technology node from 90 nm to 30 nm is around 20°C as shown in Fig. 5. This increase is not large enough to affect the adjacent cells. Furthermore, this disturbance can be suppressed by modifying the device structure and the programming method. Therefore, the thermal disturbance will not be a showstopper in the scaling of PRAM.

Another concern in the scaling of PRAM is retention characteristics. The volume of the amorphous phase in a programmed cell will decrease as the cell size scales down. The crystallization process of GST film proceeds by nucleation and subsequent crystal growth. If the crystallization kinetics does not depend on the size of the initial programmed volume in amorphous state, the transformed crystal volume fraction of a programmed cell during crystallization at a certain time should be the same, regardless of the size of the initial programmed volume. If the probability to form the percolation path depends only on the crystal fraction, the degradation of retention does not depend on the programmed volume. However, in the tail end distribution, the probability of the percolation formation increases by scaling the volume, resulting in the degradation of retention characteristics.[30] Though it remains unclear how significant the degradation of retention characteristics becomes with respect to the technology node scaling, the retention is considered to satisfy 10 years at 85°C down to the 40~50 nm node.

Table 2 summarizes the behavior of key parameters of PRAM when technology scales down. Although there are no known physical limits of

Table 2. Scaling behavior of key elements of PRAM (k is scaling factor).

|  | MOS | Diode | |
|---|---|---|---|
|  |  | Ideal | Realistic |
| Contact Size | 1/k | 1/k | 1/k |
| Programming Current | $1/k^{(1+a)}$ | $1/k^{(1+a)}$ | $1/k^{(1+a)}$ |
| On Current | 1/k | 1/k | $1/k^2$ |
| Off Current | >k | >k | 1 |
| Rset | k | K | k |
| Rreset | k | k | K |

size scaling, it will be a key success factor of future PRAM scaling to reduce the reset current while suppressing the set resistance below a tolerance limit.

## 3. Switching Speed

Since the success in silicon industry has been achieved by the above shrink technology, basic logic transistors and memory cells have been scaled down without dramatic changes in their structures. Especially, transistor scaling theory[31] tells us that power-delay product is improved to $1/k^3$. As a result, the transistor scaling not only provides high speed but also high density, which enables versatile applications and multi functional systems. However, the improvement of performance in memory has been slow compared with that in logic and even more slowly evolved compared with the increase in memory density itself. Therefore, the performance bottleneck in today's electronic system has arisen from the memory system, which drives the memory towards higher bandwidth or higher data throughput.

To understand what limits the memory performance, both intrinsic switching time of memory cell and maximum achievable total data rate which memory chip can provide should be separately reviewed and compared with each other so that we can fully take advantages of inherent properties of each memory. The incumbent memories such as DRAM, SRAM, and Flash possess similar mechanism of intrinsic cell switching time, because the switching time is mainly determined by charging and discharging times for a capacitive node. The charging time is defined as the elapsed time for the required charges to reach the capacitive node by applied charging current. By the same token, the discharging can be defined. The required charges are determined by voltage drop across node multiplied by node capacitance such as cell capacitance in DRAM, gate capacitance in SRAM, and floating gate capacitance in Flash, respectively, while charging and discharging currents are determined by on-current of cell transistor in DRAM and SRAM, and tunneling current in Flash. The intrinsic cell switching speed of DRAM and SRAM is smaller than 1ns, but intrinsic switching speed of Flash memory is very slow due to the small tunneling current. For emerging new memories, atomic movements in the perovskite structure, ordering of atoms in the chalcogenide (GST), and dipole transition in magnetic tunnel junction (MTJ) limit the switching speed in FRAM, PRAM and MRAM, respectively. The switching time in FRAM and MRAM are very fast and comparable to that in DRAM and SRAM. PRAM has faster switching time than Flash memory, but slower than DRAM, SRAM, FRAM, and MRAM. Table 3 summarizes the intrinsic switching time and maximum achievable bandwidth for each memory.

It should be noted from Table 3 that the data rates of various memories are not very closely related to their intrinsic switching times. This indicates that the data rate in conventional memories depends more strongly on other factors such as chip architectures, circuit techniques, and technology features as well as sensing delay and propagation delays of word line and bit line rather than their intrinsic switching time. Furthermore, when memory density increases, the delay owing to the word line and bit line becomes more pronounced, so that improving speed, especially random access speed, in higher density becomes more challenging. Therefore, it is highly preferred to improve the memory performance, not by just pushing the limit

Table 3. Intrinsic switching times of various memory cells and its currently achievable data rates.

|  | Intrinsic Switching | Time | Data rate/pin |
|---|---|---|---|
| DRAM | Qcell/Ion | $\sim$ ns | $\sim$2G bps/pin |
| SRAM | Qgate/Ion | $\sim$0.1ns | $\sim$0.6G bps/pin |
| NAND | $Q_{FG}/I_{tunnel}$ | $\sim$0.2ms | $\sim$20M bps/pin |
| FRAM | $To^*exp(-\gamma E/Ec)$[32] | $\sim$1ns | $\sim$100M bps/pin |
| PRAM | $f = 1 - exp(-kt^n)$[33,34] | $\sim$500ns | $\sim$2M bps/pin |
| MRAM | $P(t,H) = exp(-t/\tau)$[35,36] | $\sim$1ns | $\sim$100M bps/pin |
|  | $\tau = \tau o \; exp(E_B/K_{BT})$ |  |  |

of long wires but by other methods such as parallelism,[37] pipelining[38] and interleaving memory.[39]

Parallelism can improve the memory performance by processing the data in wide parallel data bus. Parallelism can be traded off with die size because the wide data bus needs to use a large number of data pad, giving rise to the increase of chip size. A good compromise between parallelism and cost is to use an internal wide bus, which uses 2 or more IO sense amplifiers and data buses for one data pin so that it can increases the bandwidth without appreciable increase of cost.[37] Pipelining can also provide higher bandwidth by reducing a sequential access time. Goal of pipelining is to enhance the data throughput by dividing the data path into several segments. For example, the sense amplifier unit is active during the first eighth of total access time and remains idle in the next seven-eighth of access time. However in pipelining structure, sense amplifier performs the data processing of second data in second eighth of period, third data in third eighth of period, and so on by inserting some latches in the data path. This can greatly improve the bandwidth without increasing the cost. It has been widely used in memory as well as logic in order to accelerate the operation of data paths.[38] Interleaved memory can improve the bandwidth by dividing the memory into two or more sections. The external controller can access alternate sections immediately without waiting for memory to catch up. Figure 6 shows an example for data rate improvement with aforementioned parallelism, pipelining, and interleaving. By using parallelism, pipelining, and interleaving, the data rate can be as fast as more than $\times 30$ improvements.

In summary, the most efficient way to increase the bandwidth of memory is to use parallel data processing scheme, which can be achieved by wide bus, pipelining, and interleaving memory.

## 4.    Power Consumption

Mobile devices can be uniquely characterized as diversified products, fashionable design for personal applications, short life cycles, small and light products for portability. By the nature of portability, power consumption is the critical factor for the memory in mobile devices such as high-performance PC and server where a large number of memory chips are needed to be closely packed in the limited space of memory module so that power management is of paramount importance. The power consumption consists of active mode power, idle mode power, and standby mode power consumption. Active mode power consumption arises during read and write operations and idle mode power consumption occurs when we just hold current data as it is without change, and standby mode power consumption appears when all operations stop while power supply is connected.

The analyses of active mode power consumption have been reported by many groups. In most of them, they focused power dissipation in peripheral circuit ($C_{peripheral} * V_{DD}^2 * f$), decorder ($(m+n)C_{decorder} * V_{DD}^2 f$), and cell array including core circuit ($I_{activate-row} * V_{DD}$).[40,41] (where m and n are the number of column and row address). Although considerable research has been devoted to the power dissipation in various functional blocks of chips, rather less attention has been paid to the stored energy of the cell itself and how much energy should be consumed in order to correctly read and write onto a cell. That is, storing efficiency, which is very important in minimizing the power consumption of the memory chip while maximizing the performance of memory chip. In this article, we divide the power dissipation into 4 categories — stored energy for cell, power dissipation for charging bit lines and word lines, power dissipation due to static current path, and power dissipation in core and peripheral circuits. The power consumption in each block of memory for writing process can be expressed by the following equations.

$$P_{cell,stored} = \Delta E_{cell} \times f_{cell}$$
$$\Delta E_{cell} = energy\ difference\ between\ data\ "1"\ and\ 0 \qquad (15)$$
$$f_{cell} = frequency\ of\ cell\ access$$

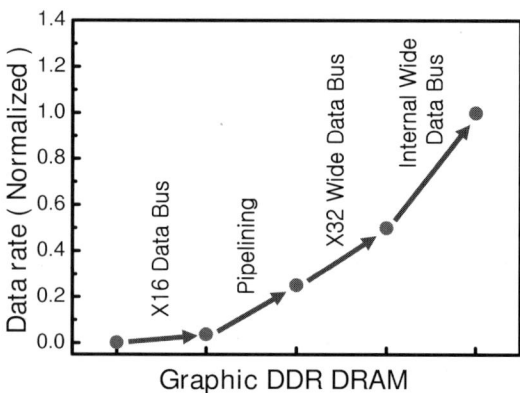

Fig. 6. Improvement of bandwidth for graphic DDR DRAM.

$$P_{BL,WL}(charging) = 1/2 C_{BL} V_{DD}^2 f_{BL} + 1/2 C_{WL} V_{DD}^2 f_{WL} \qquad (16)$$

$$f_{BL}, f_{WL} = frequency\ of\ bit\ line\ and\ word\ line\ access$$

$$P_{BL,WL}(dissipation) = I_{BL} \times V_{BL} + I_{WL} \times V_{WL} \qquad (17)$$

$$I_{CCW} = Active\ current\ for\ writing$$
$$I_{BL}, I_{WL} = static\ current\ in\ Bit\ line\ and\ Word\ line$$

$$P_{total} = I_{CCW} \times V_{DD} \qquad (18)$$

Here,

$$\Delta E_{cell} = 1/2 * C_{cell} * V_{DD}^2$$

($C_{cell}$ : *cell* capacitor capacitance in DRAM cell gate capacitance in SRAM floating gate capacitance in Flash)

$$= \Delta E(crystal - amorphous)\ (PRAM)$$
$$= 1/2 * A * d * Ec * Pr\ (FRAM)$$
$$= 1/2 * A * d * Hc * M\ (MRAM) \qquad (19)$$

where $Ec$ and $Hc$ are coercive fields, $Pr$ is remnant polarization, and $M$ is magnetization. $P_{cell,stored}$ is the intrinsic power consumption to store the data into cell, $P_{BL,WL(Charging)}$ is consumed power to charge the bit line and word line, $P_{BL,WL(dissipation)}$ is power dissipation due to the leakage current of bit line and word line during the programming time and $P_{total}$ is total power consumption in chip level.

To illustrate how much active power is consumed, typical burst writing mode of operation is selected and compared among memories. 100 MHz burst writing is referenced for RAMs, and 20Mbyte/sec programming speed is employed for NAND, and 0.2 Mbyte/sec is set for PRAM. Power consumption due to static current in bit line and word line are negligible except for PRAM and MRAM. Continuous current in the bit line is required for about 500 nsec in PRAM, and continuous current in the bit line and the digit line is required for several nano-seconds in MRAM.

The power consumption in core and peripheral circuit is higher than 90% of total power consumption for every memory. Table 4 compares the energy to store one bit data with the energy consumed in other functional blocks in order to store one bit data. The total energy required to store one bit data is almost same for DRAM, SRAM, FRAM, and MRAM. However, those of NAND Flash and PRAM are almost two orders of magnitude larger

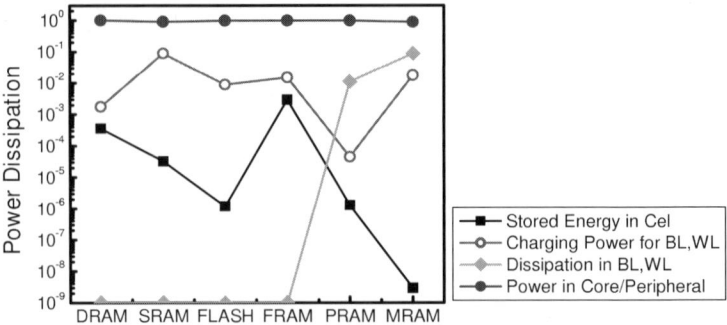

Fig. 7. Write mode active power consumption for each part of memory (normalized).

than the others. This kind of poor efficiencies are attributed to the small tunneling current in Flash and the slow crystallization kinetics in PRAM, respectively. That is, the poor efficiency is mainly due to the long programming time in NAND Flash and PRAM. Thus, the most efficient way to improve the efficiency is to reduce the programming time or perform the parallel writing. It was already realized to improve the energy efficiency per bit by parallel writing in NAND Flash memory. For example, page programming was performed with page unit of 512 bytes. Another approach to reduce the power dissipation in peripheral circuits is to use low internal voltage as long as we can achieve the required performance, which requires designing logic circuits to allow logic swing voltage as low as possible. Scaling of operating voltage in mobile era is expected to be more strongly accelerated than in EDP because of the nature of the mobility. Hence, we need to introduce dual work function gate process although introduction of this process has been postponed in commodity memory processes. In conclusion, the power consumption due to the intrinsic stored energy in a cell is negligible and the efficient way to reduce the active power consumption is to reduce the power consumption in peripheral circuits or to program the data in parallel.

Standby mode power consumption is strongly correlated with the leakage current such as transistor sub-threshold current and various junction leakage currents. Traditional approach to reduce the standby mode power consumption is to reduce the leakage current at the expense of some performances. It can also be reduced by circuit technology like DPD (Deep Power Down) mode.[42] To minimize the power consumption in DPD mode, circuit design technique is to turn each I/O pins into high impedance state and disconnect internal powers at the same time.

Table 4. Comparison of the energy to program one bit data for various memories.

| | Stored Energy in Cell | Charging Energy of BL | Dissipation Energy in BL | Total Energy with Peripheral |
|---|---|---|---|---|
| DRAM | $\sim 10^{-14}$ J | $\sim 10^{-13}$ J | $\sim 10^{-23}$ J | $\sim 10^{-10}$ J |
| SRAM | $\sim 10^{-16}$ J | $\sim 10^{-12}$ J | $\sim 10^{-23}$ J | $\sim 10^{-11}$ J |
| NAND | $\sim 10^{-16}$ J | $\sim 10^{-12}$ J | $\sim 10^{-19}$ J | $\sim 10^{-8}$ J |
| FRAM | $\sim 10^{-14}$ J | $\sim 10^{-13}$ J | $\sim 10^{-23}$ J | $\sim 10^{-10}$ J |
| PRAM | $\sim 10^{-14}$ J | $\sim 10^{-12}$ J | $\sim 10^{-10}$ J | $\sim 10^{-8}$ J |
| MRAM | $\sim 10^{-19}$ J | $\sim 10^{-12}$ J | $\sim 10^{-11}$ J | $\sim 10^{-10}$ J |

On the other hand, idle mode power consumption is different from standby mode power consumption. DRAM shows higher idle mode power consumption than non-volatile memories due to its intrinsic refresh operation. Figure 8 compares the idle mode power consumption for various memories. The idle mode power consumption of non-volatile memory is almost the same as standby mode power consumption because it is not necessary to perform the write and read operations for maintaining the stored data.

However, the idle mode power consumption is not negligible in DRAM because DRAM is required to perform the self refresh to keep the stored data. There are two categories to decrease the idle mode power consumption — circuit technology and process technology. Process technologies are already reviewed in Sec. 2, where various technologies to increase

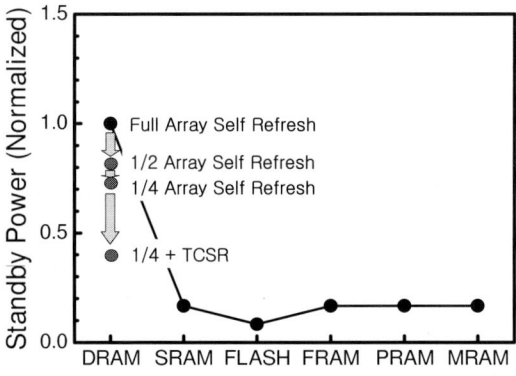

Fig. 8. Comparison of idle mode power consumptions.

the data retention time were reviewed. Examples of circuit technologies to reduce the idle mode power consumption are shown in Fig. 8. PASR (partial array self-refresh)[42] reduces power consumption by adjusting the self-refresh area according to the necessary data refresh area. Since it can be operated in several modes such as full, 1/2, and 1/4 PASR, 1/4 mode PASR can reduce the power consumption by about 30%. Another powerful technique to reduce power consumption is TCSR (temperature compensated self-refresh).[43] The TCSR function measures the chip temperature and keeps data retention by increasing refresh period when temperature is low and vice versa. Recent TCSR scheme tends to include temperature sensor in the chip. TCSR can reduce the power consumption by about 50% at below 45 °C.

## 5.    Conclusion

Critical technology barriers for future memory development and prospects of technology evolution have been reviewed to overcome the barrier. In spite of much concerns about the future development, many of the technology barriers expected in scaling down seem to be overcome with innovative breakthroughs in technology as suggested above. In addition to the simple scaling approach, nano era requires versatile memory applications, which need memory devices with multitalents such as low power consumption and ultra-high speed as well as high density. And it was reviewed that these requirements can be satisfied by the innovation of process and circuit technology. As a result, it is expected that future memory technology can continuously provide new versatile functions and cost effectiveness at the same time in future.

## References

1.    R. Sarpeshkar, J. L. Wyatt, C. Lu and P. D. Gerber, *IEEE Transactions on Circuits and Systems*, **39**(5), p 277 (1992).
2.    H. Hidaka, K. Fujishima, Y. Matsuda, M. Asakura and T. Yoshihara, *IEEE Journal of Solid-State Circuits*, **24**(1), p 21 (1989).
3.    Y. Nakagome, M. Aoki, S. Ikenaga, M. Horiguchi, S. Kimura, Y. Kawamoto and Kiyoo Itoh, *IEEE Journal of Solid-State Circuits*, **23**(5), p 1120 (1988).

4. J. Lee, Y. Ahn, Y. Park, M. Kim, D. Lee, K. Lee, C. Cho, T. Chung and K. Kim, *Dig. Tech. Papers, VLSI Technology Symposium*, pp 57–58 (2003).

5. H. Kim, S. Kim, S. Lee, S. Jang, J.-H. Kim, Y. Sung, J. Park, S. Kwon, S. Jun, W. Park, D. Han, C. Cho, Y. Kim, K. Kim and B. Ryu, *Dig. Tech. Papers, VLSI-TSA*, pp 29–30 (2005).

6. D. H. Kim, J. Y. Kim, M. Huh, Y. S. Hwang, J. M. Park, D. H. Han, D. I. Kim, M. H. Cho, B. H. Lee, H. K. Hwang, J. W. Song, N. J. Kang, G. W. Ha, S. S. Song, M. S. Shim, S. E. Kim, J. M. Kwon, B. J. Park, H. J. Oh, H. J. Kim, D. S. Woo, M. Y. Jeong, Y. I. Kim, Y. S. Lee, H. J. Kim, J. C. Shin, J. W. Seo, S. S. Jeong, K. H. Yoon, T. H. Ahn, J. B. Lee, Y. W. Hyung, S. J. Park, H. S. Kim, W. T. Choi, G. Y. Jin, Y. G. Park and K. Kim, *Electron Devices Meeting, IEDM Technical Digest. IEEE International*, p 69 (2004).

7. K. Kim and G. Jeong, *ISSCC Dig. Tech. Papers*, pp 576–577 (2005).

8. K. Kim, *Technical Digest of 2005 IEDM*, pp 333–336 (2005).

9. J. Y. Kim, C. S. Lee, S. E. Kim, I. B. Chung, Y. M. Choi, B. J. Park, J. W. Lee, D. I. Kim, Y. S. Hwang, D. S. Hwang, H. K. Hwang, J. M. Park, D. H. Kim, N. J. Kang, M. H. Cho, M. Y. Jeong, H. J. Kim, J. N. Han, S. Y. Kim, B. Y. Nam, H. S. Park, S. H. Chung, J. H. Lee, J. S. Park, H. S. Kim, Y. J. Park and K. Kim, *Dig. Tech. Papers, VLSI Technology Symposium*, pp 11–12 (2003).

10. C. H. Lee, J. M. Yoon, C. Lee, H. M. Yang, K. N. Kim, T. Y. Kim, H. S. Kang, Y. J. Ahn, D. Park and K. Kim, *Dig. Tech. Papers, VLSI Technology Symposium*, p 130 (2004).

11. J.-M. Yoon, K. Lee, S.-B. Park, S.-G. Kim, H.-W. Seo, Y.-W. Son, B.-S. Kim, H.-W. Chung, C.-H. Lee, W.-S. Lee, D.-C. Kim, D. Park, W. Lee and B.-I. Ryu, *64th DRC Digest*, pp 259–260 (2006).

12. K. Kim, J. H. Choi, J. Choi and H.-S. Jeong, *Dig. Tech. Papers, VLSI-TSA*, pp 88–94 (2005).

13. S. Mori, E. Sakagami, H. Araki, Y. Kaneko, K. Narita, Y. Ohshima, N. Arai and K. Yoshikawa, *IEEE Transactions on Electron Devices* **38**(2), 386–391 (1991).

14. S. Tiwari, F. Rana, K. Chan, H. Hanafi, W. Chan and D. Buchanan, *Electron Devices Meeting, IEDM Technical Digest. IEEE International*, pp 521–524 (1995).

15. A. Nakajima, T. Futatsugi, H. Nakao, T. Usuki, N. Horiguchi and N. Yokoyama, *Journal of Applied Physics* **84**(3), 1316–1320 (1998).

16. J.-H. Park, S.-H. Hur, J.-H. Lee, J.-T. Park, J.-S. Sel, J.-W. Kim, S.-B. Song, J.-Y. Lee, J.-H. Lee, S.-J. Son, Y.-S. Kim, M.-C. Park, S.-J. Chai, J.-D. Choi, U.-I. Chung, J.-T. Moon, K.-T. Kim, K. Kim and B.-I. Ryu, *Electron Devices Meeting, IEDM Technical Digest. IEEE International*, pp 873–876 (2004).

17. J. H. Park, H. J. Joo, S. K. Kang, Y. M. Kang, H. S. Rhie, B. J. Koo, S. Y. Lee, B. J. Bae, J. E. Lim, H. S. Jeong and K. Kim, *Electron Devices Meeting, IEDM Technical Digest. IEEE International*, p 591–594 (2004).

18. H. J. Joo, Y. J. Song, H. H. Kim, S. K. Kang, J. H. Park, Y. M. Kang, E. Y. Kang, S. Y. Lee, H. S. Jeong and K. Kim, *Dig. Tech. Papers, VLSI Technology Symposium*, p 148 (2004).

19. R. Neale, *Electronic Engineering*, pp 67–78 (2001).

20. S. Lai and T. Lowrey, *Electron Devices Meeting, IEDM Technical Digest. IEEE International*, pp 803–806 (2001).

21. S. Lai, *Electron Devices Meeting, IEDM Technical Digest. IEEE International*, pp 255–258 (2003).

22. H. Horii, J. H. Yi, J. H. Park, Y. H. Ha, I. G. Baek, S. O. Park, Y. N. Hwang, S. H. Lee, Y. T. Kim, K. H. Lee, U.-I. Chung and J. T. Moon, *Dig. Tech. Papers, VLSI Technology Symposium*, p 177 (2003).

23. Y. H. Ha, J. H. Yi, H. Horii, J. H. Park, S. H. Joo, S. O. Park, U.-I. Chung and J. T. Moon, *Dig. Tech. Papers, VLSI Technology Symposium*, pp 175–176 (2003).

24. M. G., T. Lowrey and J. Park, *Digest of Technical Papers ISSCC*, pp 202–204 (2002).

25. S. J. Ahn, Y. J. Song, C. W. Jeong, J. M. Shin, Y. Fai, Y. N. Hwang, S. H. Lee, K. C. Ryoo, S. Y. Lee, J. H. Park, H. Horii, Y. H. Ha, J. H. Yi, B. J. Kuh, G. H. Koh, G. T. Jeong, H. S. Jeong, K. Kim and B. I. Ryu, *Electron Devices Meeting, IEDM Technical Digest. IEEE International*, pp 907–910 (2004).

26. Y. J. Song, K. C. Ryoo, Y. N. Hwang, C. W. Jeong, D. W. Lim, S. S. Park, J. I. Kim, J. H. Kim, S. Y. Lee, J. H. Kong, S. J. Ahn, S. H. Lee, J. H. Park, J. H. Oh, Y. T. Oh, J. S. Kim, J. M. Shin, J. H. Park, Y. Fai, G. H. Koh, G. T. Jeong, R. H. Kim, H. S. Lim, I. S. Park, H. S. Jeong and K. Kim, *Dig. Tech. Papers, VLSI Technology Symposium*, pp 146–147 (2006).

27. F. Pellizzer, A. Pirovano, F. Ottogalli, M. Magistretti, M. Scaravaggi, P. Zuliani, M. Tosi, A. Benvenuti, P. Besana, S. Cadeo, T. Marangon, R. Morandi, R. Piva, A. Spandre, R. Zonca, A. Modelli, E. Varesi, T.

Lowrey, A. Lacaita, G. Casagrande, P. Cappelletti and R. Bez, *Dig. Tech. Papers, VLSI Technology Symposium*, p 18 (2004).

28. Y. N. Hwang, S. H. Lee, S. J. Ahn, S. Y. Lee, K. C. Ryoo, H. S. Hong, H. C. Koo, F. Yeung, J. H. Oh, H. J. Kim, W. C. Jeong, J. H. Park, H. Horii, Y. H. Ha, J. H. Yi, G. H. Koh, G. T. Jeong, H. S. Jeong and K. Kim, *Electron Devices Meeting, IEDM Technical Digest. IEEE International*, pp 893–896 (2003).

29. I. Stakgold, *A Wiley-Interscience Publication*, John Wiley and Sons, New York, p 4 (1979).

30. U. Russo, D. Ielmini, A. Redaelli and Andrea L. Lacaita, *IEEE Transactions on Electron Devices*, **53**(12) pp 3032–3039 (2006).

31. R. H. Dennard, F. H. Gaensslen, V. L. Rideout, E. Bassous and A. R. LeBlanc, *IEEE Journal of Solid-State Circuits SC-9*, p 256 (1974).

32. W. J. Merz, *Physics Review*, **95**, p 690 (1954).

33. M. Avrami, *J. Chem. Phys.* **7**, 1103 (1939).

34. M. Avrami, *J. Chem. Phys.* **8**, 212 (1940).

35. I. Zutic, J. Fabian and S. Das Sarma, *Reviews of Modern Physics*, **76**, p 323 (2004).

36. Y. Nozaki, K. Matsuyama and S. Ishii, *Journal of Applied Physics*, **93**(11), 9182 (2003).

37. http://www.samsung.com/Products/Semiconductor/common/ product_list.aspx?family_cd=GME1002

38. J. M. Rabaey, A. Chandrakasan and B. Nikolic, *A Design Perspective second edition, Prentic Hall Electronics and VLSI Series.*

39. http://www.rambus.com

40. P. Hicks, M. Walnock and R. M. Owens, *Low Power Electronics and Design, 1997. International Symposium on*, p 239 (1997).

41. C.-L. Su and A. M. Despain, *International Symposium on Low Power Electronics and Design*, p 63 (1995).

42. http://www.samsung.com/products/semiconductor/MobileSDRAM/ MobileDDRSDRAM/512Mbit/K4X51163PC/K4X51163PC.htm

43. J.-Y. Sim, H. Yoon, K.-C. Chun, H.-S. Lee, S.-P. Hong, K.-C. Lee, J.-H. Yoo, D.-I. Seo and S.-I. Cho, *IEEE Journal of Solid-State Circuits*, **38**(4), (2003).

# 8

# FeRAM and MRAM Technologies

Yoshihiro Arimoto

Fujitsu Laboratories Ltd. P.O. Box 211-8588,
4-1-1 Kamikodanaka, Nakahara, Kawasaki, Japan.

arimoto@jp.fujitsu.com

·······························

FeRAM and MRAM are promising non-volatile random access memory candidates and have been put into mass production before other new memories. The remaining scalability issues for FeRAM and MRAM are rapidly being resolved by introduction of new materials, processes, structures, memory cell circuits, and architectures. The performance of the 1T1C FeRAM was improved by optimizing its memory cell circuit and by using a ferroelectric capacitor with large, stable remanent polarization charges. The 6T4C FeRAM shows unlimited read/write endurance. The chain FeRAM has a memory cell as small as that of a DRAM. The readout margin of the MRAM was increased by using single-crystal MgO as an MTJ insulator. The half-select disturb of the MRAM has been greatly improved by the toggle writing, thermal select writing, and spin torque transfer switching schemes. The write current of the MRAM was reduced by spin torque transfer switching to 1/10. The performances of FeRAM and MRAM are improving as the demand for low power consumption devices increases.

# 1. Introduction

Ferroelectric random access memory (FeRAM) and magnetoresistive or magnetic random access memory (MRAM) are commercially available and are the most promising non-volatile memory for various electronic systems. Several hundred million FeRAM embedded chips have been shipped all over the world since the latter half of the 1990s. Mass production of MRAM chips started in 2006. A lot of excellent research is being done in the field on issues from materials to architecture.

Figure 1 shows the key components and data storing mechanisms of FeRAM and MRAM devices. The ferroelectric capacitor shown in Fig. 1 (a) is used as a memory cell for FeRAM. The two remanent polarization directions in the capacitor's ferroelectric film create the two memory states. Polarization direction is switched by applying programming voltage between the electrodes. Remanent polarization is caused by the movement of atoms that compose ferroelectricity. Stored data is read by detecting the polarization reversal or non-reversal current of a ferroelectric capacitor when the reading voltage is applied between the electrodes.

MRAM is based on the magnetic tunnel junction (MTJ) shown in Fig. 1 (b). The parallel or antiparallel magnetization of two ferromagnetic films on each side of an insulator (tunnel barrier) of magnetic tunnel junction

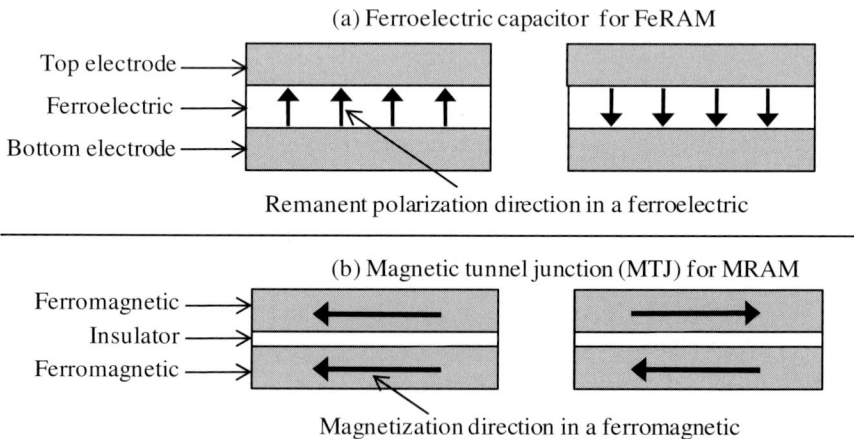

(a) Ferroelectric capacitor for FeRAM

Top electrode
Ferroelectric
Bottom electrode

Remanent polarization direction in a ferroelectric

(b) Magnetic tunnel junction (MTJ) for MRAM

Ferromagnetic
Insulator
Ferromagnetic

Magnetization direction in a ferromagnetic

Fig. 1. (a) Ferroelectric capacitor for FeRAM device and (b) Magnetic tunnel junction for MRAM device.

represents the different memory states. The magnetization direction of ferromagnetic film is switched by a magnetic field induced by the current through wiring or by spin torque transfer induced by the current through the MTJ. Tunnel resistance between ferromagnetic films is low for parallel magnetization and is high for antiparallel magnetization. Stored data is read by sensing the tunnel current change caused by the tunnel magnetoresistance effect.[1,2]

FeRAM and MRAM can read/write at high speed without losing data when the power is turned off. The power consumption of FeRAM and MRAM is smaller than that of DRAM because they do not need the refresh operation. FeRAM and MRAM have the potential to replace standard DRAMs and SRAMs, and are also used as embedded memories in various ICs because they can be fabricated in a logic chip using standard CMOS technology.[3–13]

Figure 2 shows the FeRAM and MRAM integration processes. Except for the thermal budget, the fabrication processes are compatible with the conventional CMOS process. Ferroelectric capacitors are formed above the MOSFET because the process temperature of the ferroelectric capacitor is higher than the temperature for wiring. Magnetic tunnel junctions (MTJ) are formed on the top layer of wiring because the temperature for the wiring process degrades the characteristics of the magnetic tunnel junction.

Use of FeRAM or MRAM instead of DRAM, SRAM, or other conventional memories can increases the speed and reduces the power consumption of various electronic systems, especially portable computers and

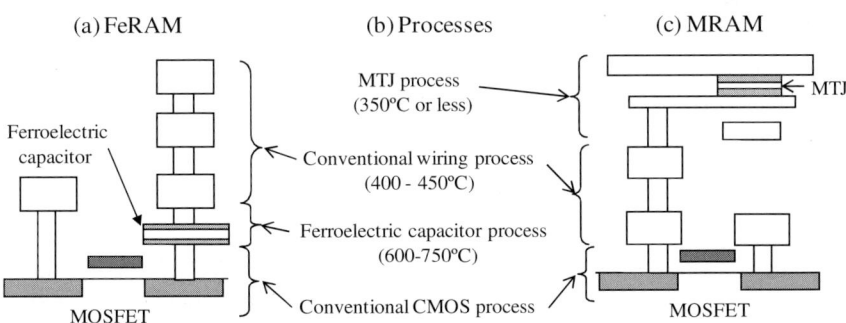

Fig. 2. (a) Ferroelectric capacitors are formed above MOSFET (b) MTJs are located in top layer of wiring.

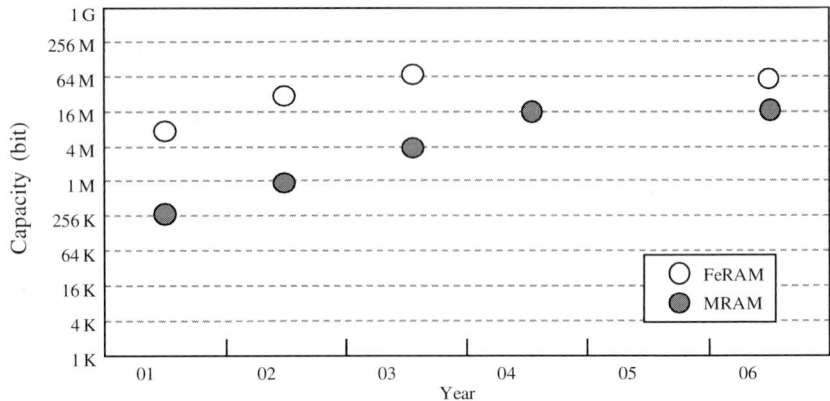

Fig. 3. Maximum capacity of FeRAM and MRAM (presentation data).

cellular phones. FeRAM- or MRAM-based systems are becoming more convenient.

However, FeRAM and MRAM have a serious scalability problem. The technology nodes of commercially available FeRAMs and MRAMs are three generations older than those of DRAM, SRAM, and Flash and have less memory capacity. The memory capacity of commercially available FeRAMs and MRAMs is 4 Mbit or less, and is 1% or less than DRAM's, as shown in Fig. 3. Furthermore, the read/write endurance of FeRAM devices is limited. In MRAMs, the programming current is too large, and the readout signal is too small. However, these problems have largely been mitigated by optimizing the materials, processes, structures, circuits, and architectures of FeRAM and MRAM devices.[8,9,14−27]

In this chapter, the current status of and advanced approaches to FeRAM and MRAM technologies are discussed.[28,29]

## 2.    FeRAM Technologies

Key characteristics of FeRAM devices are high-speed read/write operation, low power consumption, limited read/write endurance, destructive readout, and radiation hardness. 64-Mbit standard/embedded FeRAMs have been developed. The maximum memory capacity of commercially available ones is 4 Mbits for standard FeRAMs, and 512 Kbits for embedded FeRAMs. FeRAMs are commonly used in IC cards,[30,31] RFID tags,[32] MCUs, and as replacements for battery-backed up SRAM. The technical issues for FeRAMs are memory capacity (scalability) and read/write endurance.

## 2.1. Ferroelectric capacitor

Figure 4 shows a memory cell circuit schematic, the ferroelectric capacitor structure, and the materials used in an FeRAM. The memory cell circuit of the FeRAM is similar to that of a DRAM. The difference is in the capacitor materials. Ferroelectric is used in the FeRAM capacitor, and dielectric is used in the DRAM capacitor. The remanent polarization of ferroelectric makes the FeRAM non-volatile. The ferroelectric capacitor consists of a top electrode, the ferroelectric, and a bottom electrode. These materials have a strong influence on the performance and reliability of a FeRAM. PZT and SBT are used as a ferroelectric in commercially available FeRAMs. Pt was used as the electrode material FeRAM technology was first developed. In recent FeRAMs, a conductive oxide electrode is used as the electrode to prevent hydrogen from degrading the ferroelectric.

The ferroelectric layer has two directions of polarization, one for each direction of an applied electric field. In PZT, the Zr or Ti atom moves between two stable points based on the direction of an applied electric field. These two stable points correspond to data "0" and data "1", as shown in Fig. 5. Large remanent polarization charges Pr and a small coercive voltage Vc are needed to develop high-capacity and low-voltage FeRAM devices. PZT is the best ferroelectric material for large capacity FeRAM devices because PZT has a large Pr. Recently, a BFO with a larger Pr than that of PZT has been studied.[33−35] SBT, which has a low Vc, is the best ferroelectric for low-voltage FeRAM devices.[36,37]

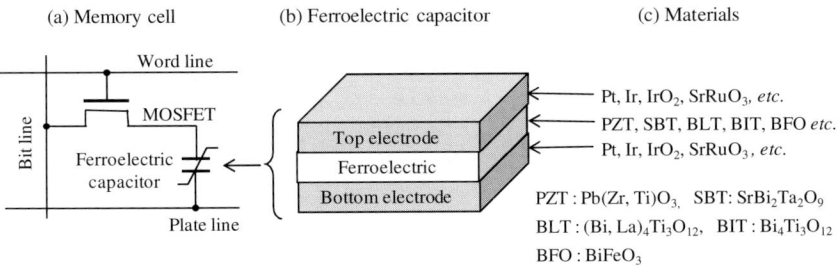

Fig. 4. (a) Memory cell circuit (b) ferroelectric capacitor, and (c) materials.

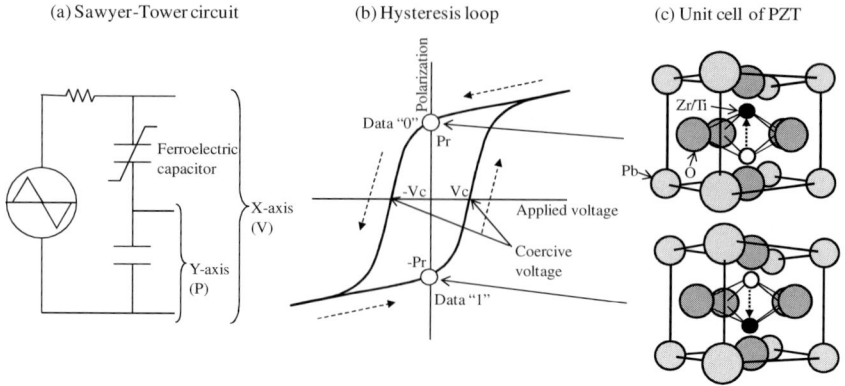

Fig. 5. (a) Sawyer-Tower circuit is used to measure polarization-voltage characteristics. (b) Polarization-voltage curve of ferroelectric capacitor shows a counter-clockwise hysteresis loop. (c) Unit cell of PZT. Ferroelectric has two directions of polarization based on direction of an applied electric field.

Many types of capacitors have been developed to miniaturize the memory cell, as shown in Table 1.[38,39] A planar structure has been used since FeRAM technology was first developed because its fabrication process is simple. A stack structure is used in more recent FeRAMs because it reduces

Table 1. Various structures of ferroelectric capacitor. Ferroelectric is formed by sputtering, spin coating, and MOCVD. 3D-stack structure makes it possible to obtain a large Pr with a small capacitor footprint. MFIS structure is used for 1T FeRAM.

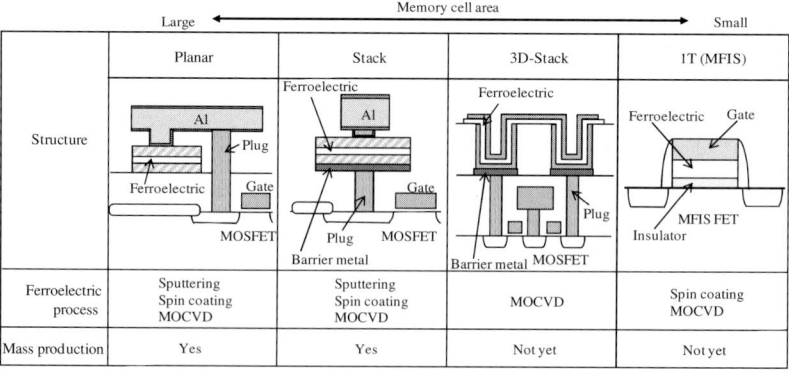

| | Planar | Stack | 3D-Stack | 1T (MFIS) |
|---|---|---|---|---|
| Structure | | | | |
| Ferroelectric process | Sputtering Spin coating MOCVD | Sputtering Spin coating MOCVD | MOCVD | Spin coating MOCVD |
| Mass production | Yes | Yes | Not yet | Not yet |

the size of the memory cell. In the stack structure, a barrier metal layer is needed to prevent the plug from oxidizing during high-temperature annealing in oxygen ambient. Many developers anticipate the introduction of a 3D-stack structure that will greatly reduce the size of the memory cell.

## 2.2. Memory cell

The many types of memory cell circuits have been developed to reduce the size of memory cells and to improve FeRAM performance are shown in Table 2. The 1T1C, 2T2C, and 6T4C cells are used in commercially available FeRAMs.

The 2T2C memory cell consists of two transistors and two ferroelectric capacitors. Data and opposite data are simultaneously written in two capacitors. Data is read by comparing the voltage of the two bit lines connected to each transistor. The 2T2C memory cell is larger and has a larger read margin than the 1T1C memory cell. The 6T4C-FeRAM has the largest memory cell of the current FeRAM devices but features unlimited read/write endurance, non-destructive readout, and sub-10 ns access time. Chain FeRAM, 1T-FeRAM, and cross point FeRAM[40,41] all have small memory cells.

Figure 6 shows the write operation of the 1T1C memory cell. Data "1" is written by setting the word, bit, and plate lines high. No voltage is applied to the capacitor in this condition. When the plate line drops, write voltage is

Table 2. Various circuits of a memory cell.

| | | 6T4C | 2T2C | 1T1C | Chain | 1T | Cross point |
|---|---|---|---|---|---|---|---|
| | | | | Memory cell area | | | |
| | Large | | | | | | Small |
| Memory cell circuit | | | | | | | |
| Access time | | >5 ns | >30 ns | >30 ns | >30 ns | >20 ns | - |
| Non-destructive readout | | Yes | No | No | No | Yes | No |
| Data retention | | >10 years | >10 years | >10 years | >10 years | ~1 month | >10 years |
| Read/Write endurance | | Unlimited | Limited | Limited | Limited | R: Unlimited ? W: Limited? | Limited |
| Mass production | | Yes | Yes | Yes | Not yet | Not yet | Not yet |

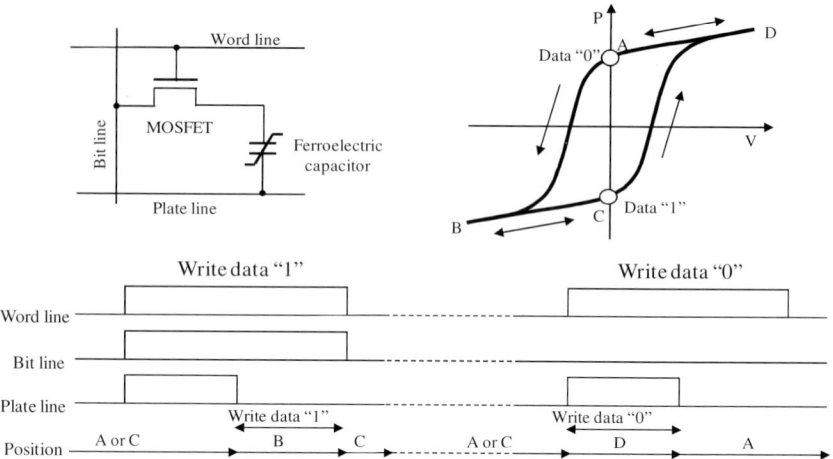

Fig. 6. Write operation of 1T1C FeRAM. During write operation, position A or C in the hysteresis curve moves to B and C for data "1', and moves to D and A for data "0".

applied to the capacitor. The position A or C in the hysteresis curve moves to B and C after the word line and bit line drop. Data "0" is written in the same manner. The position A or C in the hysteresis curve moves to D and A after operation.

Figure 7 shows the read operation of the 1T1C FeRAM. Stored data is read by applying voltage $V_{PL}$ to the plate line. $V_{PL}$ is divided into $V_{FE}$ and $V_{BL}$ by bit line capacitance $C_{BL}$ and ferroelectric capacitor capacitance $C_{FE}$. $V_{BL}$ is compared to reference voltage $V_{ref}$ by a sense amplifier. In this read operation, data "1" is lost after read (destructive readout) and is, therefore, rewritten after read. Data is read by setting the word line and plate line high. Data "1" is rewritten by setting the plate line low after read. The position C in the hysteresis curve moves to F, G, B, and C. The readout margin is determined by the amplitude of the readout signal, which depends on both $C_{BL}$ and $C_{FE}$. How $C_{BL}$ and $C_{FE}$ are designed has an extremely significant influence on FeRAM yield.[21,42−49] Various ferroelectric capacitor models and cell operation schemes have been developed for FeRAMs.

Figure 8(a) shows a simple capacitor model for simulation of readout operation. In this model, the capacitance of the ferroelectric capacitor is approximated at C0 when the polarization does not switch. The capacitance is approximated to C1 when the polarization switches. The bit line

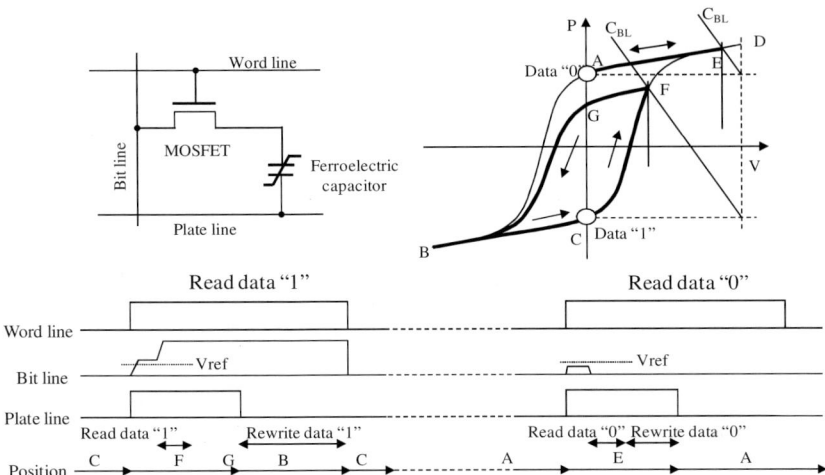

Fig. 7. Read operation of 1T1C FeRAM. Rewrite operation is executed continuously after reading operation. During read/rewrite operations, position C in hysteresis curve returns to C through F, G, and B for data "1", and position A in hysteresis curve returns to A through E for data "0".

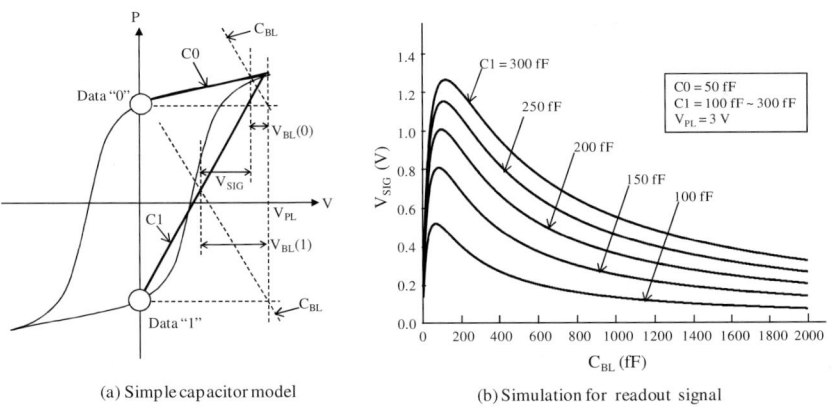

(a) Simple capacitor model      (b) Simulation for readout signal

Fig. 8. (a) Simple capacitor model and (b) simulation for readout signal.

voltage deference $V_{SIG}$ between data "0" and data "1" is approximately determined by using this model. The relations among $V_{SIG}$, C0, C1, and $C_{BL}$ are obtained by Eq. (1).

$$V_{SIG} = V_{BL}(1) - V_{BL}(0)$$
$$= C1 \times V_{PL}/(C1 + C_{BL}) - C0 \times V_{PL}/(C0 + C_{BL}) \qquad (1)$$

A large C1 and a small $C_{BL}$ are needed to obtain a large readout margin, as shown in Fig. 8(b). A large ferroelectric capacitor increases C1 and $V_{SIG}$, but the size of the memory cell also increases. Decreasing the number of capacitors connected to a bit line reduces bit line capacitance but increases the number of sense amplifiers and the size of the memory. Bit line ground sensing (BGS) architecture[50] has been developed to decrease the dependence of $V_{SIG}$ on $C_{BL}$. In BGS architecture, the bit line is kept near the ground level during read operation. Readout signal amplitude is independent of $C_{BL}$ and does not decrease even if the number of cells increases. BGS is suitable for low-voltage, large capacity FeRAMs.

The readout margin of a FeRAM increases as C1 increases and C0 decreases. Therefore, to improve the scalability of FeRAM devices, the ferroelectric capacitor should have a large remanent polarization at low voltage and a small footprint. Use of a 3D stack capacitor[51,52] with a new ferroelectric with large remanent polarization charges can resolve this issue.

## 2.3.    Reliability

The remaining issue to be solved is degradation of the ferroelectric capacitor. Remanent polarization charges are decreased by degradation of the ferroelectric capacitor due to fatigue, imprint, and retention loss, as shown in Fig. 9.[53-56] A decrease in the remanent polarization charges results in readout errors in FeRAM devices. Large capacitors must be used when the remanent polarization decreases. In other words, degradation reduces the memory capacity of FeRAM devices.

Fatigue is caused by decrease in remanent polarization, which is caused by repeated polarization switching and limits read/write endurance (the non-switching operation does not cause fatigue). Imprint leads to a shift in the hysteresis loop and causes read/write failure of the memory cell. A decrease in polarization over time in a written state is referred to as retention loss. Retention loss limits the data retention time of FeRAM devices.

Fatigue, imprint, and retention loss are related to inherent defects, process-induced crystal imperfections, and operation-induced charges of

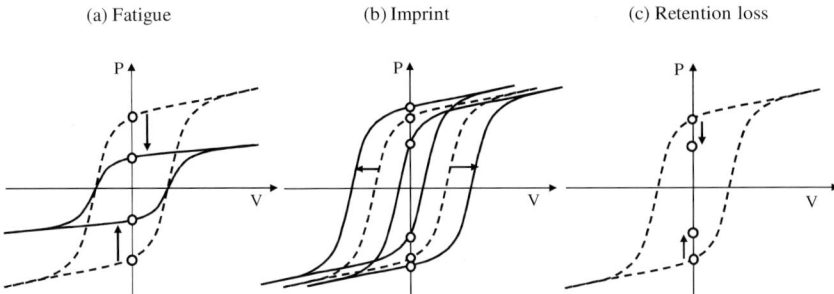

Fig. 9. Reliability issues. Remanent polarization is decreased by (a) fatigue (bi-polar cycling), (b) imprint, and (c) retention loss of ferroelectric capacitor.

the electrode/ferroelectric interface (interfacial layer with suppressed ferroelectric properties). Fatigue is caused by domain wall pinning due to charged defects, inhibition of domain nucleation by injected charges, and voltage-drop at the interfacial layer. Imprint and retention loss are related to the internal depolarizing electric field induced by charges in the interfacial layer.

Degradation of the ferroelectric capacitor was a serious problem in the early stages of FeRAM development. However, this issue has been mostly resolved by reducing the thickness of the interfacial layer. This is done by using a conductive oxide electrode, $IrO_2$ or $SrRuO_3$ (SRO), and encapsulating layers to protect against process damage.[57] A large remanent polarization with high stability is obtained, resulting in improved reliability and scalability.

## 2.4. Advanced FeRAMs

Figure 10 shows the 6T4C,[58] chain[14,17,] and 1T FeRAMs, which are being actively studied as advanced FeRAMs. The 6T4C FeRAM is composed of a six-transistor SRAM and four ferroelectric capacitors, as shown in Fig. 10(a) and is called a non-volatile SRAM (NVSRAM)[59] because it acts as an SRAM in normal operation. It features non-destructive readout, unlimited read/write endurance, and a sub-10-ns access time. The memory cell of the 6T4C FeRAM is large, but it can be reduced to the same size as that of an SRAM by placing the ferroelectric capacitors over transistors

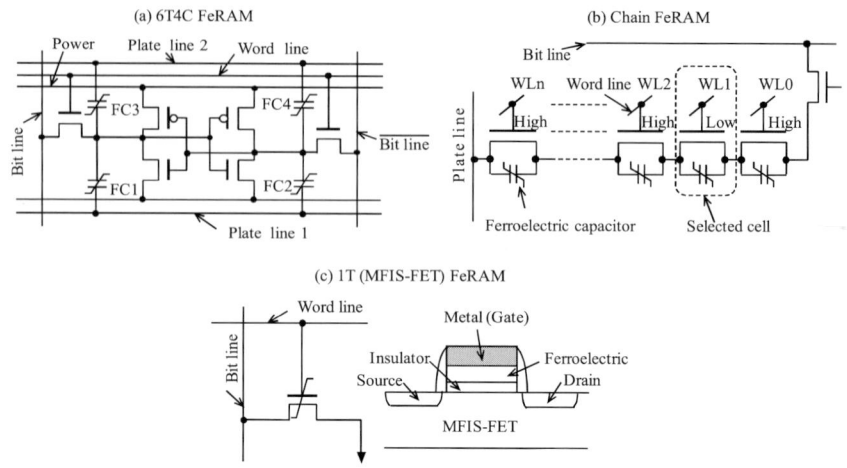

Fig. 10. (a) 6T4C FeRAM, (b) FeRAM, and (c) 1T (MFIS) FeRAM.

using a stacked structure. Ferroelectric capacitors do not fatigue because they only switch when the power is turned on or off, they operate at high speed and have unlimited read/write endurance. The recall operation is executed right after power is turned on, restoring the data stored in the ferroelectric capacitors to the SRAM cell. Data is stored before the power is turned off, and the data in the SRAM is written into the ferroelectric capacitors.

The chain FeRAM[20,60−62] has a small memory cell and a small bit line capacitance because the contacts between capacitor and transistor are shared between two unit cells. A memory cell is selected by setting unselected word lines to high and the selected word line to low. The read/write pulse is applied only to the selected cell, and the read/write operations are executed. The device's small bit line capacitance increases the read margin. A 64-Mbit chain FeRAM with excellent scalability and good reliability has been developed.

The metal-ferroelectric-insulator-semiconductor (MFIS)-FET is used in 1T FeRAM devices.[63−67] A ferroelectric is formed within the gate. The memory cell of the 1T FeRAM has excellent scalability, and can be reduced to that of Flash. Its write operation is executed by applying the write voltage to the gate. Remanent polarization charges in the ferroelectric shift the threshold voltage of the MFIS-FET. The written data is read non-destructively by the MFIS-FET's change in the drain current. Data retention time is limited by the depolarization field from charges in the insulator and the leakage current between the ferroelectric and the insulator. A retention time of over 30 days is achieved by using the $Pt/SBT/HfO_2/Si$ structure,[68] which may be adequate for many applications.

## 3. MRAM Technologies

The key characteristics of MRAM technology are very high-speed read/write operation,[69,70] low-voltage operation, nondestructive readout, unlimited read/write endurance, and radiation hardness.[71] A 16-Mbit standard MRAM and 1-Mbit embedded MRAM have been developed.[25−27,72] The maximum memory capacity of mass-produced standard MRAM is 4 Mbits.[73] It is used in standard NVRAM and to replace embedded SRAM. Technical issues include memory capacity (scalability), readout margin, and write current.

### 3.1. *Magnetic tunnel junction*

Figure 11 shows the basic structure of a magnetic tunnel junction (MTJ) for the memory cell circuit of an MRAM. An MTJ is composed of a free

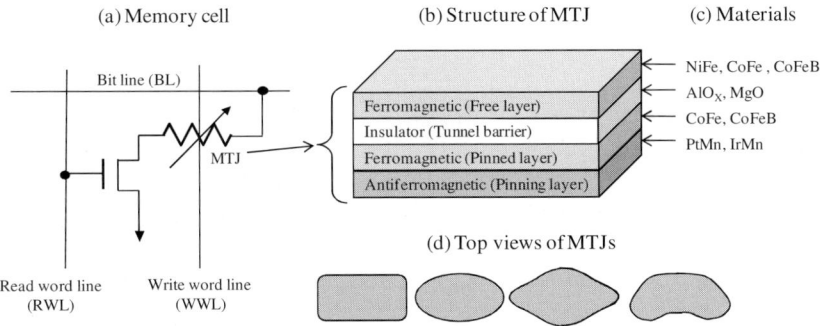

Fig. 11. (a) Memory cell, (b) magnetic tunnel junction (MTJ), (c) materials, and (d) Top views of MTJs for MRAM.

ferromagnetic layer, a thin (1∼2 nm) insulator (tunnel barrier), a pinned ferromagnetic layer, and a pinning antiferromagnetic layer.

Data is stored as a direction of magnetization of the free layer in the magnetic tunnel junction. Various kinds of MTJ have been developed to improve read/write characteristics.[74−78] Stored data is read by sensing the tunnel current change due to the tunnel magnetoresistance effect. When the magnetization directions of the free layer and the pinned layer are the same direction, namely parallel, the tunnel resistance is low. When the magnetization directions are the opposite of each other, namely antiparallel, the tunnel resistance is high. Recently, a synthetic antiferromagnet (SAF) structure has been used for the pinned layer and/or free layer of the MTJ, as shown in Fig. 12.[79] An SAF structure is formed from two ferromagnetic layers (CoFeB) separated by a non-magnetic coupling spacer layer (Ru). The size dependence of the switching field has been reduced by using a free layer with an SAF structure.[80,81] This pinned layer reduces magnetostatic coupling due to stray fields.

Figure 13 shows a schematic diagram of the band structure of a conventional MTJ. The up-spin electrons tunnel from majority band to majority band when the magnetization directions of the two magnetic layers are parallel. Therefore, the tunnel resistance is low ($R_L$), and the tunnel current is large. The up-spin electrons tunnel from majority band to minority band when the magnetization directions are antiparallel. Therefore, the tunnel resistance is high ($R_H$), and the tunnel current is small. The magnetoresistance (MR) ratio is defined by the following expression.

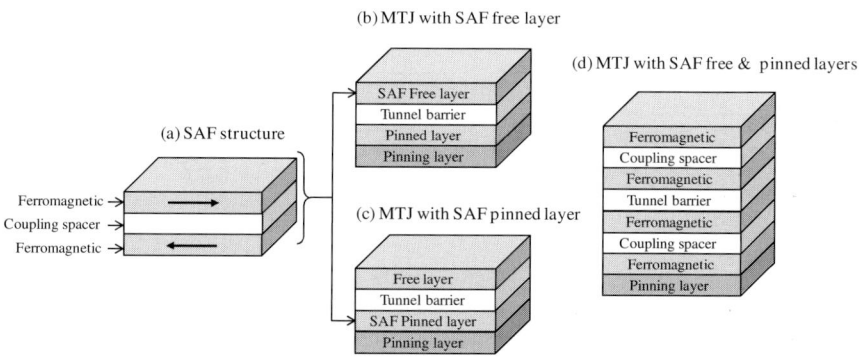

Fig. 12. (a) Synthetic antiferromagnetic (SAF) structure, (b) MTJ with SAF free layer, (c) MTJ with SAF pinned layer, and (d) MTJ with SAF free & pinned layers.

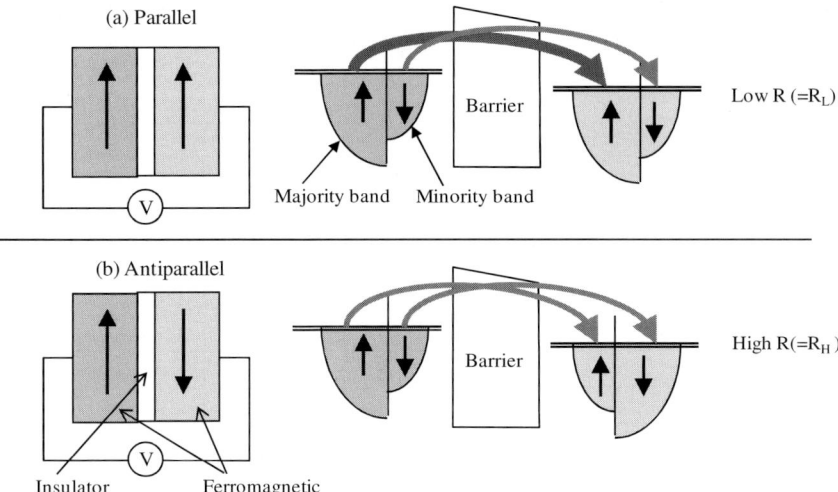

Fig. 13. Magnetoresistance effect. (a) A large current of electrons tunnels from majority band to majority band when magnetization directions are parallel. (b) A small current of electrons tunnels from majority band to minority band when magnetization directions are antiparallel.

$$R \ ratio(\%) = 100 \times (R_H - R_L)/R_L \qquad (2)$$

The MR ratio depends on both the ferromagnetic and the insulator and has a big influence on the yield, speed, and scalability of the MRAM. When an $AlO_X$ film used as the insulator (tunnel barrier) of the MTJ, the value of MR ratio is less than 100%. However, a large MR ratio has been obtained by using MgO film instead of $AlO_X$ film as an insulator.[82–84] An MTJ with a 230% MR ratio has been developed by using (100)-oriented single-crystal MgO film deposited on amorphous CoFeB by sputtering. Use of MgO film greatly improves the readout margin of an MRAM.[85]

## 3.2. Memory cell

Three basic circuits of memory cell have been developed to decrease the memory cell area, and to increase the performance of MRAM as shown in Table 3. Recently, various memory cell circuits such as 2T1MTJ[86] or 1T2MTJ[87] are proposed. The 1T1MTJ cells are used in commercially available MRAMs. 2T2MTJ memory cell consists of 2 transistors and 2 MTJs.

Table 3. 2T2MTJ consists of two 1T1MTJs, and operates at high speed. Write current of MTJ is too large for MRAM embedded SoC (system-on-a-chip) because a lot of embedded memory cells are simultaneously accessed.

| | Memory cell area<br>Large ←————————————————→ Small | | |
|---|---|---|---|
| | 2T2MTJ | 1T1MTJ | Cross point |
| Memory cell circuit | | | |
| Access time | > 2 ns | > 5 ns | > 250 ns |
| Write current/MTJ | 0.1 - 10 mA | | 4 mA |
| MR (magnetoresistance) ratio | 10 - 70% ($AlO_x$ based MTJ), >100% (MgO based MTJ) | | |
| Data retention | > 10 years | | |
| Read/Write endurance | Unlimited | | |

Data and opposite data are written in two MTJs at the same time. Date is read by comparing the current of two MTJs. 2T2MTJ memory cell has a larger memory cell area and larger read margin than 1T1MTJ memory cell as well as the relation between 2T2C and 1T1C in FeRAM. Cross point memory cell has the smallest memory cell area in all MRAM. However, data read current is disturbed by sneak current which flows across the entire array of the cross point cell. Sneak current of cross point MRAM[88,89] is reduced by using the hierarchical bit line architecture.

## 3.3. MRAM write operation

Figure 14 shows an MRAM write operation, which is executed by two orthogonal magnetic fields generated by the current flow through the bit line (BL) and write word line (WWL). Write operation is carried out properly when the magnetic fields, $H_x$ and $H_y$, are in the switching regions. Outside of these regions, the half-selected MTJs switch (this is called the half-select disturb problem). No switching occurs in the non-switching region. The BL and WWL write currents should be in the area of switching regions to prevent half-selected MTJs from switching. This means that the write margin of MRAM is small.

The switching field, which is generated by write current, strongly depends on the size of the MTJ, as shown in Fig. 15. The diamagnetic field

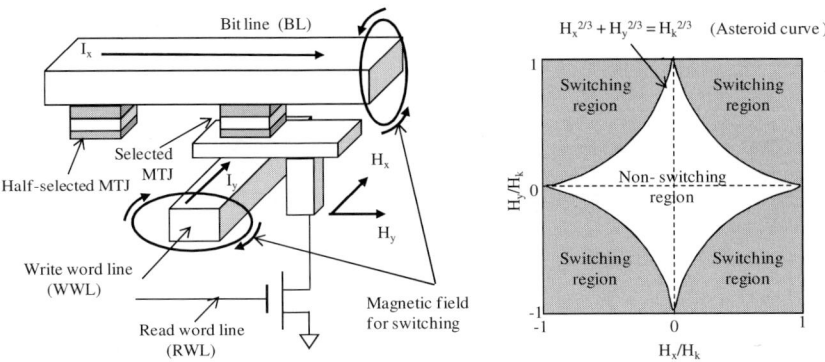

Fig. 14. Write operation to selected MTJ.

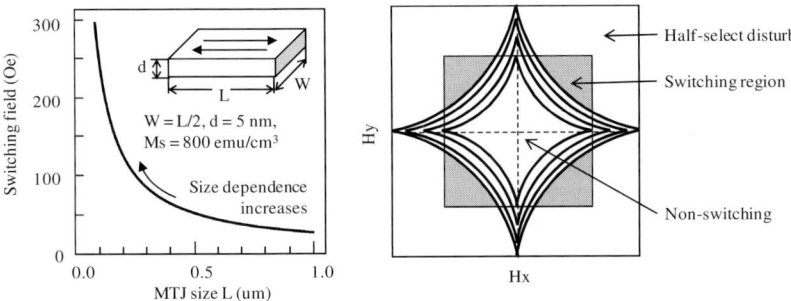

Fig. 15. Switching field (switching current) increases greatly as the size of MTJ decreases. Switching region shrinks due to the size distribution.

increases as MTJ size decreases and causes write current to increase. The MRAM device also consumes more power, the switching region shrinks, and the half-select disturb becomes more serious when MTJs vary in size. The size dependence of the switching current degrades the write margin and scalability of MRAM devices.

These issues have been mostly resolved by using cladding line, toggling architecture with a moment-balanced synthetic antiferromagnetic (SAF) free layer, thermal select architecture, MTJ design, and spin torque transfer switching. Cladding lines decrease the writing current. Cladded BL and WWL with soft ferromagnetic NiFe, which doubles the magnetic field by concentrating flux, decreases the switching current by half.

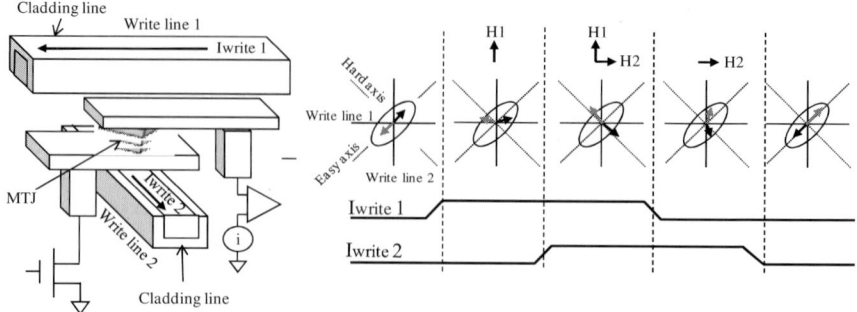

Fig. 16. Toggling architecture.

The half-select disturb has been greatly reduced by toggling architecture, as shown in Fig. 16.[81,90−92] The free layer of the MTJ is composed of a moment-balanced SAF multilayer, which is formed by two ferromagnetic layers separated by a non-magnetic coupling layer that couples the layers in antiparallel. The MTJ axis is aligned mid-angle between two orthogonal write lines (write lines 1 and 2). A two-phase writing pulse sequence is applied to rotate the magnetization direction of the SAF free layer 180 degrees. The writing pulse sequence toggles the magnetic state to the opposite state regardless of the existing state. Therefore, pre-read is used to determine if a write is required because the data stored in the MTJ changes from "0" to "1" or "1" to "0" whenever a two-phase writing pulse sequence is applied. A magnetic field generated by the current that flows through a single write line cannot switch the SAF free layer easily. Therefore, the toggling architecture is designed to prevent the half-selected MTJs from switching.

The half-selected disturb is reduced even further by using a thermal select writing scheme.[93] The selected MTJ is heated by directing the local heating current through it. When the free layer is heated above a critical temperature, the pinning vanishes, and the free layer can be set by sufficiently small magnetic fields. The half-selected disturb is suppressed because the half-selected MTJ is not heated.

## 3.4.   *Spin torque transfer switching*

Spin torque transfer switching (spin-polarized current induced switching) is a new writing method for MRAM.[94−96] MRAMs of 256 kbits and 2 Mbits

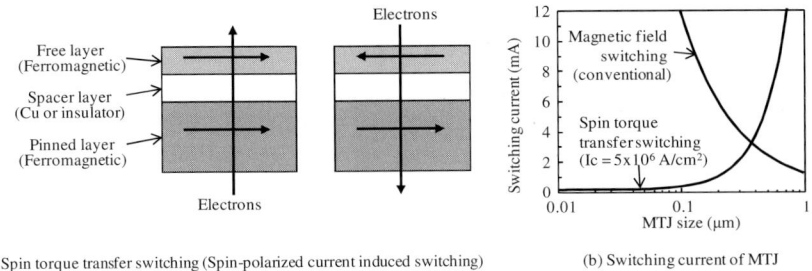

(a) Spin torque transfer switching (Spin-polarized current induced switching)   (b) Switching current of MTJ

Fig. 17. (a) Spin transfer switching (spin-polarized current induced switching) and (b) switching current of MTJ.

using the spin torque transfer switching method have been developed.[97,98] The write current is reduced to less than 1 mA. The magnetization direction of the free layer is switched by the current through the MTJ, as shown in Fig. 17(a). A high-current density write pulse results in a torque on the free layer magnetic moment due to the angular momentum carried by the spin-polarized tunneling current. The magnetization direction of the MTJ's free layer is the same as that of the pinned layer when the electrons flow from the pinned layer to the free layer. When the electrons flow from the free layer to the pinned layer, the magnetization of free layer takes the direction opposite to that of the pinned layer. In spin torque transfer switching, the switching current of a free layer depends not on the current but on the current density. Therefore, the switching current decreases as the MTJ size decreases and is smaller than that in the conventional magnetic field switching method when the MTJ size is less than $0.4 \, \mu$m and the switching current density is $5 \times 10^5$ A/cm$^2$, as shown in Fig. 17(b). When the current density of spin torque transfer switching decreases to $10^5$ A/cm$^2$, the write current drops to 0.1 mA for an MTJ with an area of $0.1 \, \mu$m$^2$. The spin torque transfer switching method is suitable for large capacity, low-power MRAM devices.

### 3.5.   *MRAM read operation*

Figure 18 shows the read operation of an MRAM device. Stored data is read by sensing the resistance of the MTJ. Amplitude of the readout signal depends on both the magnetoresistance (MR) ratio and tunnel resistance. The MR ratio also decreases as applied voltage to MTJ increases.

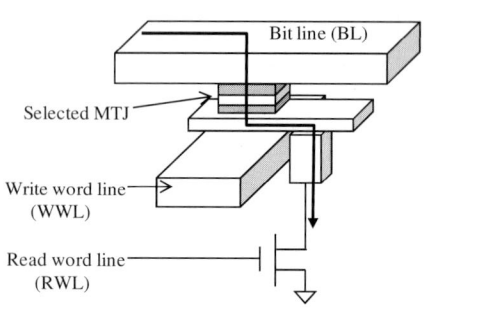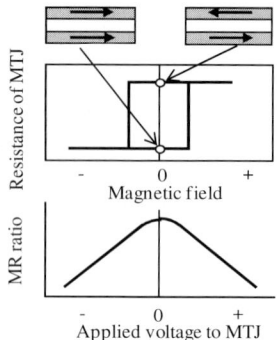

Fig. 18. Read operation to selected MTJ.

An MRAM with a large readout margin must be used to precisely control insulator thickness and reading voltage.

The readout signal is not adequate for high-speed operation even though an MTJ with an MR ratio of 230% has been developed using single-crystal MgO as an insulator. Distribution of tunnel resistance due to variations in insulator thickness or quality, MTJ size, and bias dependence is the most serious problem. A self-reference-sensing scheme[99] has been developed to eliminate the effect of tunnel resistance distribution, and has been experimentally demonstrated for a tunnel resistance variation of 100%. However, the self-reference-sensing scheme results in a destructive readout and low speed. The rewriting operation also increases power consumption.

## 4. Conclusions

FeRAM is the best memory for low-power SoC applications. The scalability and reliability of FeRAM can be greatly improved by optimizing the materials, process, structure of ferroelectric capacitor, and architecture. Conductive oxide electrodes ($IrO_2$, $SrRuO_3$ etc.) of the ferroelectric capacitor prevent the interfacial layer from growing and improve the reliability of FeRAM devices. Optimizing the memory cell circuit and architecture increases the read margin and helps reduce the size of the memory cell.

MRAM has the highest access speed and read/write endurance of all NVRAMs and is the best embedded memory for high-speed SoCs. The half-select disturb has been mostly resolved by the toggle writing, thermal select writing, and spin torque transfer switching schemes. Read margins of MRAM devices can be increased by using a single crystal MgO tunnel barrier. Spin torque transfer switching decreases the write current, and has the potential to become a key technology for sub-0.1 $\mu$m low-power and high-capacity MRAM devices. An SAF is a key structure that resolves many MRAM technical issues of and helps reduce the size of the memory cell.

A lot of research on FeRAM and MRAM is going on all over the world, and developers anticipate rapid improvements in scalability and reliability. The number of applications of FeRAM and MRAM will increase as the demand for low-power consumption increases.

## 5. Acknowledgments

I am grateful to Professor Hiroshi Ishiwara of the Tokyo Institute of Technology for supervision on advanced FeRAM technology, to Dr. Jeff Cross, Dr. Takashi Eshita, Mr. Shoichiro Kawashima, Dr. Shoichi Masui, and Dr. Tetsuro Tamura for their discussions of FeRAM reliability, process, circuitry, and modeling, and to Dr. Kazuo Kobayashi, Dr. Masaki Aoki, and Dr. Masashige Sato for discussions of MRAM technology.

## References

1.  M. Julliere, *Phys. Lett. A*, **54**, pp 225–226 (1975).
2.  T. Miyazaki and N. Tezuka, *J. Magn. Magn. Mater.*, pp L231–L234 (1995).
3.  Y. M. Kang, J.-H. Kim, H. J. Joo, S. K. Kang, H. S. Rhie, J. H. Park, D. Y. Choi, S. G. Oh, B. J. Koo, S. Y. Lee, H. S. Jeong and K. Kim, *Symposium on VLSI Technology*, pp 102–103 (2005).
4.  S. Masui, T. Ninomiya, M. Oura, W. Yokozeki, K. Mukaida and S. Kawashima, *Symposium on VLSI Circuits*, pp 200–203 (2002).
5.  H. McAdams, R. Acklin, T. Blake, J. Fong, D. Liu, S. Madan, T. Moise, S. Natarajan, N. Qian, Y. Qui, J. Roscher, A. Seshadri, S. Summerfelt, X. Du, J. Eliason, W. Kraus, R. Lanham, F. Li, C. Pietrzyk and J. Rickes, *Symposium on VLSI Circuits*, pp 175–176 (2003).
6.  T. Hanyu, H. Kimura, M. Kameyama, Y. Fujimori, T. Nakamura and H. Takasu, *ISSCC*, pp 208–208 (2002).
7.  T. Hayashi, Y. Igarashi, D. Inomata, T. Ichimori, T. Mitsuhashi, K. Ashikaga, T. Ito, M. Yoshimaru, M. Nagata, S. Mitarai, H. Godaiin,

T. Nagahama, C. Isobe, H. Moriya, M. Shoji, Y. Ito, H. Kuroda and M. Sasaki, *IEDM Tech. Dig.*, pp 543–546 (2002).

8.  T. S. Moise, S. R. Summerfelt, H. McAdams, S. Aggarwal, K. R. Udayakumar, F. G. Celii, J. S. Martin, G. Xing, L. Hall, K. J. Taylor, T. Hurd, J. Rodriguez, K. Remack, M. D. Khan, K. Boku, G. Stacey, M. Yao, M. G. Albrecht, E. Zielinski, M. Thakre, S. Kuchimanchi, A. Thomas, B. McKee, J. Rickes, A. Wang, J. Grace, J. Fong, D. Lee, C. Pietrzyk, R. Lanham, S. R. Gilbert, D. Taylor and Jun Amano, *IEDM Tech. Dig.*, pp 535–538 (2002).

9.  Y. Horii, Y. Hikosaka, A. Itoh, K. Matsuura, M. Kurasawa, G. Komuro, K. Maruyama, T. Eshita and S. Kashiwagi, *IEDM Tech. Dig.*, pp 539–542 (2002).

10. A. Itoh, Y. Hikosaka, T. Saito, H. Naganuma, H. Miyazawa, Y. Ozaki, Y. Kate, S. Mihara, H. Iwamoto, S. Mochizuki, M. Nakamura and T. Yamazaki, *Symposium on VLSI Technology*, pp 32–33 (2000).

11. Y. K. Hong, D. J. Jung, S. K. Kang, H. S. Kim, J. Y. Jung, H. K. Koh, J. H. Park, D. Choi, S. E. Kim, W. S. Ann, Y. M. Kang, H. H. Kim, J.-H. Kim, W. U. Jung, E. S. Lee, S. Y. Lee, H. S. Jeong and K. Kim, *Symposium on VLSI Technology*, pp 230–231 (2007).

12. A. R. Sitaram, D. W. Abraham, C. Alof, D. Braun, S. Brown, G. Costrini, F. Findeis, M. Gaidis, E. Galligan, W. Glashauser, A. Gupta, H. Hoenigschmid, J. Hummel, S. Kanakasabapathy, I. Kasko, W. Kim, U. Klostermann, G. Y. Lee, R. Leuschner, K.-S. Low, Y. Lu, J. Nützel, E. O'Sullivan, C. Park, W. Raberg, R. Robertazzi, C. Sarma, J. Schmid, P. L. Trouilloud, D. Worledge, G. Wright, W. J. Gallagher and G. Müller, *Symposium on VLSI Technology*, pp 15–16 (2003).

13. J. H. Park, W. C. Jeong, H. J. Kim, J. H. Oh, H. C. Koo, G. H. Koh, G. T. Jeong, H. S. Jeong, Y. J. Jeong, S. L. Cho, J. E. Lee, H. J. Kim and K. Kim, *IEDM Tech. Dig.*, pp 827–830 (2003).

14. D. Takashima, Y. Takeuchi, T. Miyakawa, Y. Itoh, R. Ogiwara, M. Kamoshida, K. Hoya, S. M. Doumae, T. Ozaki, H. Kanaya, M. Aoki, K. Yamakawa, I. Kunishima and Y. Oowaki, *ISSCC*, pp 40–41 (2001).

15. H. H. Kim, Y. J. Song, S. Y. Lee, H. J. Joo, N. W. Jang, D. J. Jung, Y. S. Park, S. O. Park, K. M. Lee, S. H. Joo, S. W. Lee, S. D. Nam and K. Kim, *Symposium on VLSI Technology*, pp 210–211 (2002).

16. H. Kanaya, K. Tomioka, T. Matsushita, M. Omura, T. Ozaki, Y. Kumura, Y. Shimojo, T. Morimoto, O. Hidaka, S. Shuto, H. Koyama, Y. Yamada, K. Osari, N. Tokoh, F. Fujisaki, N. Iwabuchi,

N. Yamaguchi, T. Watanabe, M. Yabuki, H. Shinomiya, N. Watanabe, E. Itoh, T. Tsuchiya, K. Yamakawa, K. Natori, S. Yamazaki, K. Nakazawa, D. Takashima, S. Shiratake, S. Ohtsuki, Y. Oowaki, I. Kunishima and A. Nitayama, *Symposium on VLSI Technology*, pp 150–151 (2004).

17. S. Shiratake, T. Miyakawa, Y. Takeuchi, R. Ogiwara, M. Kamoshida, K. Hoya, K. Oikawa, T. Ozaki, I. Kunishima, K. Yamakawa, S. Sugimoto, D. Takashima, H. O. Joachim, N. Rehm, J. Wohlfahrt, N. Nagel, G. Beitel, M. Jacob and T. Roehr, ISSCC, pp.282–283 (2003).

18. Y. J. Song, H. J. Joo, N. W. Jang, H. H. Kim, J. H. Park, H. Y. Kang, S. Y. Lee and K. Kim, *Symposium on VLSI Technology*, pp 169–170 (2003).

19. J -H. Kim, D. J. Jung, S. K. Kang, Y. M. Kang, H. H. Kim, J. Y. Kang, E. S. Lee, W. W. Jung, H. J. Joo, J. Y. Jung, J. H. Park, H. Kim, D. Y. Choi, S. Y. Lee, H. S. Jeong and K. Kim, *IEDM Tech. Dig.*, pp 45–48 (2006).

20. K. Hoya, D. Takashima, S. Shiratake, R. Ogiwara, T. Miyakawa, H. Shiga, S. M. Doumae, S. Ohtsuki, Y. Kumura, S. Shuto, T. Ozaki, K. Yamakawa, I. Kunishima, A. Nitayama and S. Fujii, *ISSCC*, pp 465–472 (2006).

21. M.-K. Choi, B.-G Jeon, N. Jang, B.-J. Min, Y.-J. Song, S.-Y. Lee, H.-H. Kim, D.-J. Jung, H.-J. Joo and K. Kim, *ISSCC*, pp 162–163 (2002).

22. M. Durlam, D. Addie, J. Akerman, B. Butcher, P. Brown, J. Chan, M. DeHerrera, B.N. Engel, B. Feil, G. Grynkewich, J. Janesky, M. Johnson, K. Kyler, J. Molla, J. Martin, K. Nagel, J. Ren, N.D. Rizzo, T. Rodriguez, L. Savtchenko, J. Salter, J.M. Slaughter, K. Smith, J.J. Sun, M. Lien, K. Papworth, P. Shah, W. Qin, R. Williams, L. Wise and S. Tehrani, *IEDM Tech. Dig.*, pp 995–997 (2003).

23. P. K. Naji, M. Durlam, S. Tehrani, J. Calder and M. F. DeHerrera, *ISSCC*, pp 122–123 (2001).

24. M. Durlam, P. Naji, A. Omair, M. DeHerrera, J. Calder, J. M. Slaughter, B. Engel, N. Rizzo, G. Grynkewich, B. Butcher, C. Tracy, K. Smith, K. Kyler, J. Ren, J. Molla, B. Feil, R. Williams and S. Tehrani, *Symposium on VLSI Circuits*, pp 158–161 (2002).

25. J. DeBrosse, C. Arndt, C. Barwin, A. Bette, D. Gog, E. Gow, H. Hoenigschmid, S. Lammers, M. Lamorey, Y. Lu, T. Maffitt, K. Maloney, W. Obermeyer, A. Sturm, H. Viehmann, D. Willmott, M. Wood,

W. J. Gallagher, G. Mueller and A. R. Sitaram, *Symposium on VLSI Circuits*, pp 454–457 (2004).

26. H. Hönigschmid, P. Beer, A. Bette, R. Dittrich, F. Gardic, D. Gogl, S. Lammers, J. Schmid, L. Altimime, S. Bournat and G. Müller, *ISSCC*, pp 473–482 (2006).

27. Y. Iwata, K. Tsuchida, T. Inaba, Y. Shimizu, R. Takizawa, Y. Ueda, T. Sugibayashi, Y. Asao, T. Kajiyama, K. Hosotani, S. Ikegawa, T. Kai, M. Nakayama, S. Tahara and H. Yoda, *ISSCC*, pp 483–492 (2006).

28. Y. Arimoto and H. Ishiwara, *MRS BULLETIN*, pp 823–828 (2004).

29. Y. Arimoto, *ESSDERC-ESSCIRC'05, Short Course SC4_5* (2005).

30. H. J. Joo, Y. J. Song, H. H. Kim, S. K. Kang, J. H. Park, Y. M. Kang, E. Y. Kang, S. Y. Lee, H. S. Jeong and K. Kim, *Symposium on VLSI Technology*, pp 148–149 (2004).

31. J.-H. Kim, Y. M. Kang, J. H. Park, H. J. Joo, S. K. Kang, D. Y. Choi, H. S. Rhie, B.J. Koo, S. Y. Lee, H. S. Jeong and K. Kim, *IEDM Tech. Dig.*, pp 869–872 (2005).

32. H. Nakamoto, D. Yamazaki, T. Yamamoto, H. Kurata, S. Yamada, K. Mukaida, T. Ninomiya, T. Ohkawa, S. Masui and K. Gotoh, *ISSCC*, pp 1207–1216 (2006).

33. J. Wang, J. B. Neaton, H. Zheng, V. Nagarajan, S. B. Ogale, B. Liu, D. Viehland, V. Vaithyanathan, D. G. Schlom, U. V. Waghmare, N. A. Spaldin, K. M. Rabe, M. Wuttig and R. Ramesh, *Science*, **299**, pp 1719–1722 (2003).

34. S. K. Singh and H. Ishiwara, *Jpn. J. Appl. Phys.*, **43**, pp L734–L736 (2005).

35. K. Y. Yun, D. Ricinschi, T. Kanashima, M. Noda and M. Okuyama, *Jpn. J. Appl. Phys*, **43**, pp L647–L648 (2004).

36. Y. Nagano, T. Mikawa, T. Kutsunai, S. Hayashi, T. Nasu, S. Natsume, T. Tatsunari, T. Ito, S. Goto, H. Yano, A. Noma, K. Nagahashi, T. Miki, M. Sakagami, Y. Izutsu, T. Nakakuma, H. Hirano, S. Iwanari, Y. Murakuki, K. Yamaoka, Y. Goho, Y. Judai, E. Fujii and K. Sato, *Symposium on VLSI Technology*, pp 171–172 (2003).

37. K. Yamaoka, S. Iwanari, Y. Murakuki, H. Hirano, M. Sakagami, T. Nakakuma, T. Miki and Y. Gohou, *ISSCC*, pp 50–51 (2004).

38. S, Y. Lee, H. H. Kim, D. J. Jung, Y. J. Song, N. W. Jang, M. K. Choi, B, K. Jeon, Y. T. Lee, K. M. Lee, S. H. Joo, S. O. Park and K. Kim, *Symposium on VLSI Technology*, pp 111–112 (2001).

39. S.-H. Oh, S.-K Hong, K.-H. Noh, S.-Y. Kweon, N.-K. Kim, Y.-H. Yang, J.-G. Kim, J.-Y. Seong, I.-W. Jang, S.-H. Park, K.-H. Bang,

K.-N. Lee, H-.J. Jeong, J.-H. Son, S.-S.Lee, E.-S. Choi, H.-J. Sun, S.-J. Yeom, K.-D. Ban, J.-W. Park, G.-D. Park, S.-Y. Song, J.-H. Shin, S.-I. Lee and Y.-J. Park, *IEDM Tech. Dig.*, pp 835–838 (2003).

40. T. Nishihara and Y. Ito, *ISSCC*, pp 160–161 (2002).

41. N. Sakai, Y. Ishizuka, S. Matsushita, Y. Takano, S. Ogasawara, K. Honma, T. Geshi, Y. Inoue and K. Fukase, *Symposium on VLSI Circuits*, pp 171–172 (2003).

42. J. Chow, A. Sheikholeslami, J. S. Cross and S. Masui, *Symposium on VLSI Circuits*, pp 448–449 (2004).

43. T. Tamura, Y. Arimoto and H. Ishiwara, *Jpn. J. Appl. Phys.*, pp 2654–2657 (2002).

44. Y. Eslami, A. Sheikholeslami, S. Masui, T. Endo and S. Kawashima, *Symposium on VLSI Circuits*, pp 298–301 (2002).

45. H.-B. Kang, H.-W. Kye, D.-J. Kim, G.-I Lee, J.-H. Park, J.-K. Wee, S.-S. Lee, S.-K. Hong, N.-S. Kang and J.-Y. Chung, *Symposium on VLSI Circuits*, pp 125–126 (2001).

46. H.-B. Kang, H.-W. Kye, G.-I. Lee, J.-H. Park, J.-H. Kim, S.-S. Lee, S.-K. Hong, Y.-J. Park and J.-Y. Chung, *ISSCC*, pp 38–39 (2002).

47. B.-G. Jeon, M.-K. Choi, Y. Song and K. Kim, *ISSCC*, pp 38–39 (2001).

48. C. Ohno, H. Yamazaki, H. Suzuki, E. Nagai, H. Miyazawa, K. Saigoh, T. Yamazaki, Y. Chung, W. Kraus, D. Verhaeghe, G. Argos, J. Walbert and S. Mitra, *ISSCC*, pp 36–37 (2001).

49. Y. M. Kang, H. J. Joo, J. H. Park, S. K. Kang, J.-H. Kim, S. G. Oh, H. S. Kim, J. Y. Kang, J. Y. Jung, D. Y. Choi, E. S. Lee, S. Y. Lee, H. S. Jeong and K. Kim, *Symposium on VLSI Technology*, pp 152–153 (2006).

50. S. Kawashima, T. Endo, A. Yamamoto, K. Nakabayashi, M. Nakazawa, K. Morita and M. Aoki, *Symposium on VLSI Circuits*, pp 127–128 (2001).

51. N. Nagel, R. Bruchhaus, K. Hornik, U. Egger, H. Zhuang, H.-O. Joachim, T. Röhr G. Beitel, T. Ozaki and I. Kunishima, *Symposium on VLSI Technology*, pp 146–147 (2004).

52. J.-M. Koo, B.-S. Seo, S. Kim, S. Shin, J.-H. Lee, H. Baik, J.-H. Lee, J. H. Lee, B.-J. Bae, J.-E. Lim, D.-C. Yoo, S.-O. Park, H.-S. Kim, H. Han, S. Baik, J.-Y. Choi, Y. J. Park and Y. Park, *IEDM Tech. Dig.*, pp 351–354 (2005).

53. A. K. Tagantsev, I. Stolichnov, E. L. Colla and N. Setter, *J. of Appl. Phys.*, **90**, pp 1387–1402 (2001).

54. M. Grossmann, O. Lohse, D. Bolten, U. Boettger, T. Schneller and R. Waser, *J. of Appl. Phys.*, **92**, pp 2680–2687 (2002).
55. A. K. Tagantsev, I. Stolichnov, N. Setter and J. S. Cross, *J. of Appl. Phys.*, **96**, pp 6616–6623 (2004).
56. I. Stolichnov, A. K. Tagantsev, E. Colla, N. Setter and J. S. Cross, *J Appl Phys.*, **98**, 084106 (2005).
57. D. C. Yoo, B. J. Bae, J.-E. Lim, D. H. Im, S. O. Park, H. S. Kim, U.-In. Chung, J. T. Moon and B. I. Ryu, *Symposium on VLSI Technology*, pp 100–101 (2005).
58. S. Masui, T. Ninomiya, T. Ohkawa, M. Oura, Y. Horii, N. Kin and K. Honda, *IEICE Trans. On Electron.*, **E87-C**, pp 1769–1776 (2004).
59. T. Miwa, J. Yamada, H. Koike, T. Nakura, S. Kobayashi, N. Kasai and H. Toyoshimam, *Symposium on VLSI Circuits*, pp 129–132 (2001).
60. K. Oikawa, D. Takashima, S. Shiratake, K. Hoya and H. O. Joachim, *Symposium on VLSI Circuits*, pp 169–170 (2003).
61. T. Ozaki, J. Iba, Y. Yamada, H. Kanaya, T.Morimoto, O. Hidaka, A. Taniguchi, Y. Kumura, K. Yamakawa, Y. Oowaki and I. Kunishima, *Symposium on VLSI Technology*, pp 113–114 (2001).
62. O. Hidaka, T. Ozaki, H. Kanaya, Y. Kumura, Y. Shimojo, S. Shuto, Y. Yamada, K. Yahashi, K. Yamakawa, S. Yamazaki, D. Takashima, T. Miyakawa, S. Shiratake, S. Ohtsuki, I. Kunishima and A. Nitayama, *Symposium on VLSI Technology*, pp 154–155 (2006).
63. H. Ishiwara, *IEDM Tech. Dig.*, pp 725–728 (2001).
64. H. Ishiwara, *IEDM Tech. Dig.*, pp 263–266 (2003).
65. M. Y. Yang, S. B. Chen, A. Chin, C. L. Sun, B. C. Lan and S. Y. Chen, *IEDM Tech. Dig.*, pp 795–798 (2001).
66. K. Aizawa, B.-E. Park, Y. Kawashima, K. Takahashi and H. Ishiwara, *Appl. Phys. Lett.* **85**, pp 3199–3201 (2004).
67. S. Sakai and R. Ilangovan, *IEEE Electron. Dev. Lett.* **25**, pp 369–371 (2004).
68. K. Takahashi, B.-E. Park, K. Aizawa and H. Ishiwara, *Abs. Int. Conf. Solid State Devices and Materials*, pp 52–53 (2004).
69. A. Bette, J. DeBrosse, D. Gogl, H. Hoenigschmid, R. Robertazzi, C. Arndt, D. Braun, D. Casarotto, R. Havreluk, S. Lammers, W. Obermaier, W. Reohr, H. Viehmann, W. J. Gallagher and G. Müller, *Symposium on VLSI Circuits*, pp 217–220 (2003).

70. N. Sakimura, T. Sugibayashi, T. Honda, H. Honjo, S. Saito, T. Suzuki, N. Ishiwata and S. Tahara, *Symposium on VLSI Circuits*, pp 136–137 (2006).

71. S. Tehrani, *IEDM Tech. Dig.*, pp 585–588 (2006).

72. T. Tsuji, H. Tanizaki, M. Ishikawa, J. Otani, Y. Yamaguchi, S. Ueno, T. Oishi and H. Hidaka, *Symposium on VLSI Circuits*, pp 450–453 (2004).

73. J. Nahas, T. Andre, C. Subramanian, B. Garni, H. Lin, A. Omair and W. Martino, *ISSCC*, pp 44–45 ( 2004).

74. Y. K. Ha, J. E. Lee, H.-J. Kim, J. S. Bae, S. C. Oh, K. T. Nam, S. O. Park, N. I. Lee, H. K. Kang, U.-In Chung and J. T. Moon, *Symposium of VLSI Technology*, pp 24–25 (2004).

75. M. Motoyoshi, I. Yamamura, W. Ohtsuka, M. Shouji, H. Yamagishi, M. Nakamura, H. Yamada, K.Tai, T. Kikutani, T. Sagara, K. Moriyama, H. Mori, C. Fukumoto, M. Watanabe, H. Hachino, H. Kano, K. Bessho, H. Narisawa, M. Hosomi and N. Okazaki, *Symposium of VLSI Technology*, pp 22–23 (2004).

76. M. Motoyoshi, K. Moriyama, H. Mori, C. Fukumoto, H. Itoh, H. Kano, K. Bessho and H. Narisawa, *Symposium on VLSI Technology*, pp 212–213 (2002).

77. T. Kai, M. Yoshikawa, M. Nakayama, Y. Fukuzumi, T. Nagase, E. Kitagawa, T. Ueda, T. Kishi, S. Ikegawa, Y. Asao, K. Tsuchida, H. Yoda, N. Ishiwata, H. Hada and S. Tahara, *IEDM Tech. Dig.*, pp 583–586 (2004).

78. S. Ueno, T. Eimori, T. Kuroiwa, H. Furuta, J. Tsuchimoto, S. Maejima, S. Iida, H. Ohshita, S. Hasegawa, S. Hirano, T. Yamaguchi, H. Kurisu, A. Yutani, N. Hashikawa, H. Maeda, Y. Ogawa, K. Kawabata, Y. Okumura, T. Tsuji, J. Ohtani, T. Tanizaki, Y. Yamaguchi, T. Ohishi, H. Hidaka, T. Takenaga, S. Beysen, H. Kobayashi, T. Oomori, T. Koga and Y. Ohji, *IEDM Tech. Dig.*, pp 579–582 (2004).

79. T. Suzuki, Y. Fukumoto, K. Mori, H. Honjo, R. Nebashi, S. Miura, K. Nagahara, S. Saito, H. Numata, K. Tsuji, T. Sugibayashi, H. Hada, N. Ishiwata, Y. Asao, S. Ikegawa, H. Yoda and S. Tahara, *Symposium of VLSI Technology*, pp 188–189 (2005).

80. K. Inomata, N. Koike, T. Nozaki, S. Abe and N. Tezuka, *Appl. Phys. Lett.*, **82**, pp 2667–2669 (2003).

81. J. Slaughter, J. Janesky, N. Rizzo, J. Åkerman, B. Engel, R. Dave, J. Sun, S. Pietambaram, G. Grynkewich, M. DeHerrera, M. Durlam,

K. Smith and S. Tehrani, *Cornell CNS Nanotechnology Symposium* (2004).

82.  S. S. P. Parkin, C. Kaiser, A. Panchula, P. M. Rice, B. Hughes, M. Samant and S.-H. Yang, *Nature Materials*, pp 862–867 (2004).

83.  S. Yuasa, T. Nagahama, A. Fukushima, Y. Suzuki and K. Ando, *Nature Materials*, pp 868–871 (2004).

84.  S. Ikeda, J. Hayakawa, Y. M, Lee, R. Sasaki, T. Meguro, F. Matsukura and H. Ohno, *Jpn. J. Appl. Phys.*, pp L1442–L1445 (2005).

85.  D. D. Djayaprawira, K. Tsunekawa, M. Nagai, H. Maehara, S. Yamagata, N. Watanabe, S. Yuasa, Y. Suzuki and K. Ando, *Appl. Phys. Lett.*, **86**, 092502 (2005).

86.  H. Numata, T. Suzuki, N. Ohshima, S. Fukami, K. Nagahara, N. Ishiwata and N. Kasai, *Symposium on VLSI Technology*, pp 232–233 (2007).

87.  M. Aoki, H. Iwasa and Y. Sato, *Symposium on VLSI Circuits*, pp 171–171 (2005).

88.  Y. Asao, T. Kajiyama, Y. Fukuzumi, M. Amano, H. Aikawa, T. Ueda, T. Kishi, S. Ikegawa, K. Tsuchida, Y. Iwata, A. Nitayama, K. Shimura, Y. Kato, S. Miura, N. Ishiwata, H. Hada, S. Tahara and H. Yoda, *IEDM Tech. Dig.*, pp 571–574 (2004).

89.  N. Sakimura, T. Sugibayashi, T. Honda, S. Miura, H. Numata, H. Hada and S. Tahara, *ISSCC*, pp 278–279 (2003).

90.  L. Savtchenko, A. A. Korkin, B. N. Engel, N. D. Rizzo, M. F. Deherrera and J. A. Janesky, US patent 6,545,906.

91.  J. M. Slaughter, R. W. Dave, M. Durlam, G. Kerszykowski, K. Smith, K. Nagel, B. Feil, J. Calder, M. DeHerrera, B. Garni and S. Tehrani, *IEDM Tech. Dig.*, pp 893–896 (2005).

92.  M. Durlam, T. Andre, P. Brown, J. Calder, J. Chan, R. Cuppens, R. W. Dave, T. Ditewig, M. DeHerrera, B. N. Engel, B. Feil, C. Frey, D. Galpin, B. Garni, G. Grynkewich, J. Janesky, G. Kerszykowski, M. Lien, J. Martin, J. Nahas, K. Nagel, K. Smith, C. Subramanian, J. J. Sun, J. Tamim, R. Williams, L. Wise, S. Zoll, F. List, R. Fournel, B. Martino and S. Tehrani, *Symposium on VLSI Technology*, pp 186–187 (2005).

93.  R. Leuschner, U. K. Klostermann, H. Park, F. Dahmani, R. Dittrich, C. Grigis, K. Hernan, S. Mege, C. Park, M. C. Clech, G. Y. Lee, S. Bournat, L. Altimime and G. Mueller, *IEDM Tech. Dig.*, pp 165–168 (2006).

94. F. J. Albert, J. A. Katine, R. A. Buhrman and D. C. Ralph, *Appl. Phys. Lett.*, pp 3809–3811 (2000).

95. W. H. Butler, X.-G. Zhang and T. C. Schulthess, *Physical. Review* B, **63**, 054416 (2001).

96. Y. Huai, F. Albert, P. Nguyen, M. Pakala and T. Valet, *Appl. Phys. Lett.*, **84**, 3118 (2004).

97. K. Miura, T. Kawahara, R. Takemura, J. Hayakawa, S. Ikeda, R. Sasaki, H. Takahashi, H. Matsuoka and H. Ohno, *Symposium on VLSI Technology*, pp 234–235 (2007).

98. T. Kawahara, R. Takemura, K. Miura, J. Hayakawa, S. Ikeda, Y. Lee, R. Sasaki, Y. Goto, K. Ito, T. Meguro, F. Matsukura, H. Takahashi, H. Matsuoka and H. Ohno, *ISSCC*, pp 480–481 (2007).

99. G. Jeong, W. Cho, S. Ahn, H. Jeong, G. Koh, Y. Hwang and K. Kim, *ISSCC*, pp 280–281 (2003).

# Advanced Charge Storage Memories: From Silicon Nanocrystals to Molecular Devices

Barbara De Salvo* and Gabriel Molas

Department of Nanotechnology, CEA-LETI/Minatec,
17, rue des Martyrs, 38054 Grenoble, France.

*barbara.desalvo@cea.fr

..................................

In this paper, we will present a general overview of different technological approaches suitable for charge storage memories. Several solutions to extend the floating gate Flash memory technology to the 32 nm, and possibly 22 nm nodes, are presented. In particular, new modules (discrete traps memories, and more specifically silicon nanocrystal memories), new materials (high-k materials for the interpoly layer) and innovative architectures (FinFlash memories) are discussed. Moreover, hybrid approaches which make use of organic molecules as storage sites will be also introduced. Finally the main theoretical limits of ultra-scaled charge storage memories (i.e. reliability issues linked to few electron phenomena) will be analyzed, opening the path to the introduction of disruptive technologies based on new storage mechanisms.

## 1.    Introduction

Driven by the IC Industry, the International Technology Roadmap for Semiconductor[1] states that the 22 nm Flash technology node will be required in industrial production from the year 2016, for application ranging from high-density data storage to high-performance code storage. Nevertheless, it is widely believed that the scaling of the standard planar Flash beyond the 45 nm node will be extremely difficult.[2] In particular: (1) the scaling of tunnel and control dielectric thickness is limited by concerns for data-retention, especially in the presence of defects (SILC) in the dielectrics. This results in high operating voltages. (2) Drain voltage scaling in NOR memories is also limited by the need for maintaining coupling and program voltage for channel hot electron injection. This phenomenon gives rise to the drain turn-on phenomenon, which limits the channel length, and consequently the cell area, of NOR devices. (3) Moreover, the scaling of ultra-dense NAND devices is limited by the parasitic floating gate (FG) interferences, a lower coupling ratio and less tolerant charge loss.

In order to further scale the standard Flash architecture, at least to the 32 nm and possibly 22 nm nodes, evolutionary solutions based on the floating-gate memory concept, which involve new modules, materials and/or architectures must be investigated. For sub-22 nm memory nodes, we believe that disruptive technologies should finally be adopted (Fig. 1).

## 2.    Silicon Nanocrystal Memories

The basic idea of discrete traps memories is to replace the standard continuous poly-Si layer of the floating gate by discrete storage nodes, which can be made by natural traps in an appropriate insulator (like the nitride

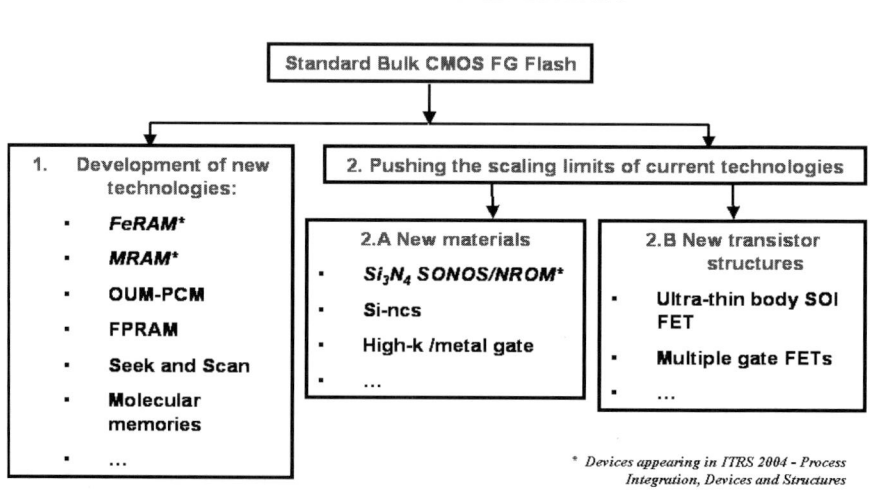

Fig. 1. Organization of research activities on advanced NVMs.

layer in SONOS, MONOS and NROM memories or composed of semiconductor nanocrystals.[3] Silicon nanocrystal (Si-NC) memories are one of the most promising solutions to push the scaling limits of Flash memories at least to the 32–20 nm technology nodes.[4−7] Due to their discrete nature, Si-NCs are robust to defects in the oxide. Thinner tunnel dielectrics and lower operating voltages can be used without compromising data-retention, especially after cycling. Cells with abnormally short retention times ("erratic bits") are suppressed. Moreover, due to the decreased capacitance coupling ratio, floating gate interferences in ultra-dense NAND memories are eliminated. Recently, it has been shown that optimized Chemical Vapor Deposition (CVD) process results in partially self-organized nucleation and growth of Si-NCs,[5] mitigating the impact of fluctuations on memory array characteristics. Finally, thanks to the use of a single poly-Si, Si-NC memories require a simple and low cost device fabrication process which make them particularly interesting in view of embedded memory applications.[8]

Recent works[6,7] have demonstrated the discrete storage node concept on a 32 Mb Si-NC NOR Flash memory product, fabricated in a 130nm technology platform. To integrate the Si-NCs in a 32 Mb NOR Flash memory array (see Fig. 2), two main key integration challenges were faced: (1) Si-NC robustness to strong oxidation steps and (2) Si-NC removal in logic periphery. To solve these issues, the integration strategy was the following: firstly, the periphery devices (i.e. CMOS logic, High Voltage, and I/Os) were produced using a SASTI (Self Aligned Silicon Trench Isolation) approach. Secondly, the memory bitcells were defined in a conventional flow (non SASTI) and thirdly, the memory gate stack was removed by dry etch in the periphery of the arrays. The remaining process steps (gate patterning, halo implants, LDDs, Source/Drain implants and back end) closely followed conventional 130 nm process flow. As shown in Fig. 2, the gate length and width of the Si-NC memory bitcells are 0.23 $\mu$m and 0.16 $\mu$m, respectively. The memory gate stack consists in 5nm-thick thermal $SiO_2$ tunnel dielectric covered by the Si-NC storage layer, the 10nm-thick High Temperature top Oxide and the $n^+$ poly-Silicon control gate. Nanocrystals were deposited following a two-step LPCVD process, deeply described in Ref. 4. Several nanocrystal deposition conditions (yielding similar densities, $N_{dot}$, and different dot sizes, $\Phi_{dot}$) have been explored (see Fig. 2). Subsequent to deposition, Si-NCs were properly passivated (giving rise to a thin nitrided oxide shell) to definitely avoid any parasitic oxidation.

(a)

(b)

(c)

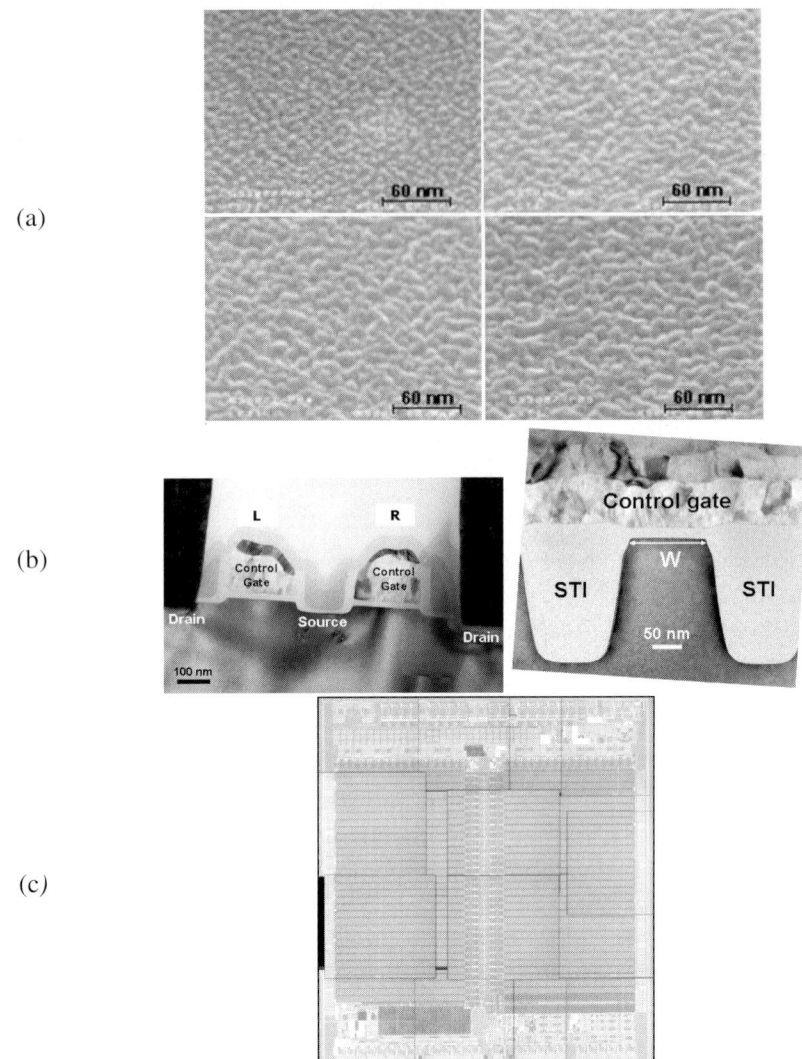

Fig. 2. (a) SEM images of Si-NCs with same nucleation step (yielding similar densities: $N_{dot} \approx 1E12/cm^2$) and increasing dot diameter ($\Phi_{dot}$); (b) Si-NC memory bitcell (up: cross-section along cell length; down: cross-section along channel width); (c) Image of the 32 Mb Si-NC array. After Refs. 6 and 7.

Concerning the memory cell results, devices are programmed by Channel Hot Electron (CHE) injection and erased by Fowler-Nordheim tunnelling (FN). Figure 3(a) shows the $I_d$-$V_g$ curves of a memory bitcell corresponding to the sample with 9 nm Si-NC diameter and $1E12/cm^2$ Si-NC

(a)

(b)

(c)

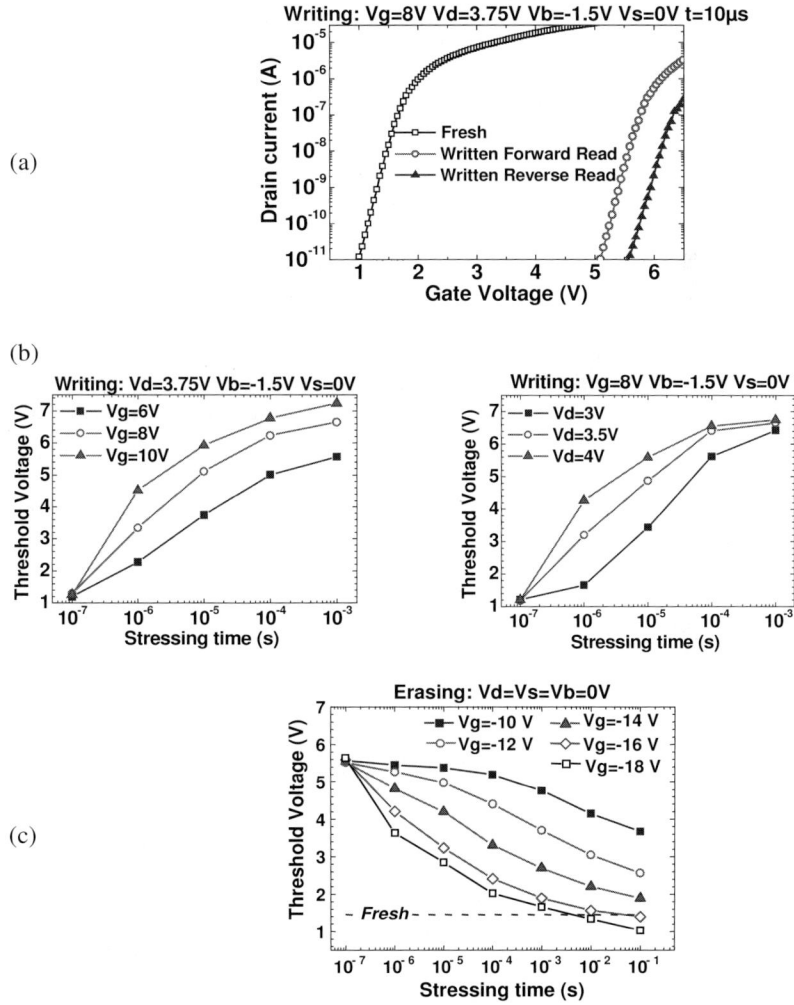

Fig. 3. (a) $I_d$-$V_g$ of a memory bitcell. (b) Writing by Channel Hot Electron. (c) Erasing by Fowler-Nordheim. After Refs. 6 and 7.

density. Writing and erasing dynamics for different bias conditions are also shown in Figs. 3(b) and (c). A very large programming window of 4 V is achieved in 10 $\mu$s with drain (Vd), gate (Vg) and substrate (Vb) biases equal to 3.75 V, 8 V, $-1.5$ V, respectively. Moreover, the asymmetry in the written curves (Fig. 3(a)), read in the forward ($V_{ds} = 1$ V) and in the reverse mode ($V_{sd} = 1$ V), clearly states the discontinuity of the Si-NC layer. Looking at Fig. 3(c), we can also observe that fast erase operations can be achieved ($\Delta V_{th} = -3$ V with Vg $= -16$ V, 100 $\mu$s) in the FN regime.

Concerning the memory array results, the 32 Mb array is divided in 64 sectors of 512 Kb. Programming and erasing of the memory sectors are achieved by using the internal voltages regulated by the charge pumps of the 32 Mb Flash product. In order to compare the Si-NC processes, single program and erase pulses have been issued instead of using the usual algorithms of the embedded logic on the die. In particular the distributions after sector erasing are recorded without the soft-programming step, which is usually required to individually recover the bits which have been over-erased. Sector distributions have been obtained by using a 4 $\mu$A margin test mode supported by the product. The distributions of fresh, erased and programmed threshold voltages of a 32 Mb array of the sample with 9 nm Si-NC diameter and 1E12/cm$^2$ Si-NC density are shown in Fig. 4(a). We can see that the average threshold voltage shift is higher than 3 V, the separation between the less programmed cell and the less erased cell being of the order of 500 mV. We think that a large margin of improvement exists concerning this value in view of future products, in particular by developing intelligent write/erase algorithms suitable for Si-NC arrays. One solution to separate more the erased and written distributions is to push the written distribution towards higher values as shown in Fig. 4(b), where the separation between the less programmed cell and the less erased cell reaches about 1.7 V. Finally, data retention has been measured for a 512 Kb sector of the same sample at 150°C, before (see Fig. 5(a)) and after 5 K write/erase cycles (see Fig. 5(b)). After 5 K cycles and 1 week of data-retention at 150°C, the programmed threshold voltage reduces of about 500 mV, which is close to the charge loss before cycling. Once again, we note that no extrinsic bits are observed following cycling and data retention.

However, it should be stated that Silicon nanocrystal memories still suffer from some inherent weaknesses. One of the main limitations resides in the low nanocrystal/control gate coupling ratio value, so that the introduction of high-k dielectrics as top oxide in order to obtain effective Fowler-Nordheim program/erase is mandatory.[9] Another issue is the

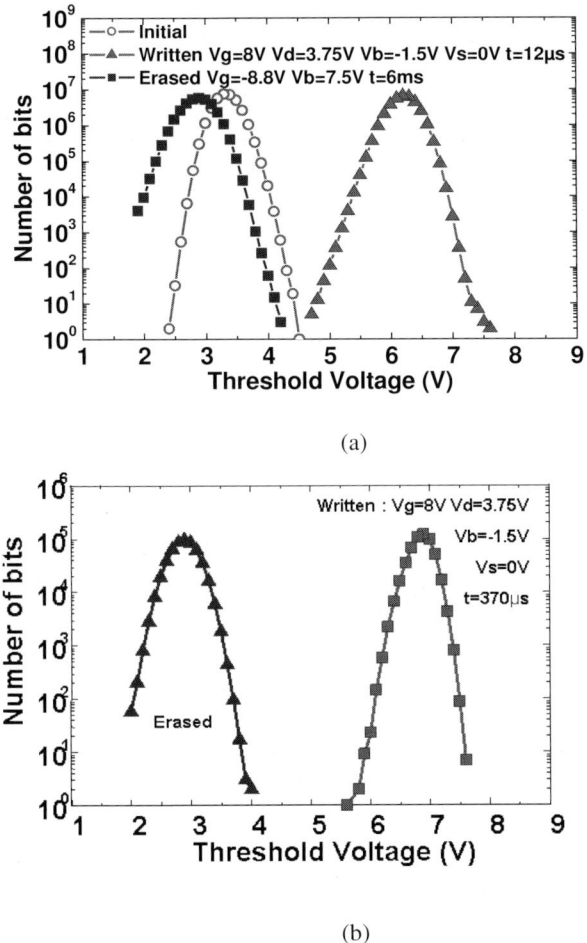

(a)

(b)

Fig. 4. (a) Threshold voltage distributions of erased and written states of a Si-NC 32 Mb array (9 nm Si-NC diameter and 1E12/cm$^2$ Si-NC density) produced using a 130 nm technology. (b) Threshold voltage distributions of erased and written states of a Si-NC 512 Kb sector obtained with different writing conditions. After Refs. 6 and 7.

relatively limited threshold voltage shift value, especially in view of multi-level NAND devices. To fit these applications, today different technologies are under study, essentially based on metal nanocrystals[10,11] and ordered nanocrystal matrixes.[12]

(a)

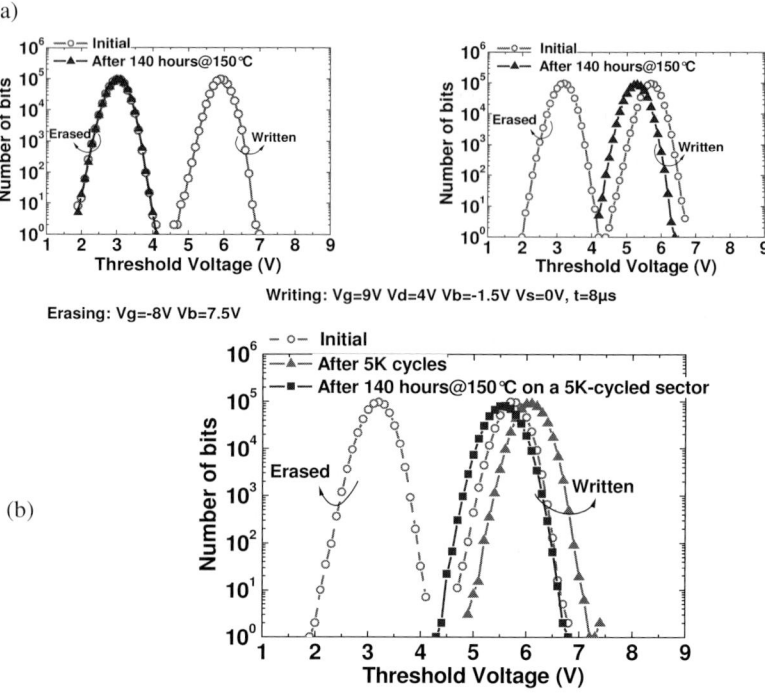

Writing: Vg=9V Vd=4V Vb=-1.5V Vs=0V, t=8µs

Erasing: Vg=-8V Vb=7.5V

(b)

Fig. 5. Data retention at 150°C on (a) two different uncycled 512 Kb sectors and (b) on a 5K-cycled 512 Kb sector. After Refs. 6 and 7.

## 3. High-k Based Memories

One of the nearest major changes of non volatile memories will concern the engineering of the Interpoly Dielectric (IPD) stack. In fact, according to the ITRS,[1] from the 45 nm-32 nm node, the IPD should be drastically reduced due to the loss of the vertical sidewalls of the poly-Si floating gate.[2,13,14] It is forecasted that the coupling ratio will be dropped to 0.3 for the 32 nm node instead of the value of 0.6 fixed by the roadmap.[14,15] On the other hand, standard interpoly Oxide-Nitride-Oxide (ONO) dielectrics reach their lower thickness limit and cannot be scaled in theory beyond 15 nm without dramatically compromising the reliability of the memory. Consequently, high-k dielectric materials (as $HfO_2$, $Al_2O_3$, HfAlO, HfSiO...) are envisaged to replace the standard ONO IPD stack of Flash memories, allowing for a high coupling ratio while maintaining good data-retention.

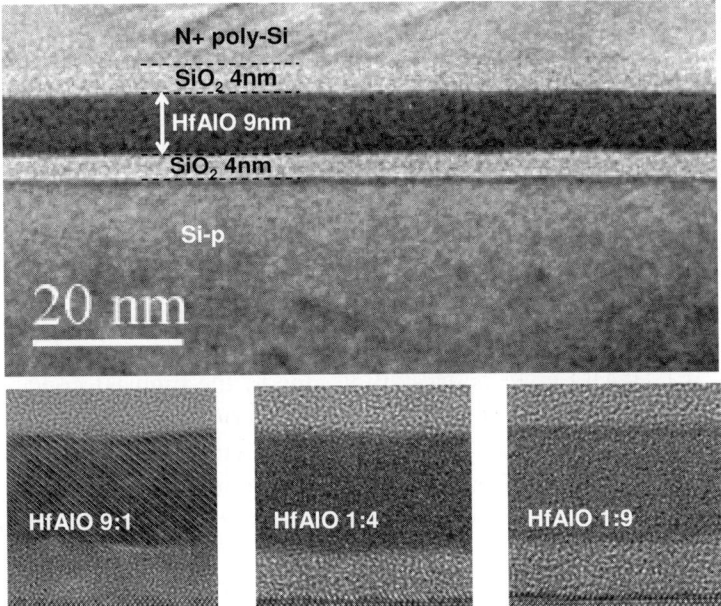

Fig. 6. High Resolution TEM images of the studied HfAlO-based triple-layer capacitors. After Ref. 16.

Demonstrations of AlO blocking oxides combined to a SiN trapping layer have been already presented in the literature for multi-level NAND applications targeting the 40 nm technology node.[15]

It has been also demonstrated that the integration of $HfO_2$ as interpoly dielectrics (instead of standard ONO stacks) in poly-Si floating gate cells gives rise to a reduction of the programming voltages, due to the better coupling coefficient between the control and the floating gates.[17] Moreover, some works have been recently presented where the discrete FGs (SiN layer or a high-k material layer) and high-k based IPDs are associated to metal control gate, to reduce the parasitic electron back tunneling from the control gate during the erase operation.[17–23] Among the different studied materials, a strong interest is given to Hafnium Aluminate (HfAlO) compounds, as they have the potentials to combine the high dielectric constant of $HfO_2$ and the elevated energy barrier height and good thermal stability of $Al_2O_3$. In our recent works,[16,24] we deeply investigated the coupling properties, insulating capabilities, electron conduction modes and parasitic trapping phenomena of HfAlO layers. HfAlO compounds are deposited by

ALD, using $H_2O$ and $HfCl_4$ as precursors for $HfO_2$ deposition and $H_2O$ and $Al(CH_3)_3$ for $Al_2O_3$ deposition. Three compositions are investigated, designed by the $HfCl_4{:}Al(CH_3)_3$ deposition cycle ratio: 9:1 (Hf-rich), 1:4, and 1:9 (Al-rich), corresponding to the following Hf concentrations: 94%, 31% and 27%, respectively. Pure $HfO_2$ and $Al_2O_3$ based samples are also processed as reference. Oxide/HfAlO/Oxide (OHO) triple-layer stacks with n+ poly-Si control gate were processed. The HfAlO thickness ranges between 3 nm and 9 nm. Both the bottom and top dielectrics of the triple layer stacks are 4 nm-thick Silane-based high thermal oxides (HTO). Figure 6 presents the cross section images of the triple-layer capacitors made by High Resolution Transmission Electron Microscopy. It appears that the 9:1 HfAlO layer is crystalline, due to the high Hf concentration in the alloy, while the 1:9 HfAlO layer is amorphous. This agrees with results previously reported in the literature, which stated that the crystallisation temperature of the HfAlO alloy increases monotonically as the Al percentage increases, the Al acting as a stabilizer of the amorphous phase.[25]

The modifications of optical properties of HfAlO layers are obtained from spectroscopic ellipsometry (Fig.7(a)). A vacuum ultraviolet (VUV) ellipsometer (Jobin Yvon phase modulated ellipsometer, 1.5 eV to 8 eV) has been used to assess the complex dielectric function $\varepsilon = \varepsilon_1 - i\varepsilon_2$, where $\varepsilon_1$ and $\varepsilon_2$ are the real and imaginary part of $\varepsilon$.[26] The thickness and the bandgap of the film are determined from the analysis of the raw data using a Tauc Lorentz model with two oscillators.[27,28] The bandgap is extracted by considering a linear variation of $\sqrt{\alpha(E).E}$ vs the photon energy E, where $\alpha$ is the absorption coefficient. The bandgap (Fig. 7) is correlated with the hafnium content of the layer and is ranging from 6.4 eV for pure $Al_2O_3$ to 5.6 eV for pure $HfO_2$, confirming the intermixing of $HfO_2$ and $Al_2O_3$ during the ALD process. The obtained values are in very good agreement with data reported in the literature[28,29] also obtained on ALCVD HfAlO films. Concerning the coupling properties, based on $C(V_G)$ measurements, and taking into account quantum effects into both the n+ poly-Si gate and the Si substrate, we extracted the Equivalent Oxide Thickness (EOT) of the different stacks and the dielectric constants "k" of the HfAlO compounds (Table 1). One can notice that the dielectric constant of HfAlO progressively increases as the Hf concentration increases, varying between the value of $HfO_2$ and that of $Al_2O_3$, which is in agreement with previous results published in the literature.[25] It is thus possible to adjust the HfAlO dielectric constant, and consequently the EOT of the IPD by tuning the

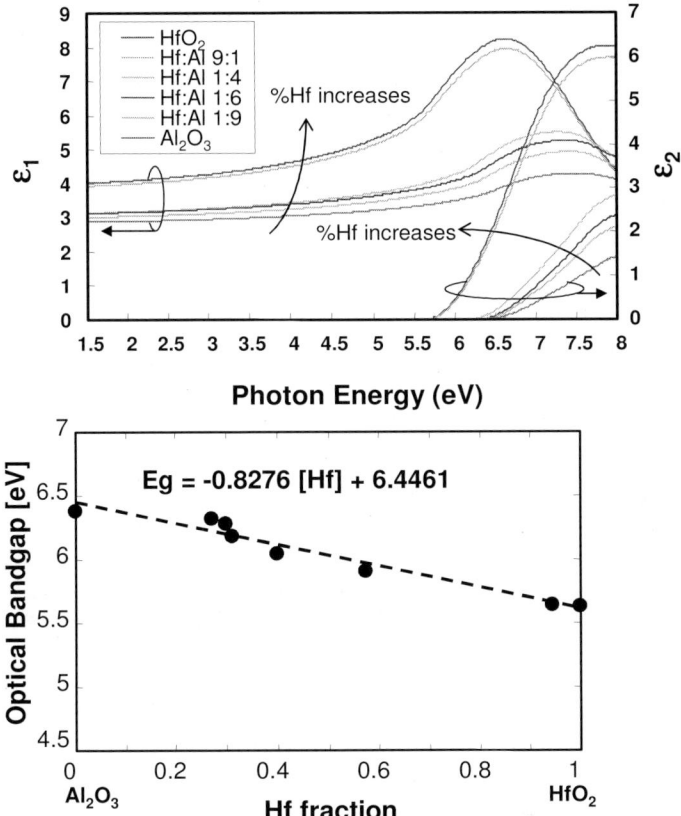

Fig. 7. High Resolution TEM images of the studied HfAlO-based triple-layer capacitors. After Ref. 16.

Table 1. Dielectric Constant of the HfAlO Stacks.

| $HfCl_4:Al(CH_3)_3$ deposition cycle ratio | HfAlO dielectric constant k |
|---|---|
| 1:0 ($HfO_2$) | 20 |
| 9:1 | 17 |
| 1:4 | 15 |
| 1:9 | 11.5 |
| 0:1 ($Al_2O_3$) | 8 |

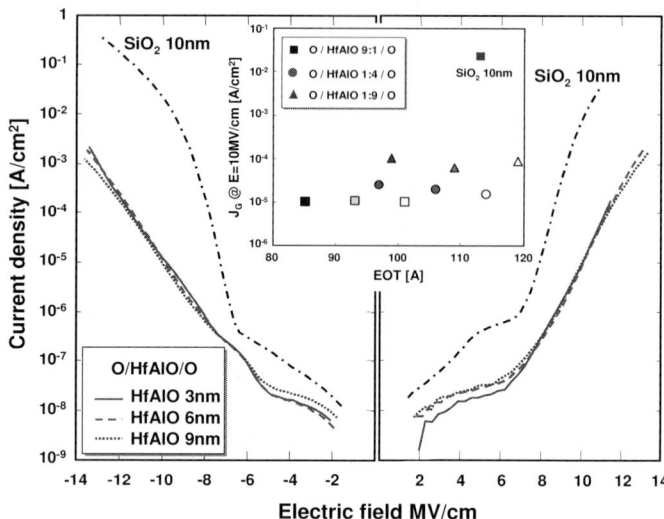

Fig. 8. Leakage currents of O/HfAlO/O stacks with fixed HfAlO composition (9:1) and different HfAlO thickness. Inset: Leakage currents for different HfAlO compositions (at 10 MV/cm) as a function of EOT (results correspond to p-Si substrate. After Ref. 24.)

Hf and Al content of the compound. Figure 8 represents the gate current densities of different OHO samples with various HfAlO compositions. We can observe that the insulating capabilities at a fixed electric field increase with the Hf concentration of the HfAlO layer, which may be linked to the reduced EOT of the stack. Indeed, for a given electric field in the $SiO_2$ layer, the electric field in the high-k layer is more important in an Al-rich HfAlO based stack than in an Hf-rich HfAlO based stack, resulting in an increase of the leakage current. In order to investigate the electron conduction mechanisms governing the leakage currents of OHO triple layer stacks, $J_G(V_G)$ measurements were performed at high temperatures (up to 250°C). The leakage currents of the OHO samples were found to be strongly activated in temperature, as illustrated in Fig. 9. The Arrhenius plots, extracted at 10 MV/cm, are shown in Fig. 10 for different OHO samples with various HfAlO compositions. Assuming at the first order that the gate current is proportional to $\exp(-qE_A/k_BT)$ where q is the elementary charge, $k_B$ the Boltzmann constant and T the temperature (in Kelvin), it is possible to extract a parameter $E_A$ which is the activation energy (eV). It clearly appears that this activation energy $E_A$ increases as the Hf concentration increases (see Table 2).

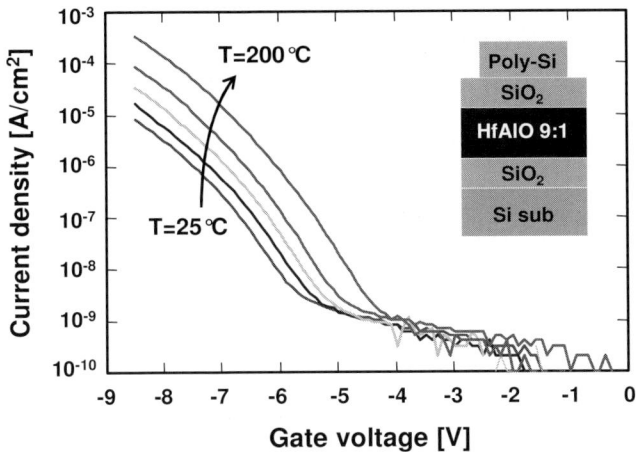

Fig. 9. Leakage currents of OHO samples at different temperatures varying between 25°C and 200°C. The 9:1 HfAlO layer is 3 nm-thick. After Ref. 24.

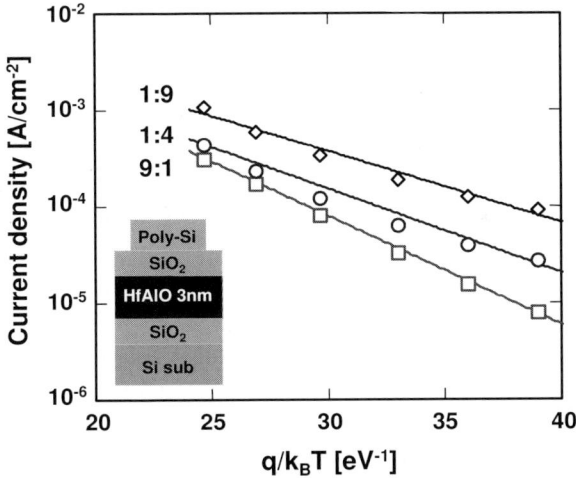

Fig. 10. Arrhenius plots of OHO samples with various compositions of HfAlO. The HfAlO layer is 3 nm thick. The current density is extracted at 10MV/cm. After Ref. 24.

However, one should note that even at 200°C, the Hf-rich alloy still presents the lowest leakage current. In order to clearly identify the conduction modes involved in the triple layer stacks, we plotted in Fig.11 the Hill diagrams, starting from previous high temperature measurements. Indeed,

Table 2. Activation Energies for OHO Samples.

| HfAlO composition | EA |
|---|---|
| HfO$_2$ | 340 meV |
| HfAlO 9:1 | 260 meV |
| HfAlO 1:4 | 200 meV |
| HfAlO 1:9 | 170 meV |

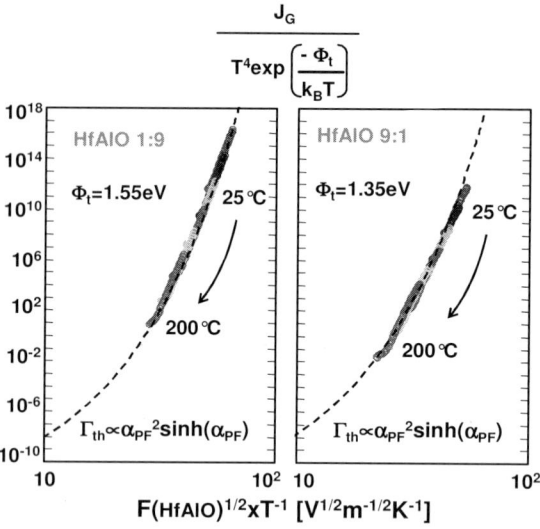

Fig. 11. (a) Hill diagrams of OHO samples with two different HfAlO compositions (1:9 and 9:1). The HfAlO layer is 3-nm-thick. After Ref. 24.

a Poole-Frenkel conduction, probably assisted by the traps in the HfAlO layer, is put in evidence. The extracted trap depth (referenced with the conduction band of the high-k) varies typically between 1 eV and 1.5 eV, depending on the HfAlO composition. This consideration well agrees with results of OHO stacks already reported in the literature.[30] The Poole-Frenkel model allows matching correctly our experimental data of OHO samples at strong voltages, for all the tested HfAlO thicknesses, from 3 nm to 9 nm. Traps in HfAlO layers were related to both oxygen vacancies, and oxygen-interstitial-related defects states of the HfO$_2$,[31] based on XPS low loss spectra and ab-initio studies. Other works also reports two local-ized electron traps in HfAlO alloys, based on electrical data obtained on capacitors.[32] In this latter case, the defects are respectively assigned to AlO$^-$ bounding groups deriving from a breaking of the network compo-nent, and to antibounding Hf atom d states that form the lowest conduction bands of the alloys.

To evaluate more precisely the trapping capabilities of the interpoly stacks, we monitored the evolution of the flatband voltages as a function of time when the devices were submitted to different gate stresses (Fig. 12). A continuous $V_{FB}$ shift is observed, showing the progressive electron trapping in the stack as the stress time increases. It clearly appears that for a given stress condition, the trapping capability increases with the Hf concentration. This result could be correlated with the crystalline structure of the high-k materials: the larger the Hf concentration, the more crystalline the layer, and hence, the higher the trapping capability.[33]

Finally HfAlO IPDs were integrated in memory transistors, using sili-con nanocrystals (Si-ncs) as floating gate.[9] Si-ncs were deposited by CVD (with a diameter of 6 nm and a density of d $=$ 9E11/cm$^2$) on a 4 nm thick thermally grown tunnel oxide and then passivated by a nitridation process (750°C, NH$_3$) to protect them from the following oxidizing steps. As IPD a 8-nm-thick HfAlO layer (with $\sim$30% of Hf, named 1:4 in the previous sections) sandwiched between two 4-nm-thick HTOs was fabricated, with

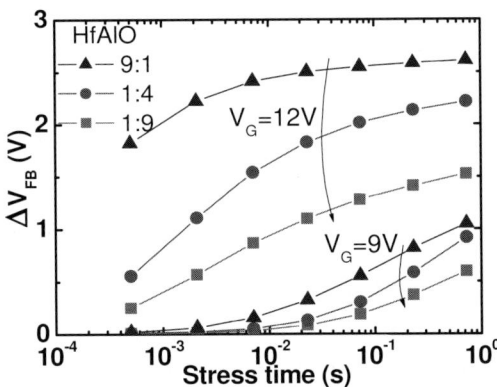

Fig. 12. Trapped charges in OHO samples, with various compositions and thicknesses of the HfAlO layer, as extracted from the programming characteristics. The stressing conditions are performed at constant VG/EOT. After Ref. 24.

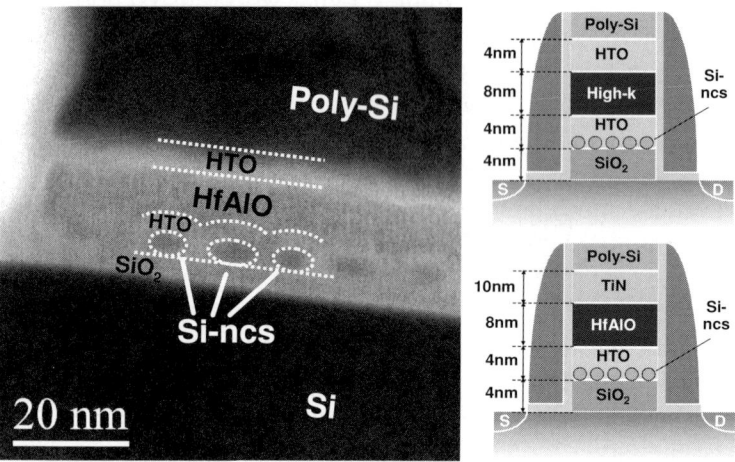

Fig. 13. TEM cross section of a silicon nanocrystal memory with HfAlO based IPD. Right: schematics of the processed samples. After Ref. 9.

a final EOT of 10.5 nm. Poly-Si was used as a control gate. Other samples were also processed, where the interpoly HTO top oxide was skipped to reduce the EOT of the stack. In this case, a TiN control gate was deposited. The gate length was defined by electron beam lithography, down to 90 nm. TEM cross section and schematics of the samples are given in Fig. 13. Fig. 14 shows the program erase characteristics of the memory devices in FN/FN mode. A $\Delta$Vth of 3 V can be achieved with a programming time of 1 ms for triple layers IPDs. On the other hand double layer IPDs allow reducing the programming voltages of several volts due to the lower IPD EOT.

## 4. FinFlash Devices

As already said, it is widely believed that the scaling of Flash memories down to the 32 nm technological node and beyond will face major issues, due to the high electric fields required for the programming and erasing operations and the stringent leakage requirements for long term charge storage.[2] In this context, new transistor architectures such as tri-gate Fin-Flash memory devices[34] coupled with the discrete storage node approaches (i.e. nitride storage layer or Silicon Nanocrystals, Si-NC[3]) offer the possibility of scaled gate dielectrics, implying scaled operating voltages, along

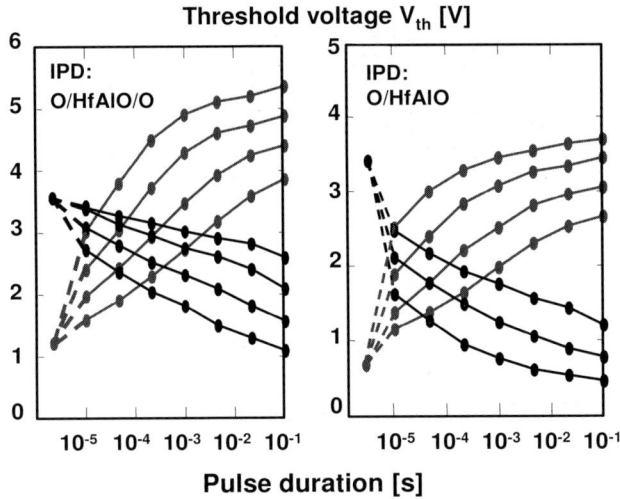

**Threshold voltage V$_{th}$ [V]**

**Pulse duration [s]**

Fig. 14. Program erase characteristics of Si-ncs memories. Left: the programming (resp. erasing) voltages vary between 14 V and 17 V (resp. $-12$ V and $-15$ V). Right: the programming (resp. erasing) voltages vary between 10 V and 13 V (resp. $-11$ V and $-13$ V). After Ref. 9.

with short channel effect immunity and higher sensing current drivability. The idea of the FinFlash memory[46] is to take advantage of the FinFET architecture for memory applications. In particular: (1) The electrostatic control of the channel will be improved, with reduced DIBL and improved short channel effects. Moreover, the use of a narrow fin channel eliminates sub-surface leakage paths, allowing the reduction of the memory gate length. (2) The drive currents will be increased due to the multi-channel conduction, improving the memory access time and programming speed. In such innovative architectures, the charge trapping in the floating gate may be affected by the three-dimensional character of the structure, leading in particular to corner effects. Currently, many FinFlash demonstrations are presented in the literature, with SiN trap layer, on bulk substrate or SOI substrate, showing results that demonstrate the high interest of these structures. Hereafter, we will present the recent results obtained in our group on FinFlash devices. The schema of the structure fabricated is shown in Fig. 15, where the critical dimensions of the device are also reported. The fabrication of our FinFlash devices is based upon a standard Finfet process flow.[35] E-beam lithography and resist trimming are used to pattern both the fin and the gate. Sidewall oxidation is carried out to round fin corners and

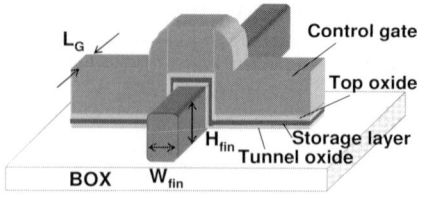

$W_{fin}$ = 10 ... 40nm

Hfin = 35 nm

$L_G$= 30 ... 130nm

Tunnel oxide: 5nm-$SiO_2$

Storage: Si-NCs, $Si_3N_4$

Top oxide: 8nm-HTO

Control gate: Poly-Si N+

Fig. 15. Schema of the FinFlash memory cells. After Ref. 46.

decrease fin width. After fin patterning and boron channel implantation, gate stack deposition is performed, i.e. the 5-nm thermal $SiO_2$, the storage layer made of Si-NC (directly deposited by LPCVD or obtained by Silicon Rich Oxide annealing) or 6 nm-thick LPCVD $Si_3N_4$, the blocking dielectric (8 nm-thick HTO) and, finally, the 100 nm $N^+$ Poly-Si control gate.

After the gate etching, nitride spacers are deposited and etched. Raised Source/Drain are epitaxially grown in order to decrease the series resistance. After the completion of source/drain implantation, the flow is terminated by standard Back-End-Of-Line. TEM images of the FinFlash devices are reported in Fig. 16, demonstrating fin widths $W_{FIN}$ and gate lengths $L_G$ down to 10 nm and 30 nm, respectively. Figure 17(a) shows the transfer characteristics ($I_D$–$V_G$) of virgin Si-NC FinFlash cells with $W_{FIN}$ = 10 nm and different gate lengths. The enhanced electrostatic control of the gate over the channel at very small fin widths clearly appears. In particular, in the Inset, we can see that the threshold voltage $V_{TH}$ roll-off disappears

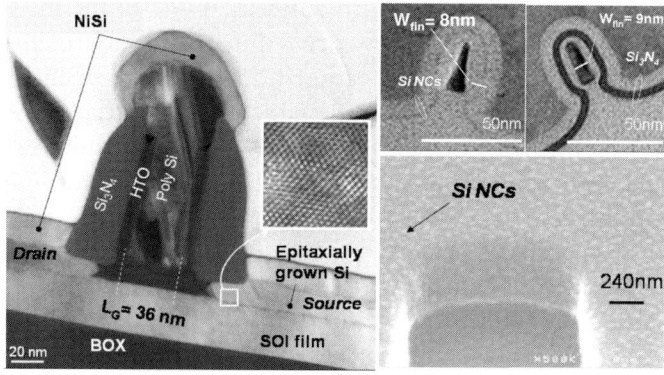

Fig. 16. TEM views of FinFlash devices with different storage nodes. After Ref. 46.

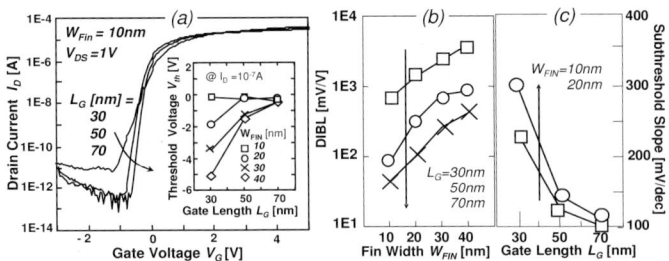

Fig. 17. (a) $I_D$-$V_G$ of Si-NCs FinFlash (in the virgin state) with $W_{FIN} = 10$ nm and different $L_G$. Inset: VTH versus $L_G$, for devices with different $W_{FIN}$. (b) DIBL versus $W_{FIN}$, for devices with different $L_G$. (c) Subthreshold slope versus $L_G$, for devices with different $W_{FIN}$. After Ref. 46.

in narrow fins. Figures 17(b) and (c) show that in the smallest devices ($W_{FIN}/L_G = 10/30$ nm), 220 mV/dec Subthreshold Slope ($SS@V_D = 1$ V) and 0.7 V/V Drain Induced Barrier Lowering (DIBL) are achieved.

Ultra-scaled Si-NC FinFlash devices are first studied in NOR configuration (i.e. Channel Hot Electron writing & Fowler-Nordheim erasing).

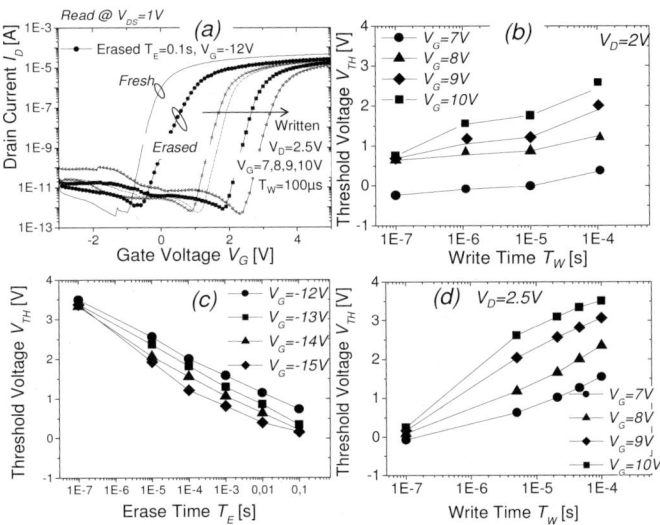

Fig. 18. CHE/FN characteristics of Si-NC FinFlash with $W_{FIN} = 10$ nm, $L_G = 30$ nm. (a) $I_D$ − $V_G$ in virgin, written (CHE) and erased (FN) states. (b,d) Write and Erase (c) dynamics. After Ref. 46.

Fig. 18 shows that, in scaled devices ($W_{FIN}/L_G = 10/30$ nm), CHE yields large programming window with low $V_D$ biases (lower than the $Si/SiO_2$ conduction band difference, i.e. $\sim3.2$ V). In particular, $\Delta V_{TH} \sim 3$ V can be achieved when $V_D = 2.5$ V, $V_G = 9$ V, and $t_{stress} = 100\,\mu s$. We can also observe that Fowler-Nordheim erasing can be achieved in Silicon Nanocrystal FinFlash devices even with a 5 nm-thick tunnel oxide (nevertheless, a saturation of the erase $V_{th}$ occurs in the smallest device). Si-NC FinFlash devices can also be programmed in the NROM operating scheme (i.e. Channel Hot Electron writing & Hot Hole Injection erasing). The W/E dynamics are reported in Fig. 19, with the programmed threshold voltages read either in the forward mode ($V_{DS} = 1$ V) or in the reverse mode ($V_{SD} = 1$ V).[36] Indeed, we can clearly observe the asymmetry between the forward/reverse $V_{th}$s, clearly suggesting that even for such strongly scaled devices the charges injected at the drain do not spread over to the source. Moreover,

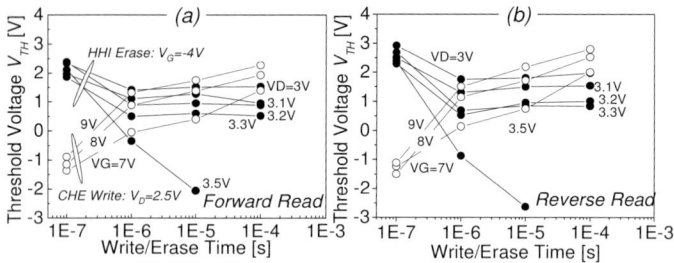

Fig. 19. CHE/HHI characteristics of Si-NC FinFlash with $W_{FIN} = 10$ nm, $L_G = 30$ nm. Programmed threshold voltages are read (a) in the forward mode ($V_{DS} = 1$ V) or (b) in the reverse mode ($V_{SD} = 1$ V). After Ref. 46.

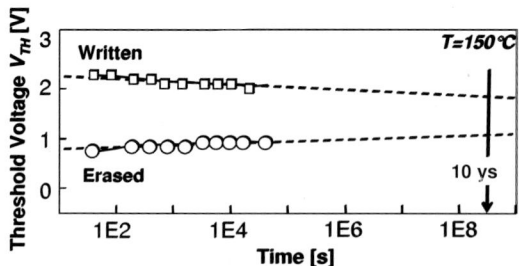

Fig. 20. Data retention @ T = 150°C of Si-NC FinFlash with $W_{FIN} = 20$ nm, $L_G = 30$ nm (CHE/FN Written/Erased). After Ref. 46.

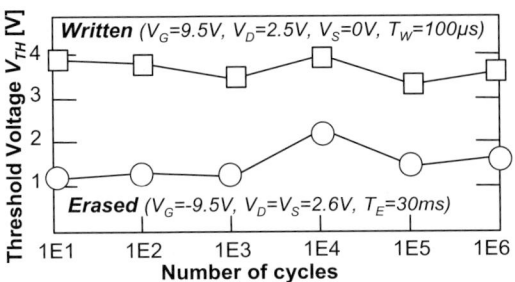

Fig. 21. Endurance of Si-NC FinFlash with $W_{FIN} = 20\,nm$, $L_G = 30\,nm$. After Ref. 46.

it can be noticed that erasing by hot holes is effective at a very low drain bias (lower than the $Si/SiO_2$ valence band difference, i.e. $\sim 4.5\,V$). Data-retention of Si-NC device with $W_{FIN}/L_G=10/30\,nm$ is reported in Fig. 20, showing small charge loss at high temperature ($150°C$). Good endurance (up to 1E6 cycles) of Si-NC device with $W_{FIN}/L_G = 20/30\,nm$ also appears in Fig. 21. Nevertheless, it should be stated that a slight degradation of the $I_D$-$V_G$ characteristics appeared after 1E5 cycles.

Ultra-scaled nitride FinFlash devices are studied in NROM configuration (i.e. Channel Hot Electron writing & Hot Hole Injection erasing), the Fowler-Nordheim erasing of charged nitride memories being not effective with 5 nm-thick tunnel oxide. As we previously observed in Si-NC devices, strongly scaled $Si_3N_4$ devices can be efficiently written with $V_D$ biases lower than 3.2 V and erased by HHI with $V_D$ biases lower than 4.5 V (Fig. 22), while these low-voltage stresses are not effective for long devices. Moreover, even in nitride devices with ultra reduced cell lengths, a good threshold voltage difference between the reverse and forward states appears. In Fig. 22 we can remark that the nitride storage layer gives rise to a larger programming window than the Si-NC storage layer, probably due to the higher trap density of amorphous nitride compared to crystalline Si-NCs. Data-retention of $Si_3N_4$ devices with $W_{FIN}/L_G = 10/30\,nm$ is reported in Fig. 23, showing small charge loss at high temperature ($150°C$) and still detached forward and reverse threshold voltages after 10 years.

## 5.    Molecular Memories

The device scaling in Silicon (Si) technologies, and namely in memory applications, is starting to face important issues. In the few-nanometer

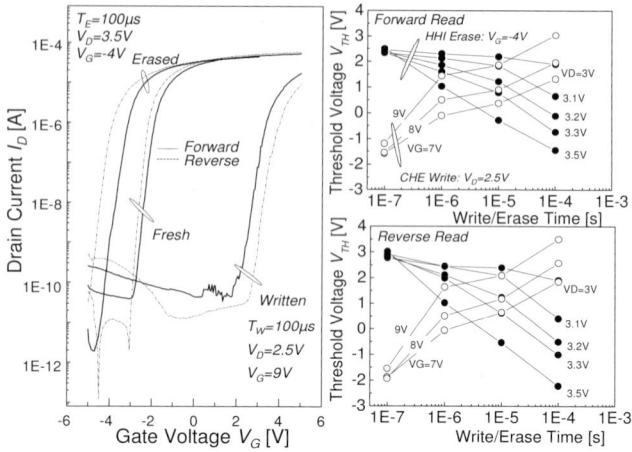

Fig. 22. CHE/HHI characteristics of Nitride FinFlash with ($W_{FIN} = 10$ nm, $L_G = 30$ nm). Left: $I_D - V_G$ characteristics. Right: W/E dynamics, with Vth read in forward and reverse mode. After Ref. 46.

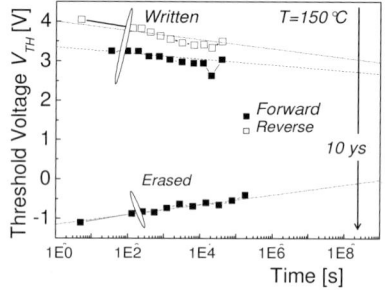

Fig. 23. Data retention @ T = 150°C of Nitride FinFlash with ($W_{FIN} = 10$ nm, $L_G = 30$ nm) (CHE/HHI Written/Erased). After Ref. 46.

range, device performance/reliability will be governed by few electron phenomena,[37] being strongly sensitive to the unavoidable fabrication spreads. Moreover, the exponentially growing fabrication costs will be one of the main critical factors. In this context, molecular electronics are of growing interest. Such a technology uses low-cost "bottom-up" approaches (i.e. chemical synthesis, molecular self-assembly), and the behaviour of devices is governed by the properties of specifically designed molecular species. In view of tera-bit memories, several concepts of hybrid semiconductor/molecular "crossbar" systems have been suggested.[38,39] Recent works have demonstrated an electron transfer between Si and redox-active monolayers in a transistor-like structure.[40] Such an approach seems to be the most suitable starting point for the experimental understanding of memories based on molecular layers, due to the robust signal readout and fewer new process technology steps. Nevertheless, it should be stated that, today, the work in this field is at a starting point. Great challenges in terms of device fabrication and integration still remain, and an extensive set of proof-of-concept experiments should still be provided. In a recent paper,[41] we propose an electrical investigation of hybrid molecular/Si memory capacitor structures, where redox active Ferrocene molecules act as storage medium. Different characterization techniques (cyclic-voltammetry, impedance spectroscopy) allow to show the strong impact of the engineering of the redox molecules and their linker on the electron transfer properties. In particular, redox-active two-state Ferrocene (Fc) organic molecules have been anchored as a monolayer on Si surface (p-type, (100) Si), either directly or with a $N_3(CH_2)_{11}$ linker, using combined hydrosilylation-cycloaddition reactions. In both cases, the monolayer formation affords a covalent attachment between the Si surface and the organic molecules. X-ray Photoelectron Spectroscopy (XPS) has been performed in order to control the chemical composition of the monolayer (Figs. 24(a) and (b)). As reference sample, structures with redox inert 1-octadecene molecules grafted on Si have also been prepared. Preliminary Cyclic-Voltammetry (CyV) measurements, using molecule-grafted Si working electrodes, were performed under an Argon atmosphere (Fig. 25(a)). Note that the voltage in these experiments is referred to the working electrode. Electrochemical capacitors, with an active area of $150 \times 300 \, \mu m^2$, were also fabricated (Fig. 25(b)). A 2-nm-thick sacrificial oxide was grown on the Si substrate and removed before molecular attachment. The walls which contain the electrolyte are made of 500-nm-thick thermal $SiO_2$ plus 10-$\mu$m-thick PECVD $SiO_2$. After molecular grafting, an electrolyte solution (1.0 M tetrabutylammonium hexafluorophosphate in propylene carbonate), acting as a conducting gate, was contacted with the molecular monolayer. Capacitance-Voltage (C-V) and Conductance-Voltage (G-V) characteristics were measured with standard equipment, in a nitrogen atmosphere. The gate voltage in these experiments was applied on the Ag tip. Cyclic-Voltammetry (CyV) tests are shown in Figs. 26 and 27.

In Fig. 26, the oxidation wave (corresponding to the transfer of electrons from the molecules to the Si) and the reduction wave (corresponding to the

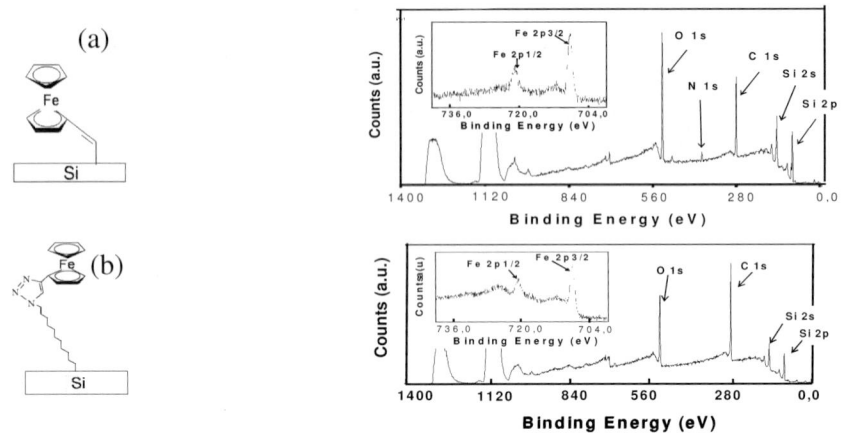

Fig. 24. Chemical structures and XPS spectra of Ferrocene functionalized on Silicon with (a) direct grafting and (b) grafting with linker. Insets: High-resolution XPS spectra of Fe 2p regions. After Ref. 41.

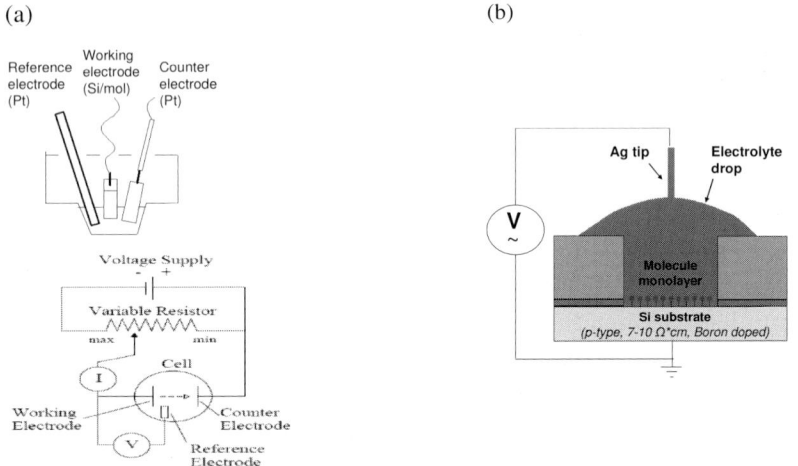

Fig. 25. (a) Electrical schema of the Cyclic-Voltammetry (CyV) experiment. (b) Electrochemical capacitors used for Capacitance-Voltage (C-V) and Conductance-Voltage (G-V) measurements. After Ref. 41.

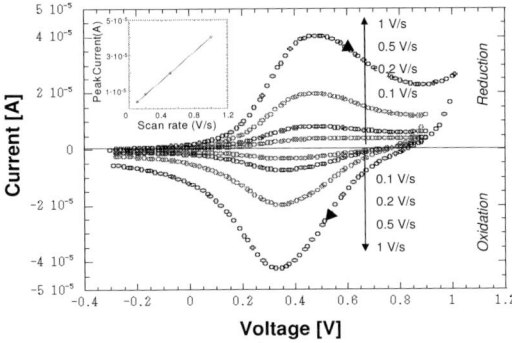

Fig. 26. CyV of Fc directly grafted on Si (p-type) at different scan rates. Inset: Linear dependence between the intensity of reduction peak and scan rate. After Ref. 41.

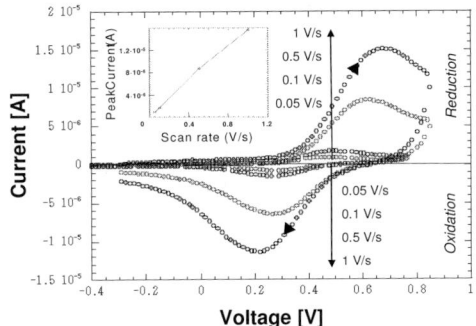

Fig. 27. CyV of Fc grafted with linker on Si (p-type) at different scan rates. Inset: Linear dependence between the intensity of reduction peak and scan rate. After Ref. 41.

electrons tunneling back to the molecules from the Si) of Fc molecules without linker clearly appear. The monolayer exhibits a reduction peak at 0.34 V and an oxidation one at 0.47 V, with a 0.5 V/sec scan rate. The peak amplitude is proportional to the amount of molecules on the Si surface which undergo the redox reactions. A high molecular density can been extracted equal to $6.38 \times 10^{13}$ molecules/cm$^2$. CyV results of Fc grafted with linker are shown in Fig. 27. In this case, the monolayer exhibits a reduction peak at 0.28 V and an oxidation one at 0.46 V, with a 0.05 V/sec scan rate. The extracted surface coverage is here equal to $7.64 \times 10^{13}$ molecules/cm$^2$. Note that the larger redox peak separation in the case of Fc molecules grafted on Si with a linker indicates that the electron transport to/from

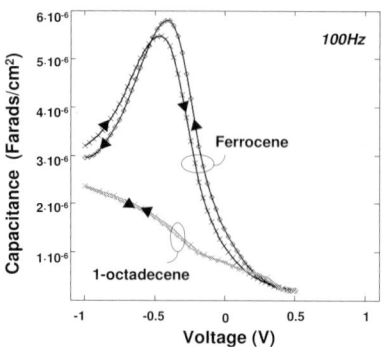

Fig. 28. C-V characteristics of redox-active Fc directly grafted on Si and of redox-inert 1-octadecene molecule. After Ref. 41.

the molecules is lower than in the case of Fc directly grafted on Si, the linker acting as a tunneling barrier for electrons. Then, electrical properties of molecules/Si systems have been studied through Capacitance-Voltage (C-V) and Conductance-Voltage (G-V) measurements. Fig. 28 shows the C-V curves of the capacitor cells either with Fc directly grafted on Silicon or with the redox-inert molecule. When the gate voltage sweeps up and down, the C-V curve of the Fc cell shows a peak at −0.45 V. These peaks are due to the charging/discharging transient currents associated with the oxidation/reduction of molecules (note that no peak appears on the redox-inert cell curve). We also studied the Fe/Si electron transfer rate behaviour by varying the measurement frequency from 100 Hz to 1 kHz (Fig. 29). An attenuation of the peak intensity on the C-V curve is observed with increasing frequencies, while the G-V peak intensity increases. Indeed, at low frequencies the charge movement can occur at a rate comparable to the measurement signal and is reflected by the presence of the peak, while at high frequencies the electron transfer process becomes rate limited and no capacitance peaks appear.[40] C-V and G-V experiments have been also carried out on a Fc with linker (Fig. 30). A peak on the C-V curves appears at −0.8 V, at a frequency of 20 Hz. The higher peak voltage value and the lower threshold frequency denote a slower electron transfer in Fc grafted on silicon with a linker compared to directly grafted Fc, in agreement with the results obtained from CyV measurements. In this work, we have shown electrical tests on hybrid Ferrocene organic molecules/Silicon capacitors clearly demonstrating that the charge transfer properties from/to the redox-active monolayer is tuned by the used linker. Indeed, this indicates

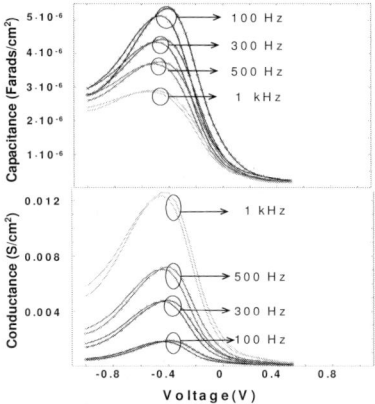

Fig. 29. C-V and G-V characteristics of Fc directly grafted on Si, performed at different frequencies. After Ref. 41.

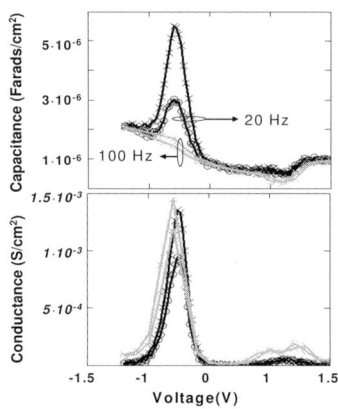

Fig. 30. C-V and G-V characteristics of Fc grafted on Si with a linker, performed at two different frequencies. After Ref. 41.

that the engineering of the molecular linker, which acts as a tunneling barrier for electrons, could be the key to control the retention properties of future molecular memory devices. Moreover, an original electrical model has been proposed, where Ferrocene molecules grafted on the Silicon substrate are considered as interface trap states, the trap characteristics directly depending on the redox molecule properties. Finally, we think that the single-electron functionality provided by properly engineered redox-active molecules has enormous potential for application to future tera-bit memories, allowing to reduce the feature sizes to molecular dimensions and to achieve high-density circuits.

## 6.    Effects of Few Electron Phenomena

Following the ITRS rules, NAND and NOR Flash memory devices are aggressively scaled down for high performance applications and high density integration. Currently, extensive studies are in progress in IC companies in order to scale further the memory cell and to solve the extrinsic reliability concerns (process related variations, ionic contamination) of future floating gate devices.[42] In this context, it becomes also urgent to address the intrinsic fundamental limits that FG memories will face once in the deca-nanometer range, even before reaching the ultimate Single Electron Memory.[43] In particular, as the dimensions of flash memories are scaled down, the number of electrons representing one bit N dramatically reduces, enhancing the effects of single electron phenomena. In a recent work,[37] we study the impact of these single electron phenomena on the performances of floating gate memory devices. We demonstrate that the charging and the discharging of scaled floating gate memories should no longer be considered as a continuous phenomenon, but as a sum of discrete stochastic events. This leads to an intrinsic dispersion of both the retention time and of the memory programming window. In Fig. 31, we have represented the number of electrons per bit N as a function of the technological node, for NAND and NOR devices.

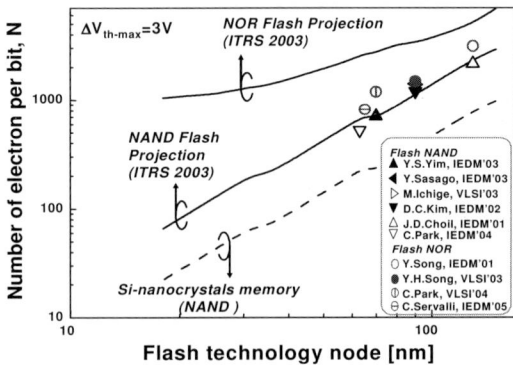

Fig. 31. Number of electrons representing one bit as a function of the Flash technology node according to the ITRS 2003 edition. After Ref. 37.

First of all, we can see that floating gate memory devices use less and less electrons and naturally become few electron devices. Moreover, it appears that the number of electrons per bit is more critical in the case of NAND memory devices than in the case of NOR memory cells, given the smaller cell active area: the number of electrons per bit reduces by a factor $\sim 0.77$ for each NAND Flash generation, each generation being defined as a 0.9 size reduction. So, in other words, the number of electrons which should be stored in the FG in order to set correctly the state of the memory cell dramatically decreases as the dimensions of Flash memory devices will be reduced. For example, the number of electrons per bit for the 35 nm NAND technology node will be equal to 200. It should also be considered that these calculations have been done assuming only one bit per cell, while the use of multi-bit or multi-level cell memory technologies[44,45] will result in an even more reduced number of electrons per bit. These theoretical calculations can be validated by advanced NOR and NAND devices in the literature, calculated from the described structures, for technology nodes going from 130 nm to 65 nm. Finally, this trend will further strengthen if new technologies are introduced, using limited charge storage sites, such as in Si-ncs memories.[4,8] In Fig. 32, we represented the calculated retention time distribution for various numbers of electrons per bit N. This figure shows that decreasing N implies a strong evolution of the retention time probability density, evolving from a Gaussian-like distribution (when $N \sim 250$) to a pure exponential/Poisson-like distribution (when $N \sim 5$). We can also see that the dispersion around the mean value increases as N is reduced. Note that if

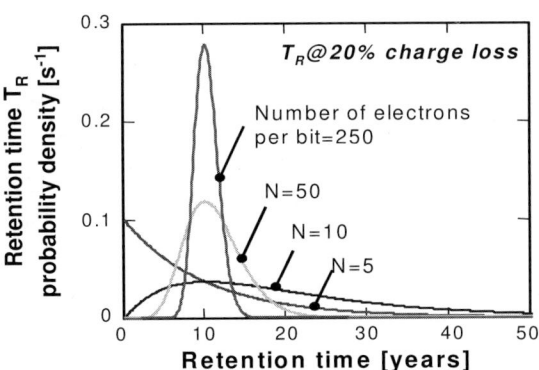

Fig. 32. Probability density of the retention time TR for memories with reduced number of electrons per bit, N. The mean TR is fixed at 10 years. After Ref. 37.

the memory working includes over-erase process, i.e. if the erased Vth is smaller than Vth0, the same single electron phenomena could occur during retention for the erased state. Indeed, charge gain could also take place, following stochastic behaviors. The widening of the retention time distribution, by scaling the number of electrons per bit, yields to an increase of the relative dispersion of the retention time following a $1/\sqrt{N}$ law, which is consistent with the central limit theorem. It is also important to notice that the relative dispersion of the retention time does not depend on the mean retention time. For a retention criterion of 20% of charge loss, we reach a relative dispersion of 10% when the number of electrons involved in one bit is equal to 500, which corresponds to the 55 nm NAND technology node according to ITRS 2003. We can thus understand the difficulties and theoretical limits of few electron memories, extremely sensitive to the stochastic discharging behavior of the storage node. Fig. 33 reports the retention time relative dispersion as a function of the number of electrons per bit. As the number of electrons per bit N is reduced, we measured an increase of the retention time relative dispersion, with a factor ~2 when we pass from 100 to 10 electrons. Finally, one should note that poly-Si and Si-nc based memories follow comparable dispersion laws, in an experimental and a theoretical way. In conclusions, at the first order, the retention time relative dispersion simply depends on the number of electrons per bit, and is slightly dependant on the nature of the floating gate. One should also note that while the increasing of the measurement temperature accelerates

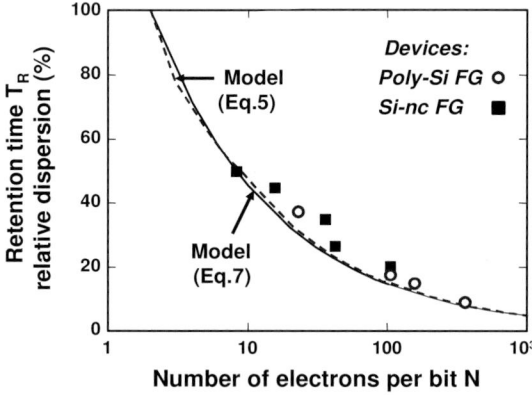

Fig. 33. Experimental and theoretical evolution of the relative dispersion of the retention time as a function of the number of electrons per bit. After Ref. 37.

the mean retention time, the retention time relative dispersion shows no temperature dependence. Indeed, experiments were performed on the same memory sample at 30°C and at 200°C, and it was found that the retention time relative dispersion remained unchanged, being respectively equal to 52% and 49%.

## 7. Conclusions

Evolutionary solutions, still based on variations of the well-proven floating-gate architecture, essentially consist in the integration of new materials (as nanocrystals or nitride traps for the floating gate, and kigh-k materials for the cell active dielectrics) and in the use of new device architecture (as multi-gate transistors). Through theses solutions, it seems possible to extend current floating gate technologies to the 45 nm and possibly 32 nm nodes. Other important emerging concepts (with high potential for low cost application to the 22 nm and smaller IC generations) make use of bottom-up approaches (i.e. chemical synthesis, self-assembly and template self-assembly) either as promising precise fabrication techniques of device structures, or even for the entire functional entity.

Nevertheless, it should be stated that as the dimensions of flash memories scale down, the number of electrons representing one bit dramatically reduces, enhancing the effects of single electron phenomena. Moreover, the number of electrons per bit further reduces in multi-level memories which will thus become extremely sensitive to the stochastic discharging behavior of the storage node. This means that the charging and the discharging of ultra-scaled floating gate memories should no longer be considered as a continuous phenomenon, but as a sum of discrete stochastic events. This leads to an intrinsic dispersion of both the retention time and of the memory programming window. Finally we argue that few electron phenomena are the intrinsic ultimate scaling limit of charge storage memory devices.

For this reason, it is widely believed that some disruptive technologies will be required beyond the 32 nm node. Possible solutions are based on the introduction of new storage mechanisms, like magnetic storage (MRAM), ferroelectric storage (FeRAM), phase-change materials (PCM memories). Nevertheless, today, all these technologies have a limitation on the cell size and, moreover, they cost several times more than DRAM and Flash. So that, the question if and when one of the above mentioned technologies will gain the position to take over the standard technologies is still open, but adequation of cost and application requirement will drive adoptions.

# References

1.  http://www.itrs.net/Links/2006Update
2.  K. Kim, *IEEE IEDM 2005 Tech. Dig.*, p 333.
3.  B. De Salvo, C. Gerardi, R. van Schaijk, S. Lombardo, D. Corso, C. Plantamura, S. Serafino, G. Ammendola, M. van Duuren, P. Goarin, W.Y. Mei, K. van der Jeugd, T. Baron, M. Gely, P. Mur and S. Deleonibus, *IEEE Tr. on Dev. and Mat. Reliab.*, **4**(3), p 377 (2004).
4.  B. DeSalvo, C. Gerardi, S. Lombardo, T. Baron, L. Perniola, D. Mariolle, P. Mur, A. Toffoli, M. Gely, M. N. Semeria, S. Deleonibus, G. Ammendola, V. Ancarani, M. Melanotte, R. Bez, L. Baldi, D. Corso, I. Crupi, R. A. Puglisi, G. Nicotra, E. Rimini, F. Mazen, G. Ghibaudo, G. Pananakakis, C. Monzio Compagnoni, D. Ielmini, A. Spinelli, A. Lacaita, Y. M. Wan and K. van der Jeugd, *IEEE IEDM 2003 Tech. Dig.*, p 597, (2003).
5.  S. Lombardo, R. A. Puglisi, I. Crupi, D. Corso, G. Nicotra, L. Perniola, B. DeSalvo and C. Gerardi, *Proc. IEEE NVSMW*, p 69, (2004).
6.  S. Jacob, L. Perniola, G. Festes, S. Bodnar, R. Coppard, J. F. Theiry, T. Pedron, E. Jalaguier, F. Boulanger, B. De Salvo and S. Deleonibus, *Proc. of IEEE NVSMW*, p 71, (2007).
7.  S. Jacob, G. Festes, S. Bodnar, R. Coppard, J. F. Thiery, T. Pate-Cazal, T. Pedron, B De Salvo, L. Perniola, E. Jalaguier, F. Boulanger and S. Deleonibus, *Proc. of ESSDERC*, p 410, (2007).
8.  R. Muralidhar, R.F. Steimle, M. Sadd, R. Rao, C.T. Swift, E.J. Prinz, J. Yater, L. Grieve, K. Harber, B. Hradsky, S. Straub, B. Acred, W. Paulson, W. Chen, L. Parker, S. G. H. Anderson, M. Rossow, T. Merchant, M. Paransky, T. Huynh, D. Hadad, K.-M. Chang and B. E. White Jr., *IEEE IEDM Tech. Dig.*, p 601, (2003).
9.  G. Molas, M. Bocquet, J. Buckley, J. P. Colonna, L. Masarotto, H. Grampeix, F. Martin, V. Vidal, A. Toffoli, P. Brianceau, L. Vermande, P. Scheiblin, M. Gély, A. M. Papon, G. Auvert, L. Perniola, C. Licitra, T. Veyron, N. Rochat, C. Bongiorno, S. Lombardo, B. De Salvo and S. Deleonibus, *IEEE IEDM Tech. Dig.*, p 453, (2007).

10.   Z. Liu, C. Lee, V. Narayanan, G. Pei and E. C. Kan, *IEEE Tr. on El. Dev*, **49**(9), p 1606 (2002).

11.   Z. Liu, C. Lee, V. Narayanan, G. Pei and E. C. Kan, *IEEE Tr. on El. Dev.*, **49**(9), p 1614 (2002).

12.   K. W. Guarini, C. T. Black, Y. Zhang, I. V. Babich, E. M. Sikorski and L. M. Gignac, *IEEE IEDM 2003 Tech. Dig.*, p 541 (2003).

13.   M. Alessandri, R. Piagge, S. Alberici, E. Bellandi, M. Caniatti, G. Ghidini, A. Modelli, C. Wiemer, S. Spiga and M. Fanciulli, *ECS Transactions* **1**(91), 2006.

14.   J. V. Houdt, *IRPS 2005 Tech. Dig.*, pp 234–239 (2005) .

15.   Y. Park, J. Choi, C. Kang, C. Lee, Y. Shin, B. Choi, J. Kim, S. Jeon, J. Sel, J. Park, K. Choi, T. Yoo, J. Sim and K. Kim, *IEEE IEDM 2006 Tech. Dig.*, 29 (2006).

16.   G. Molas, H. Grampeix, J. Buckley, M. Bocquet, X. Garros, F. Martin, J. P. Colonna, P. Brianceau, V. Vidal, M. Gély, B. De Salvo, S. Deleonibus, C. Bongiorno and S. Lombardo, *Proc. of ESSDERC 2006*, 242 (2006).

17.   M. Van Duuren, R. V. Schaijk, M. Slotboom, P. G. Tello, N. Akil, A. H. Miranda and D. S. Golubovic, *Proc. of ICICDT 2006*, p 36 (2007).

18.   C. H. Lai, A. Chin, H. L. Kao, K. M. Chen, M. Hong, J. Kwo and C. C. Chi, *Tech. Dig. of VLSI Technology 2006*, p 210 (2005).

19.   Y. Q. Wang, P. K. Singh, W. J. Yoo, Y. C. Yeo, G. Samudra and A. Chin, *IEEE IEDM 2005 Tech. Dig.*, p 169 (2005).

20.   C. H. Lai, A. Chin, K. C. Chiang, W. J. Yoo, C. F. Cheng and S. P. McAlister, *Tech. Dig. of VLSI Technology 2005*, p 210.

21.   A. Chin, C. C. Laio, C. Chen, K. C. Chiang, D. S. Yu, W. J. Yoo, G. S. Samudra, T. Wang, I. J. Hsieh, S. P. McAlister and C. C. Chi, *IEEE IEDM 2005 Tech. Dig.*, pp 165–168.

22.   C.H. Lee, C. Kang, J. Sim, J. S. Lee, J. Kim, Y. Shin, K. T. Park, S. Jeon, J. Sel, Y. Jeong, B. Choi, V. Kim, W. Jung, C. I. Hyun, J. Choi and K. Kim, *Proc. NVSMW 2006*.

23.   M. V. Duuren, R. V. Schaijk, M. Slotboom, P. Tello, P. Goarin, N. Akil, F. Neuilly, Z. Rittersma and A. Huerta, *Proc. of NVSMW 2006*, p 48 (2006).

24.   G. Molas, M. Bocquet, J. Buckley, H. Grampeix, M. Gély, J. P. Colonna, C. Licitra, N. Rochat, T. Veyront, X. Garros, F. Martin, P. Brianceau, V. Vidal, C. Bongiorno, S. Lombardo, B. De Salvo and S. Deleonibus, *Solid State Electronics,* **51**(11–12), p 1541 (2007).

25.   W. J. Zhu, T. Tamagawa, M. Gibson, T. Furukawa and T. P. Ma, *IEEE Elec. Dev. Lett.*, **23**(11), p 649 (2002).
26.   C. Licitra, E. Martinez, N. Rochat, T. Veyron, H. Grampeix, M. Gely, J. P. Colonna and G. Molas, *Proc. of NIST 2007*.
27.   P. Boher, P. Evrard, O. Condat, C. Dos Reis, C. Defranoux and E. Bellandi, *Thin Solid Films*, 455–456, p 798 (2004).
28.   N. V. Nguyen, S. Sayan, I. Levin, J. R. Ehrstein, I. J. R. Baumvol, C. Driemeier, C. Krug, L. Wielunski, P. Y. Hung and A. Diebold, *J. Vac. Sci. Technol. A*, **23**, p 1706 (2005).
29.   H. Y. Yu, M. F. Li, B. J. Cho, C. C. Yeo, M. S. Joo, D. L. Kwong, J. S. Pan, C. H. Ang, J. Z. Zheng and S. Ramanathan, *Appl. Phys. Lett.*, **81**(2), p 376 (2002).
30.   D. Deleruyelle, B. De Salvo, M. Gely, V. Vidal, X. Garros, N. Buffet, F. Martin, S. Lombardo, G. Nicotra and S. Deleonibus, *Proc. of Wodim 2004*, p 73, (2004).
31.   Q. Li, K. M. Koo, W. M. Lau, P. F. Lee, J. Y. Dai, Z. F. Hou and X. G. Gong, *Appl. Phys. Lett.*, **88**, 182903 (2006).
32.   R. S. Johnson, J. G. Hong, C. Hinkle and G. Lucovsky, *J. Vac. Sci. Technol. B* **20**(3), p 1126 (2002).
33.   C. Leroux, J. Mitard, G. Ghibaudo, X. Garros, G. Reimbold, B. Guillaumot and F. Martin, *IEEE IEDM 2006 Tech. Dig.*, p 737 (2004).
34.   S. H. Lee, J. J. Lee, J. D. Choe, E. S. Cho, Y. J. Ahn, W. Hwang, T. Kim, W. J. Kim, Y. B. Yoon, D. Jang, J. Yoo, D. Kim, K. Park, D. Park and B.-I. Ryu, *Tech. Dig. of IEDM 2006*, p 33 (2006).
35.   C. Jahan, O. Faynot, M. Cassé, R. Ritzenthaler, L. Brévard, L. Tosti, X. Garros, C. Vizioz, F. Allain, A. M. Papon, H. Dansas, F. Martin, M. Vinet, B. Guillaumot, A. Toffoli, B. Giffard and S. Deleonibus, *VLSI Tech. Dig. 2005*, p 112 (2005).
36.   B. Eitan, P. Pavan, I. Bloom, E. Aloni, A. Frommer and D. Finzi, *IEEE Electron Dev. Lett.*, **21**(11), pp 543–545, (2000).
37.   G. Molas, D.Deleruyelle, B. De Salvo, G. Ghibaudo, M. Gely, L. Perniola, D. Lafond and S. Deleonibus, *IEEE Trans. on El. Dev.*, **53**(10), p 2610 (2006).
38.   D. B. Strukov and K. K. Likharev, *Nanotechnology*, **16**, p 137, (2005).
39.   J. E. Green, J. Wook Choi, A. Boukai, Y. Bunimovich, E. Johnston-Halperin, E. DeIonno, Y. Luo, B. A. Sheriff, K. Xu, Y. Shik Shin, H.-Rong Tseng, J. F. Stoddart and J. R. Heath, *Nature*, **445**, p 414.

40. S. Gowda, G. Mathur, Q. Li, S. Surthi and V. Misra, *IEEE Trans. on Nanotech.*, **5**(3), p 258 (2006).

41. T. Pro, J. Buckley, M. Gely, K. Huang, G. Delapierre, G. Ghibaudo, F. Duclairoir, E. Jalaguier, B. De Salvo and S. Deleonibus, *Proc. of SIlicon Nanoelectronics Workshop*, 2007.

42. S. Lai, *MIT-Stanford-UC Berkeley Nanotechnology Forum*, Feb. 26, 2004.

43. H. Silva, M. K. Kim; A. Kumar, U. Avci and S. Tiwari, *IEEE IEDM 2003 Tech. Dig.*, p 271 (2003).

44. G. Atwood, A. Fazio, D. Mills and B. Reaves, *Intel Technol. J., Q4-97*, Online. http://www.intel.com/design/flash/ isf/overview.pdf

45. B. Cambou, *Mirrorbit*, Online. http://www.spansion.com/ flash_memory_products/mirrorbit.html

46. J. J. Razafindramora, L. Perniola, C. Jahan, P. Scheiblin, M. Gély, C. Vizioz, C. Carabasse, F. Boulanger, B. De Salvo, S. Deleonibus, S. Lombardo, C. Bongiorno and G. Iannaccone, *Proc. of ESSDERC, 2007.*

**Section 2**

...........................................

# New Concepts for Nanoelectronics. New Paths Added to CMOS Beyond the End of the Roadmap

# 10

# Single Electron Devices and Applications

Jacques Gautier*,‡, Xavier Jehl†, and Marc Sanquer†

*CEA-LETI, 17 avenue des Martyrs,
38054 Grenoble cedex 9, France

†CEA-INAC, 17 avenue des Martyrs,
38054 Grenoble cedex 9, France

‡jacques.gautier@cea.fr

·····································

Single electron devices have specific characteristics and properties, particularly the existence of periodic Coulomb blockade oscillations, a high charge sensitivity and an operation based on charge quantization. Many investigations have been done to take advantage of them. However, there are several challenging issues to face before any concrete application. Here, we overview the potential of these devices and discuss their merits and drawbacks. Although some niche applications exist, it is concluded that their future should be thought in hybrid association with CMOS.

## 1.    Introduction

Whereas the outstanding progress in microelectronics has resulted mainly from the scaling down of CMOS technology, detrimental effects are playing an increasing role, leading to a difficult and costly miniaturization of MOS-FETs and even more to a future end of the classic scaling. This is why many different approaches have been conceived, either to alleviate the problems, making possible an extension of the conventional top-down route, or to go beyond CMOS. They are known as evolutionary or disruptive solutions. For information processing, they are related to the different levels of the hierarchy: devices, state variables, architectures and computational models. Among the corresponding emerging solutions,[1] it is not clear to date which ones will be really implemented in future products, because they all have advantages and drawbacks.

This paper is focused on the use of single or few electron devices for nanoelectronic applications, a domain which is known as single electronics.[2] The first part is a brief description of single-electron transistors, from main characteristics to modeling, including their fabrication. Then we overview the potential of these devices, exploiting their specific features in comparison to CMOS, and discuss the important issues that could limit their interest to niche applications.

## 2.  Specific Features of Single Electron Devices

In a Single Electron Device, SED, the flow of current between electrodes is quantized by the electron charge. This is the main difference to conventional electronic devices, MOSFET or BJT, where this flow is continuous. Of course, to obtain such a behavior, there are conditions to meet. Especially, at each time, it is required to have an integer number of electron, or hole, in the body of the device, which implies a localization of the electron wave function. The most obvious way to achieve that is to implement two tunneling junctions, or potential barriers, which define an island in between. Their equivalent resistance $R_T$ should be higher than the quantum of resistance $R_Q = h/2e^2 \sim 13\,k\Omega$.[2] When a third electrode is added for an electrostatic control of the island potential, a Single Electron Transistor, SET, is obtained (Fig. 1).

### 2.1.  *Characteristics of SETs*

The characteristics of SETs are very different from those of MOSFETs. In both of them, electrostatic effects are dominant, but, due to the existence of tunneling junctions in SETs, electrons are not so free to move from source to drain. The Coulomb blockade effect, that is the electrostatic repulsion experienced by an electron approaching a small negatively charged region, limits the number of electrons in the island. As a result, for given values of gate and drain voltages, only a range of charge is possible and for some

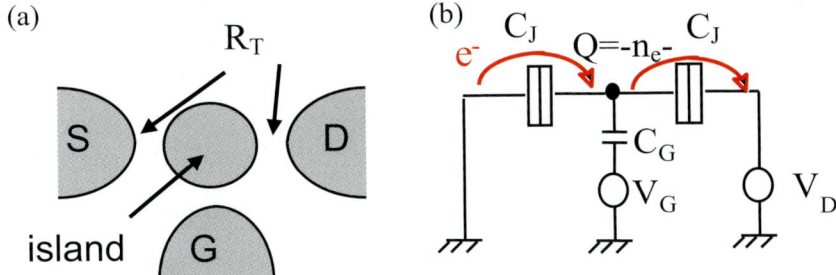

Fig. 1. (a) Schematic image of a Single Electron Transistor; (b) equivalent circuit.

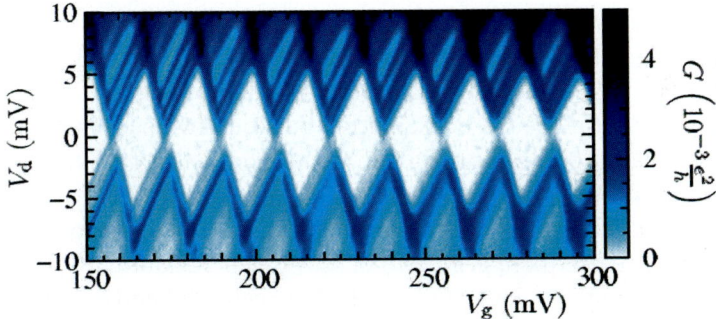

Fig. 2. Measured drain-source conductance versus gate and drain voltages of a SET. White areas correspond to Coulomb blockade regions (no detectable current) known as Coulomb diamonds. From diamond to diamond, the number of electrons in the island increases one by one.[3]

of them this is even more restricted to a single value. Due to the shape of the corresponding domains in the ($V_G$, $V_D$) plan, they are called Coulomb diamonds (Fig. 2).

The current in the device is due to the sequential tunneling of electrons through the source and drain junctions. According to the orthodox theory,[2,4] the tunneling rate $\Gamma$ is a function of the transparency of the barriers and of the drop of electrostatic energy $\Delta W$ corresponding to a single electron transition:

$$\Gamma = \frac{\Delta W}{e^2 R_T \left(1 - \exp\left(\frac{-\Delta W}{kT}\right)\right)} \tag{1}$$

where $\Delta W = E_{el.(before\ transition)} - E_{el.(after\ transition)} > 0$. The amount $E_c$ $= e^2/2C_{eff}$ is the charging energy of one extra electron, where $C_{eff}$ is the effective capacitance, the sum of gate and junction capacitances ($C_{eff} = C_g + 2C_j$). Note that here the I(V) characteristics of the junctions, in the absence of single-electron charging effects, has been replaced by the Ohmic approximation $I(V) = V/R_T$. From this equation, we infer that the Coulomb blockade regime can only exist for $E_c \gg kT$, which implies a maximum $C_{eff}$ close to 0.3aF for RT (room temperature) operation.

This condition is quite difficult to satisfy for RT, because it requires device geometries in the nanometer range, but for larger sizes Coulomb blockade effects can still be observed at a sufficiently low temperature, provided the localization condition is fulfilled. An example of gate voltage characteristics is given in Fig. 3.

For a low drain voltage, the device is alternatively on and off, the characteristics displaying periodic Coulomb blockade oscillations, CBOs. In the valleys there is a stable state of charge in the island, whereas at the peaks of current the charge is oscillating between two successive integers, the average charge being a half integer. Consequently the gate voltage period is equal to $e/C_g$, where $C_g$ is the gate to island capacitance.

The output characteristics of SETs are also very different from those of MOSFETs, since there is a more or less linear variation of their current beyond a Coulomb blockade threshold which is modulated by the gate bias. Another important feature for applications is their quite low level of current:

$$I_{Dpeak} \approx \frac{V_D}{2R_\Sigma} \ll \frac{V_D}{4R_Q} \tag{2}$$

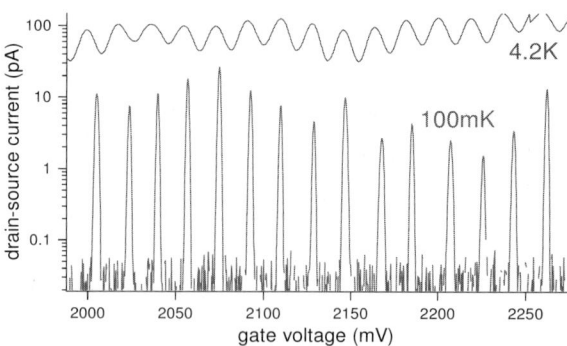

Fig. 3. Measured CBOs on a nanowire SET on SOI (W = 50 nm, L = 40 nm).[5]

where $R_\Sigma$ is the sum of both tunneling junction resistance ($=2R_T$ for symmetrical device) and $R_Q$ is the quantum of resistance.

## 2.2.   Fabrication of SETs

Concerning the design and the fabrication of SETs, there is a great flexibility about materials and processes. For the island, any conductive material can be used, metal, semiconductor, carbon nanotubes and molecules. The localization of electrons can be obtained with tunneling junctions, potential barriers and even resistors, provided they have a resistance higher than the quantum of resistance. As a result a semiconductor is not mandatory, opening the possibility to put SETs on top of other devices in the frame of 3D integration. However, from a practical point of view, implementation on silicon or SOI for dielectric isolation could be advantageous, to take benefit of the huge amount of knowledge and process control acquired on that substrate and also for hybrid integration with MOSFET.

The archetype of SETs was fabricated by Al evaporation at two large angles through a shadow-mask, tunnel $Al_2O_3$ junctions being defined at the overlap of Al layers.[6] Since this technique is simple and does not need costly equipment, it has been widely used in condensed matter research labs. To improve it and relax lithography requirements, several modifications have been reported: step-like tunnel junctions,[7] stencil masks,[8] anodization,[9] self-alignment,[10] etc. There were also attempts to replace the single island by arrays of metallic nano particles.[11] A Coulomb gap is measured but in general CBOs are not reported or there is a multiple periodicity. To increase the charging energy, SETs have been fabricated by STM/AFM nano-oxidation of thin metallic films (Ti, Nb). RT operation was achieved using SWCNT AFM cantilever, but to the detriment of the level of current, due to the thickness of tunnel junctions.[12] Instead of such barriers, it has been demonstrated the possibility to confine electrons with high-ohmic ($R \gg R_Q$) metallic microstrips,[13] in agreement with theoretical work performed by Nazarov.[14]

With semiconductors, there is still more flexibility to design SETs. Similarly to metallic SETs, it is possible to localize electrons with tunnel oxide[15] or highly resistive material,[16] however their lower density of states offers other attractive features from engineering viewpoints. Using lateral split-gates or Schottky wrap-gates, quantum dots devices and SETs have been fabricated in III-V 2DEG.[17] In silicon, it is also possible to define junctions with sidewall depletion gates[18] or by modulation of doping along a nanowire on SOI.[19] An interesting feature of this approach is the possibility to tune the transparency of junctions, contrary to dielectric tunnel barriers.[20] Another way, is to exploit quantum mechanical effects taking place in semiconductors. This has been done by NTT in the PADOX process, combining confinement effects in a narrow Si wire and opposite band-gap reduction induced by oxidation stress.[21] The operation of single-hole transistor with high PVCR (peak-to-valley current ratio) has also been explained by quantum effects, but here the mechanism is the quantization of energy levels in naturally formed dots along a Si ultra-narrow-wire.[22]

In comparison with metallic SETs, another important advantage of Si SETs is their excellent stability. Whereas a detrimental charge offset noise problem has been reported in Al-based SETs,[23] the long-term drift is better by several orders of magnitude in the case of Si.[24] This is also demonstrated in Fig. 4, showing no visible evolution of CBOs for more than 10 hours.

Despite these essential features of Si, for RT operation it is also crucial to master nanometer size fabrication while reaching a level of reproducibility compatible with the complexity of applications. This is still far to be achieved. A promising approach is to exploit naturally formed nanostructures, like carbon nanotubes (CNT) or molecules, although the issues related to their controlled localization. CBOs have already been observed at RT on CNT devices[25] or close to RT on nano-gap devices combining ultra small gold islands and bridging molecules.[26] Also, on single-molecule transistors, well defined Coulomb diamonds have been measured at low temperature.[27]

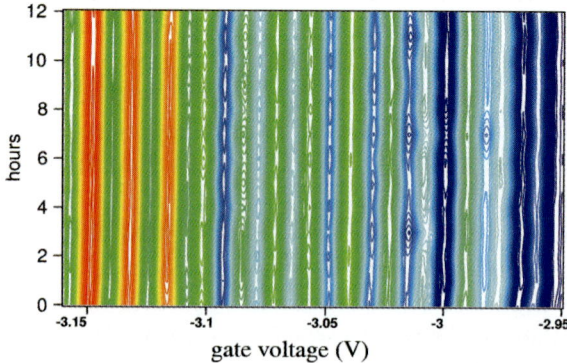

Fig. 4.  Stability of CBOs measured on a SOI constriction.

## 2.3.    Modeling and simulation of SETs

To design circuits including SETs, CAD models are essential. They are based on the resolution of the master equation[2,4] which gives the time variation of the probability $p_n$ of state n, *i.e.* to have n electrons in the island:

$$\dot{p}_n = \Gamma_{n,n+1} p_{n+1} + \Gamma_{n,n-1} p_{n-1} - (\Gamma_{n+1,n} + \Gamma_{n-1,n}) p_n \qquad (3)$$

where $\Gamma_{k,i}$ is the rate of transition from state i to state k, as a result of electron tunneling from/to the source or the drain. For example $\Gamma_{n+1,n}$ is the sum of the tunneling rates from the source and the drain (given by Eq. (1), when there are initially n electrons in the island:

$$\Gamma_{n+1,n} = \vec{\Gamma}_1(n) + \overleftarrow{\Gamma}_2(n)$$

The current in the branch i of the device is given by:

$$I(V) = e \left( \sum_{n=-\infty}^{+\infty} p_n(\vec{\Gamma}_i(n) - \overleftarrow{\Gamma}_i(n)) \right) \qquad (4)$$

Quasi analytical solutions of Eqs. (5)–(7) have been reported for the stationary case $\dot{p}_n = 0$, considering only the relevant values of $n$.[28-30] A compact SET transient model has been developed in Ref. 31.

These models have been integrated in SPICE or other circuit simulators to study hybrid SET/CMOS circuits, taking advantages of both of these devices.[32,33]

For more complex single-electron devices, and in general for circuits comprising several nodes coupled by single-electron effects, CAD models are not available. In this case, the approach is to perform numerical simulation with a Monte Carlo solver such as the one described in Ref. 34.

## 3.    Digital Low Power Electronics

An important challenge for microelectronic applications is the reduction of energy consumption while increasing performance. In fact this is not new, but just becoming more critical for microprocessors, since they have reached a level of $\sim 100$ W above which heat extraction becomes expansive, as well as for nomadic products or for self-powered or wireless remote-powered autonomous micro-systems. Thanks to the downsizing of CMOS technology, the amount of energy required to perform an elementary operation has been dramatically reduced, but this gain was not sufficient in comparison to the combined effects of complexity and clock frequency increases in the case of microprocessors. For the future, up to the end of the CMOS roadmap, ~2020, the scaling trend will continue and the complexity will also rise at more or less the same pace, but the clock frequency will stay saturated at roughly the current value. Nevertheless, as pointed out by several people, performance will still improve at least by innovation at the system level and by holistic design.[35] For instance, instead of increasing the processing power by frequency, it is now more advantageous to combine multi processing elements in the same chip. Furthermore, an alternative technology roadmap for semiconductors has been proposed recently, suggesting a reduction of the frequency of such processing elements, and of $I_{Dsat}$, to maintain the power consumption at a low level, while improving the processing power.[36]

From the previous remarks, and looking at the specific characteristics of SEDs, several paths can be considered for low power applications.

## 3.1.  Conventional digital circuit

One of the features of SETs with size in the nanometer range is the very low number of electrons located in the island, which results from low gate capacitance. In the case of CMOS, for the most aggressive generation of the roadmap corresponding to a double-gate transistor with a length of 5 nm, there will be only 20–30 electrons in the channel, depending on parasitic capacitances (here it is supposed a gate width of $3 \times Lg$). So, from this simple consideration, using SETs instead of MOSFETs, it can be expected a further reduction in the number of electrons implied in switching operations by a factor of 10, which would be very attractive for energy saving. However, the replacement of MOSFETs by SETs is not so easy due to some severe constraints or drawbacks as discussed in the following section. In addition, one has to take the parasitic and interconnect capacitances into account.

A first point is the requirement of two types of device to obtain complementary actions in CMOS-like logic gates. This is not directly available with SETs. The only possibility is to play with the CBO characteristics where SETs behave as NMOS in the rising parts, while they behave as PMOS in the falling parts. Nevertheless in basic SETs there is no way to control the phase of CBOs, contrary to MOSFETs where the threshold voltage can easily be adjusted by doping. In fact it is necessary to modify the device architecture, either adding another control gate, or introducing a given amount of charges in proximity, to shift CBOs by electrostatic effect on the island potential. Both approaches are also useful to balance the impact of parasitic charges which offset CBOs. An example of the second solution will be presented in section 4. With control gates, it is possible to design inverter or more complex digital circuits, provided there is a careful optimization of SET parameters.[37−41] One has to note that with basic twin SETs it is also possible to obtain an inverter, but at the drawback of a lost of degree of freedom on the value of the supply voltage, to get complementary pull-up and pull-down actions ($V_{DD} = e/(C_g + C_j)$) at low temperature.

The requirement of control gates has *a priori* detrimental effects on packing density, however it can be viewed as beneficial to design compact digital circuits, thanks to the increased functionality it provides. The key point is the existence of CBOs and the possibility to play with them. For instance the current in a symmetrical double-gate SET is an X-OR function of the inputs, provided the high-level logic is defined by the voltage $e/2C_{in}$, where $C_{in}$ is the capacitance of the input gate. When only one input is high, the device is biased at a peak of current, whereas the device is off when both inputs are low or high. With the same principle, adding a control gate $C_c$ biased at $e/2C_c$ to a SET, in a logic circuit, transforms its inputs to complementary values. This is very powerful in the frame of Pass-Transistor-Logic,[42] but the peak position of CBOs should not be shifted by random offset charges. Redundant design and reconfigurability have been suggested for defect tolerance, despite an area overhead.

The previous approach takes advantage of only the first period of CBOs, leading to design flexibility not achievable with CMOS. Going further, these periodic oscillations can be exploited in multiple-valued logic.[43,44] Concepts of circuits based on SET or hybrid CMOS-SET architectures have been reported, emphasizing their potential to alleviate the interconnect problem between modules in a chip.[45] For example 50% reduction in global line is possible by using quaternary logic instead of binary logic. Concepts of hybrid multiple-valued SRAM have also been proposed.[46,47]

Coming back to the design of SET circuits, it should be noted that several constraints limit the value of device parameters. For room temperature operation, supposing also that $C_g$ is the dominant contribution to $C_{eff}$, which is required for voltage gain since $G = C_g/C_j$, the period of CBOs cannot be lower than ∼0.5 V as shown in Ref. 48. This implies a quite large gate voltage swing to exploit several periods of CBOs at RT, e.g. about 2 V for a quaternary logic. On the contrary, the drain voltage should be as small as possible, in comparison to the Coulomb blockade gap, to achieve a sufficiently high peak-to-valley-current-ratio, PVCR, knowing that the maximum blockade voltage is e/$C_{eff}$ and because the subthreshold swing

$$S = \frac{C_{eff}}{C_g} kT/eLn10 \qquad (5)$$

is not better than 60 mV/dec at RT. As a result, hybrid-SET architectures are currently implemented for current biasing of SET, cascode stage or output voltage amplification.[45,47,49] Besides, this is useful to dynamic performance, due to the poor drive ability of SETs given by Eq. (2).

In the case of logic tree built with SETs,[50] the node capacitance between SET devices should be higher than $C_{eff}$, to avoid shot noise effects. This would be easily fulfilled in practice due to the dominant contribution of local interconnects or input capacitance of MOSFET buffer. However load capacitances should not be too large, otherwise the energy saving discussed at the beginning of 3.1 will be degraded, as well as the speed. Nevertheless, the operation at very low $V_{DD}$ for the SET based logic tree remains an attractive feature for low power operation.

## 3.2. *Disruptive architectures*

The previous section was related to SET, or hybrid SET-CMOS based circuits, taking advantage of the specific features of SETs. Single electron effects were accounted for device operation, but from the circuit point-of-view SETs were considered as black box. For other architectures, especially Quantum Cellular Automata (QCA), such a decoupling between inside device operation and circuit topology does not exist, due to the direct Coulomb interaction between the electrons of a cell and those of proximity cells.[51] Since there is no flow of carrier from cell to cell, very low power consumption is expected. In QCA, there are no conventional interconnects, data are transmitted along arrays of cells and binary functions result from their topology. To ensure a direction of propagation, a clock is required, coming with the benefit of adiabatic operation for low power.[52] However there are very challenging issues about the fabrication and reproducibility of cells. Tiny structures are essential for RT operation and the localization of cells is critical. In addition, if offset charges are not avoidable, a tolerant concept is required.[53] Current investigations address experimental demonstrations of clocked QCA through relatively large structures, at low temperature, but real applications are only expected in the long term, at molecular level.[54]

Artificial neural network is another architectural model which has the inherent merit of being defect tolerant, a key feature for nanoscale implementation. The use of neuromorphic networks is attractive to solve problems like pattern recognition or classification for which conventional digital CMOS circuits are not really efficient. Several conceptual approaches have been proposed to design or fabricate single electron based neural network.[55,56] In the latter, there is an interesting hybrid combination of a bottom CMOS layer for the soma function and of a single-electron device layer for the synapses. These last ones are latching switches designed with a SET and a single-electron trap. Their molecular implementation has been suggested for ultra high density and also to achieve RT operation.

## 4.    Single Electron Memories

Due to the importance of memory in circuits and applications, single electron phenomena have been considered to design new memory devices, either taking advantage of the Coulomb blockade effect to store electrons, or exploiting charge quantization effects in structures embedding discrete traps or nano-size dots. In the latter, charge retention results mainly from confinement barriers, but Coulomb repulsion can also play a role through self-limited charging process.

In a first approach to design new memory devices, the starting point was the concept of single-electron trap which can be viewed as a generalization of the single-electron box.[2] The idea is to replace the single-island single-tunnel-junction structure by a one-dimensional array of N islands and N tunnel junctions, i.e. a MTJ (Multiple Tunnel Junction), to obtain a hysteretic effect (Fig. 5). If N is large enough, the energy barrier provided by the array may suppress thermal and macroscopic tunneling rates along the array, leading to trapping of one or a few electrons at the storage node. To sense this charge, a SET or a MOSFET is combined with the single–electron trap, resulting in a single-electron memory, SEM. Similarly to the case of SETs, achieving RT and fast operation is a serious issue, which has not been resolved yet, mainly due to the difficulty in the fabrication of reproducible nanometer size structure. Metallic devices have been made by the shadow deposition technique,[57] gold island deposition[58] or AFM nano-oxidation process.[59] Hysteretic characteristics at RT, storage of

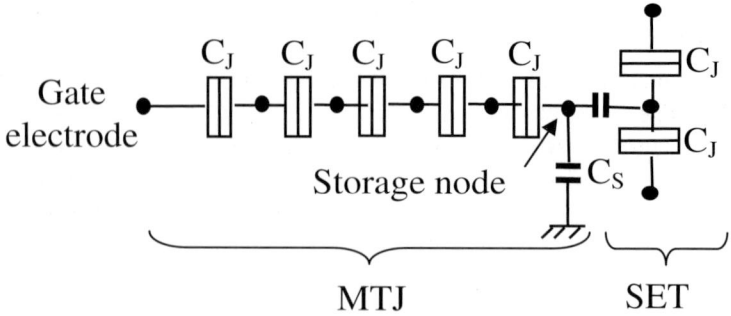

Fig. 5. Equivalent circuit of a Single-Electron-Trap.

several electrons and retention time of 600 s have been reported with the latter approach, however the current in the readout device is in the range of pA, which is detrimental for the speed of operation. In the case of Si based MTJ SEM, a $3 \times 3$ array of hybrid SET/MOSFET cells has been integrated on SOI.[60] A right operation has been demonstrated above 30 K and the output current is in the $\mu$A range, but for RT and retention of 1s, comparable to DRAM, it has been estimated that $\sim$1 nm-size islands would be required.

In addition to this serious issue, the use of SET for sensitive readout can be strongly impaired by the shift of characteristics due to background charges. A solution to this problem has been proposed by Likharev and Korotkov.[61] Instead of sensing any absolute change in charge, the idea is to sense the relative change in charge which occurs in the storage node when a voltage ramp is applied to the control gate of the device. The resulting current oscillations in the SET are detected by a FET sense amplifier. The only drawback of this reading mode is the erasing of the stored information, implying its rewriting like in DRAM. This method has been demonstrated experimentally in aluminum single-electron floating gate (FG) memory cells, where a cancellation gate voltage is applied to balance the electrostatic influence of the ramp control gate voltage on the potential of the island of the SET.[62] Under this condition the SET is only sensitive to charge change in the storage node.

Nevertheless, due to stochastic effects, small retention time, slow write and high soft error rate, that kind of memory was not fitting any application categories of the ITRS[1] and has been withdrawn from the list of emerging research memory in the 2005 edition.

Another approach to design SEM is based on the ultimate scaling of FG non-volatile memories, in which a trap charge shifts the threshold voltage of a readout transistor. The first RT operation of a real SEM was demonstrated in 1993.[63] The trick was to replace the conventional stacking of Si channel and floating gate by an ultra-thin poly-Si film in which a percolation channel and nano-grain storage node are naturally formed for a range of biasing. The signature of the charge quantization in the storage node is the abrupt shift of gate voltage characteristics. The merit of this approach is the simplicity of the process to get tiny Si dots, but since storage node and current path are located in the same granular film, it is not possible to independently tune their characteristics. A 128 Mb early prototype was fabricated but, despite design techniques to overcome inherent stochastic variations, it did not resulted in industrial product.[64]

Most of other attempts to fabricate single or few-electron memories are based on stacked structures embedding nano-size dot[65−68] to trap electrons, leading to the so-called floating dot memory or quantum dot flash memory. An important difference with conventional flash memories is that charge quantization effects can be observed on gate voltage characteristics.[69] During charging or discharging, they exhibit current discontinuities related to quantized threshold voltage shifts. For a single dot on narrow Si wire, as well as for several dots, the threshold voltage shift induced by one trapped electron can be estimated as

$$\Delta Vth = \frac{e}{C_{gd} + \left(C_{gd} + C_{dc}\right)\frac{C_{gc}}{C_{dc}}} \tag{6}$$

where $C_{gc}$ is the gate to channel capacitance, $C_{gd}$ is the gate to dot capacitance and $C_{dc}$ is the dot to channel capacitance. It has been shown[70] that this relationship can be further simplified to:

$$\Delta Vth \approx \frac{e \cdot t_{gd}}{A\epsilon_{ox}} \tag{7}$$

where $t_{gd}$ is gate to dot distance and A is the total active area, emphasizing the importance of small size to observe single electron effects. A large $t_{gd}$ is not desirable since it would imply a high writing voltage.

Instead of using the channel of a MOSFET as readout device, it is possible to implement a SET. Also, electrons can be trapped in discrete defects instead of dots. This has been demonstrated in a multiple-valued memory device.[71] In order to detect easily the electron exchange occurring between SET and silicon nitride traps during writing and reading, a cancellation method was implemented, similar to the one described previously and used in Ref. 62. With one current oscillation in the SET corresponding to a change of a few tens of electrons in SiNx, more than 10-value operation was achieved at 77 K, one value per oscillation.

For any memory device based on the coding of bits through single or a few electrons, the fluctuation in the number of electron can induce a high probability of error which should not be incompatible with targeted applications. The confinement energy should be much greater than kT to guarantee a low failure rate.[72] In addition, due to the stochastic nature of the tunnel process, the exact time of tunneling of an electron is not known, which could result in retention time and programming window dispersions.[73]

In the case of the neuromorphic architecture concept[56] discussed in 3.2, the situation is different since neural networks offers natural default tolerance.

Besides the previous memory applications, the storage of a few electrons, to shift and control CBOs, can be very useful to programmable logic.[49,74]

## 5.    Analog Applications

The unique functionality of SET devices can also be exploited in analog applications. For metrology and implementation of current or capacitance standards, the possibility to accurately control a flow of electron one by one, or to count them, is very interesting. Especially in a single electron pump, the current is linearly proportional to the frequency of the clock signal applied to the circuit, only one electron being transferred through it, thanks to the Coulomb blockade effect. Such a circuit has been designed using multiple islands SET structure[4,75] or a combination of SET and MOSFETs.[76] A challenging issue is to avoid leakages and cotunneling which have detrimental effects on the accuracy. Besides that, operation at cryogenic or low temperature is acceptable for metrology applications, which alleviates small size requirements. Devices fabricated on Si exhibit an excellent stability as shown in Fig. 4.

Other useful applications of SETs have been reported, especially the design of a flash ADC (Analog-to-Digital Converter)[47] and of a random-number generator.[77] As for logic applications, FETs are integrated with SETs as current load or for amplification.

Another feature of SETs is their extreme charge sensitivity.[78] It can be exploited in instrumentation, for instance to measure the displacement of nano mechanical resonator,[79,80] or to perform quantum measurements on a charge qubit for future quantum computers.[81−83] Single-electron spin manipulation is also under investigation.[84]

## 6. Summary

SEDs have specific features that can be exploited in useful applications, although they are impeded by challenging issues. Advantages and drawbacks have been discussed in comparison to CMOS and it is concluded that, in addition to some niches, SEDs should be thought in hybrid association with CMOS.

## References

1.    ITRS 2006 Edition, http://www.itrs.net/
2.    K. K. Likharev, *Proc. IEEE*, **87**(4), pp 606–632 (1999).
3.    M. Hofheinz, X. Jehl, M. Sanquer, G. Molas, M. Vinet and S. Deleonibus, *European Physical Journal*, **B54**, pp 299–307 (2006).
4.    H. Grabert and M. Devoret, *NATO ASI Series*, Volume 294, Plenum Press, New York and London (1992).
5.    M. Bohm, M. Hofheinz, X. Jehl, M. Sanquer, M. Vinet, G. Molas, B. De Salvo, B. Previtali, D. Fraboulet, D. Mariolle and S. Deleonibus, *LITHO 2004*, Agelonde, France.
6.    T. A. Fulton and G. J. Dolan, *Phys. Rev. Lett.*, **59**(1), pp 109–112 (1987).
7.    K. Hofmann, B. Spangenberg and H. Kurz, *J. Vac. Sci. Technol. B* **20**(1), pp 271–273 (2002).
8.    R. Dolata, H. Scherer, A. B. Zorin and J. Niemeyer, *J. Vac. Sci. Technol. B* **21**(2), pp 775–780 (2003).
9.    Y. Nakamura, C. D. Chen and J.-S. Tsai, *Jpn. J. Appl. Phys.* **35**, pp L1465–L1467 (1996).
10.   Y. Chen, P. Hadley, C. Harmans, J. E. Mooij, G. I. Ng and S. F. Yoon, *SPIE* **3183**, pp 138–145 (1997).
11.   C. T. Black, C. B. Murray, R. L. Sandstrom, and S. Sun, *Science* **290**, pp 1131–1134 (2000).

12. Y. Gotoh and K. Matsumoto, *Jpn. J. Appl. Phys.*, **41**(1)(4B), pp 2578–2582 (2002).
13. V. A. Krupenin, A. B. Zorin, M. N. Savvateev, D. E. Presnov and J. Niemeyer, *JAP* **90**(5), pp 2411–2415 (2001).
14. Y. V. Nazarov, *Phys. Rev. Lett.*, **82**(6), pp 1245–1248 (1999).
15. Y. Ito, T. Hatano, A. Nakajima and S. Yokoyama, *APL*, **80**(24), pp 4617–4619 (2002).
16. X. Jehl, M. Sanquer, G. Bertrand, G. Guégan, S. Deleonibus and D. Fraboulet, *IEEE Trans. Nanotech.*, **2**(4), pp 308–313 (2003).
17. S. Kasai and H. Hasegawa, *Jpn. J. Appl. Phys.*, **40**(1)(3B), pp 2029–2032 (2001).
18. D. H. Kim, S.-K. Sung, K. R. Kim, J. D. Lee and B.-G. Park, *IEEE Trans. Nanotech.*, **1**(4), pp 170–175 (2002).
19. M. Hofheinz, X. Jehl, M. Sanquer, G. Molas, M. Vinet and S. Deleonibus, *Appl. Phys. Lett.* **89**, pp 143504 (2006).
20. A. Fujiwara, H. Inokawa, K. Yamazaki, H. Namatsu and Y. Takahashi, *Appl. Phys. Lett.*, **88**, 053121 (2006).
21. M. Nagase, S. Horiguchi, A. Fujiwara and Y. Takahashi, *Jpn. J. Appl. Phys.*, **42**(1)(4B), pp 2438–2443 (2003).
22. M. Kobayashi, M. Saitoh and T. Hiramoto, *Jpn. J. Appl. Phys.*, **45**(8A), pp 6157–6161 (2006).
23. N. M. Zimmerman and J. L. Cobb, *Dig. of IEEE Conf. on Precision Elec. Meas.*, pp 138–139 (1998).
24. N. M. Zimmerman, W. H. Huber, A. Fujiwara and Y. Takahashi, *Dig. of IEEE Conf. on Precision Elec. Meas.*, pp 124–125 (2002).
25. C.K. Hyon, A. Kojima, T. Kamimura, M. Maeda and K. Matsumoto, *Jpn. J. Appl. Phys.*, **44**(4A), pp 2056–2060 (2005).
26. T. Goto, K. Degawa, H. Inokawa, K. Furukawa, H. Nakashima, K. Sumitomo, T. Aoki and K. Torimitsu, *Jpn. J. Appl. Phys.*, **45**(5A), pp 4285–4289 (2006).
27. J. Park *et al.*, *Nature*, **417**, pp 722–725 (2002).
28. K. Uchida *et al.*, *Jpn. J. Appl. Phys.*, **39**(4B), pp 2321–2324 (2000).
29. S. Mahapatra, A. Ionescu and K. Banerjee, *IEEE Elec. Dev. Lett.*, **23**(6), pp 366–368 (2002).
30. H. Inokawa and Y. Takahashi, *IEEE Trans. on Elec. Dev.*, **50**(2), pp 455–461 (2003).
31. Y. S. Yu, S. W. Hwang and D. Ahn, *IEE Proc. Circuits Dev. Syst.*, **152**(6), pp 691–696 (2005).
32. S. Mahapatra and A. Ionescu, *Integrated Microsystems Series*, Artech House, Boston and London.

33.   M. Kirihara, K. Nakazato and M. Wagner, *Jpn. J. Appl. Phys.*, **38**(1)(4A), pp 2028–2032 (1999).
34.   C. Wasshuber, *IEEE Trans. on CAD*, **16**(9), pp 937–944 (1997).
35.   B. Meyerson, *In-Stat/MDR Fall Processor Forum*, San Jose, (2004).
36.   J.-P. Schoellkopf, *ASYNC'2006* (2006).
37.   J. R. Tucker, *J. Appl. Phys.*, **72**(9), pp 4399–4413 (1992).
38.   A. N. Korotkov, R. H. Chen and K. K. Likharev, *J. Appl. Phys.* **78**(4), pp 2520–2530 (1995).
39.   K. Liu *et al.*, *Jpn. J. Appl. Phys.*, **41**(2A), pp 458–463 (2002).
40.   M.-Y. Jeong, B.-H. Lee and Y.-H. Jeong, *Jpn. J. Appl. Phys.*, **40**(1)(3B), pp 2054–2057 (2001).
41.   J. Gautier, *Ultra Low Power and Design*, ed. by E. Macii, Kluwer Academic Publishers (2004).
42.   Y. Ono, H. Inokawa and Y. Takahashi, *IEEE Trans. on Nanotechnology,* **1**(2), pp 93–99 (2002).
43.   M. Akazawa, K. Kanaami, T. Yamada and Y. Amemiya, *IEICE Trans. Elec.*, **E82-C**(9), pp 1607–1614 (1999).
44.   K.-W. Song *et al.*, *Jpn. J. Appl. Phys.*, **44**(4B), pp 2616–2622 (2005).
45.   S. Mahapatra and A. M. Ionescu, *IEEE Trans. Nano.*, **4**(6), pp 705–714 (2005).
46.   Y. S. Yu, H. W. Kye, B. N. Song, S.-J. Kim and J.-B. Choi, *Elec. Lett.*, **41**(24), pp 1316–1317 (2005).
47.   H. Inokawa, A. Fujiwara and Y. Takahashi, *IEEE Trans. on Elec. Dev.*, **50**(2), pp 462–470 (2003).
48.   H. Harata, M. Saitoh and T. Hiramoto, *JJAP*, **44**(20), pp L640–L642 (2005).
49.   K. Uchida, J. Koga, R. Ohba and A. Toriumi, *IEEE TED* **50**(7), pp 1623–1630 (2003).
50.   K. Uchida, J. Koga, R. Ohba and A. Toriumi, *IEICE Trans. Elec.*, **E84-C**(8), pp 1066–1070 (2001).
51.   G. H. Bernstein, *DAC 2003*, pp 268–273 (2003).
52.   J. Timler and C. Lent, *JAP*, **94**(2), pp 1050–1060 (2003).
53.   A. Fijany and B. N. Toomarian, *J. Nano. Res.* **3**, pp 27–37 (2001).
54.   C. S. Lent, B. Isaksen and M. Lieberman, *JACS* **125**, pp 1056–1063 (2003).
55.   T. Oya, A. Schmid, T. Asai, Y. Leblebici and Y. Amemiya, *IEICE Elec. Exp.*, **2**(3), p 76 (2005).
56.   K. Likharev and D. Strukov, *Introduction to Molecular Electronics*, ed. by G. Cuniberti *et al.*, Springer, Berlin (2005).

57.   S. V. Lotkhov, H. Zanerle, A. B. Zorin and J. Niemeyer, *APL*, **75**(17), pp 2665–2667 (1999).
58.   A. Pépin, C. Vieu, H. Launois, M. Rosmeulen, M. Van Rossum, H. O. Mueller, D. Williams, H. Mizuta and K. Nakazato, *Microelectronic Engineering* **53**, pp 265–268 (2000).
59.   K. Matsumoto, Y. Gotoh, T. Maeda, J. A. Dagata and J. S. Harris, *IEDM Tech. Digest*, pp 449-452 (1998).
60.   Z. A. Durrani, A. C. Irvine and H. Ahmed, *IEEE Trans. on Elec. Dev.*, **47**(12), pp 2334–2339 (2000).
61.   K. K. Likharev and A. N. Korotkov, *VLSI Design*, **6**(1-4), pp 341–344 (1998).
62.   K. K. Yadavalli, A. O. Orlov, G. L. Snider and A. N. Korotkov, *J. Vac. Sci. Technol. B* **21**(6), pp 2860–2864 (2003).
63.   K. Yano, T. Ishii, T. Hashimoto, T. Kobayashi, F. Murai and K. Seki, *IEDM Tech. Dig.*, pp 541–544 (1993).
64.   K. Yano, T. Ishii, T. T. Sano, T. Mine, F. Murai, T. Kure and K. Seki, *ISSCC Tech. Dig.*, pp 344–345 (1998).
65.   L. Guo, E. Leobandung and S. Y. Chou, *Science*, **275**, pp 649–651 (1997).
66.   T. Futatsugi, A. Nakajima and H. Nakao, *Fujitsi Sci. Tech. J.*, **34**, pp 142–152 (1998).
67.   S.-K. Sung, D. H. Kim, J.-S. Sim, K. R. Kim, Y. K. Lee, J. D. Lee, S. D. Chae, B. M. Kim and B.-G. Park, *Jpn. J. Appl. Phys.* **41**(1)(4B), pp 2606–2610 (2002).
68.   X. Tang, N. Reckinger, V. Bayot, C. Krzeminski, E. Dubois, A. Villaret and D.-C. Bensahel, *IEEE Trans. Nanotech.*, **5**(6), pp 649–656 (2006).
69.   S. Tiwari, J. A. Wahl, H. Silva, F. Rana and J.-J. Welser, *Appl. Phys. A 71*, pp 403–414 (2000).
70.   G. Molas, B. De Salvo, G. Ghibaudo, D. Mariolle, A. Toffoli, N. Buffet, R. Puglisi, S. Lombardo and S. Deleonibus, *IEEE Trans. Nanotech.*, **3**(1), pp 42–48 (2004).
71.   H. Sunamura, H. Kawaura, *Si Nanoelectronics Work.*, pp 69–70 (2002).
72.   C. Wasshuber, H. Kosina and S. Selberherr, *IEEE Trans. on Elec. Dev.*, pp 2365–2371 (1998).
73.   G. Molas, D. Deleruyelle, B. De Salvo, G. Ghibaudo, M. Gely, L. Perniola, D. Lafond and S. Deleonibus, *IEEE Trans. on Elec. Dev.*, pp 2610–2619 (2006).

74.    G. Molas, X. Jehl, M. Sanquer, B. De Salvo, D. Lafond and S. Deleonibus, *IEEE Trans. Nano.* **4**, pp 374–379 (2005).

75.    H. E. van den Brom *et al.*, *IEEE TIM*, **52**(2), pp 584–588 (2003).

76.    Y. Ono, A. Jujiwara, Y. Takahashi and H. Inokawa, *MRS Symp. Proc.*, **864**, pp 259–270 (2005).

77.    K. Uchida, T. Tanamoto, R. Ohba, S.-I. Yasuda and S. Fujita, *IEDM Tech. Digest*, pp 177–180 (2002).

78.    M. H. Devoret and R. J. Schoelkopf, *Nature* **406**, pp 1039–1046 (2000).

79.    R. G. Knobel and A. N. Cleland, *Nature* **424**, pp 291–293 (2003).

80.    A. Naik, O. Buu, M. D. LaHaye, A. D. Armour, A. A. Clerk, M. P. Blencowe and K. C. Schwab, *Nature*, **443**, pp 193–196 (2006).

81.    A. Aasime, G. Johansson, G. Wendin R. J. Schoelkopf and P. Delsing, *Phys. Rev. Lett.*, **86**(15), pp 3376–3379 (2001).

82.    T. M. Buehler, D. J. Reilly, R. Brenner, A. R. Hamilton, A. S. Dzurak and R. G. Clark, *Appl. Phys. Lett.*, **82**(4), pp 577–579 (2003).

83.    D. Williams, J. Gorman and D. G. Hasko, Si Nano. *Workshop*, pp 21–22 (2006).

84.    F. H. L. Koppens *et al.*, *Nature*, **442**, pp 766–771, (2006).

# 11

# Electronic Properties of Organic Monolayers and Molecular Devices

Dominique Vuillaume

Molecular Nanostructures & Devices group,
Institute for Electronics, Microelectronics and Nanotechnologies,
CNRS & university of Lille
BP60069, avenue Poincaré, F-59652cedex, Villeneuve d'Ascqn France

dominique.vuillaume@iemn.univ-lille1.fr

........................................

We propose a review on the electronic properties of devices comprising from a single to a monolayer of organic molecules. After a brief description on how to make such molecular devices, the properties of several functional devices (e.g. tunnel barrier, diode, switch, memory, etc.) are presented and discussed. Actual device performances, limitations, challenges and perspectives are highlighted for each of these molecular devices.

## 1.     Introduction

Since the first measurement of electron tunneling through an organic monolayer in 1971,[1] and the *gedanken* experiment of a molecular current rectifying diode in 1974,[2] molecular-scale electronics have attracted a growing interest, both for basic science at the nanoscale and for possible applications in nano-electronics. In the first case, molecules are quantum object by nature and their properties can be tailored by chemistry opening avenues for new experiments. In the second case, molecule-based devices are envisioned to complement silicon devices by providing new functions or already existing functions at a simpler process level and at a lower cost by virtue of their self-organization capabilities, moreover, they are not bound to von Neuman architecture and this may open the way to other architectural paradigms.

Molecular electronics, i.e. the information processing at the molecular-scale, becomes more and more investigated and envisioned as a promising candidate for the nanoelectronics of the future. One definition is "information processing using photo-, electro-, iono-, magneto-, thermo-, mechanico- or chemio-active effects at the scale of structurally and functionally organized molecular architectures" (adapted from Ref. 3). In the following, we will consider devices based on organic molecules with size ranging from a single molecule to a monolayer. This definition excludes devices based on thicker organic materials referred to as organic electronics. Two works paved the foundation of this molecular-scale electronics field. In 1971, Mann and Kuhn were the first to demonstrate tunneling transport through a monolayer of aliphatic chains.[1] In 1974, Aviram and Ratner theoretically proposed the concept of a molecular rectifying diode where an acceptor-bridge-donor (A-b-D) molecule can play the same role as a semiconductor p-n junction.[2] Since that, many groups have reported on the electrical properties of molecular-scale devices from single molecules to monolayers.

After a brief overview of the nanofabrication of molecular devices, we review in this chapter, the electronic properties of several basic devices, from simple molecules such as molecular tunnel junctions and molecular wires, to more complex ones such as molecular rectifying diodes, molecular switches and memories.


## 2.    Nanofabrication for Molecular Devices

To measure the electronic transport through an organic monolayer, we need a test device as simple as possible. The generic device is a metal/monolayer/metal or metal/molecules/metal (MmM) junction (for simplicity, we will always use this term and acronym throughout the paper even if the metal electrode is replaced by a semiconductor). Organic monolayers and sub-monolayers (down to single molecules) are usually deposited on the electrodes by chemical reactions in solution or in gas phase using molecules of interest bearing a functional moiety at the ends which is chemically reactive to the considered solid surface (for instance, thiol group on metal surfaces such as Au, silane group on oxidized surfaces,

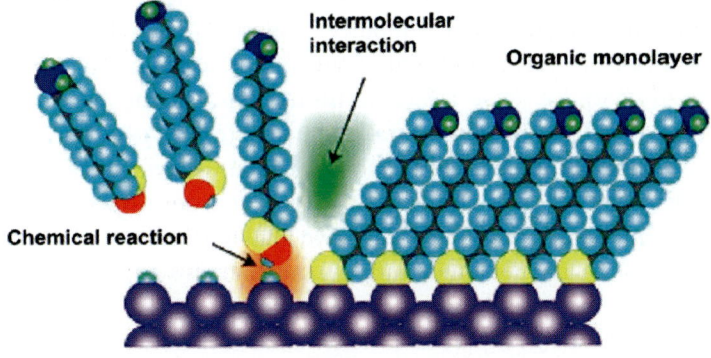

**Fig. 1.** A schematic description of the formation of an organic monolayer on a solid substrate, showing the chemical reaction between a functionalized end of the molecule and the substrate, and the interactions between adjacent molecules (from www.mtl.kyoto.u.ac.jp/groups/sugimura-g/index-E.html).

etc ...) – Fig. 1. However, Langmuir-Blodgett (LB) monolayers have also been used for device applications early in the 70s (see a review in a textbook[4]). Some important results are, for instance, the observation of a current rectification behavior through LB monolayers of hexadecylquinolinium tricyanoquinodimethanide[5−11] and the fabrication of molecular switches based on LB monolayers of catenanes.[12,16] The second method deals with monolayers of organic molecules chemically grafted on solid substrates, also called self-assembled monolayers (SAM).[4] Many reports in the literature concern SAMs of thiol terminated molecules chemisorbed on gold surfaces, and to a less extend, molecular-scale devices based on SAMs chemisorbed on semiconductors, especially silicon. Silicon is the most widely used semiconductor in microelectronics. The capability to modify its surface properties by the chemical grafting of a broad family or organic molecules (e.g. modifying the surface potential[17−19]) is the starting point for making almost any tailored surfaces useful for new and improved silicon-based devices. Between the end of the silicon road-map and the envisioned advent of fully molecular-scale electronics, there may be a role played by such hybrid-electronic devices.[20,21] The use of thiol-based SAMs on gold in molecular-scale electronics is supported by a wide range of experimental results on their growth, structural and electrical properties (see a review by F. Schreider[22]). However, SAMs on silicon and silicon dioxide surfaces were less studied and were more difficult to control. This has resulted in an irreproducible quality of these SAMs with large time-to-time and lab-to-lab variations. This feature may explain the smaller number of attempts to use these SAMs in molecular-scale electronics than for the thiol/gold system. Since the first chemisorption of alkyltrichlorosilane molecules from solution on a solid substrate (mainly oxidized silicon) introduced by Bigelow, Pickett and Zisman[23] and later developed by Maoz and Sagiv,[24] further detailed studies[25-28] have lead to a better understanding of the basic chemical and thermodynamical mechanisms of this self-assembly process. For a review on these processes, see Refs. 4 and 22.

In their pioneering work, Mann and Kuhn used a mercury drop to contact the monolayer,[1] and this technique is still used nowadays[29-32] at the laboratory level as an easy technique for a quick assessment of the electrical properties. Several types of MmM junctions have been built. The simplest structure consists of depositing the monolayer onto the bottom electrode and then evaporating a metal electrode on top of the monolayer through a masking technique. These shadow masks are fabricated from metal or silicon nitride membranes and the dimensions of the holes in the mask may range from few hundreds of $\mu$m to few tens of nanometers. Chen and coworkers[33,34] have used nanopores (about 30 nm in diameter in a silicon nitride membrane), in which a small numbers of molecules are chemisorbed to fabricate these MmM junctions. From $\sim 10^{10}$ to $\sim 10^2$ molecules can be measured in parallel with these devices. The critical point deals with the difficult problem of making a reliable metal contact on top of an organic monolayer. Several studies[35-40] have analyzed (by X-ray photoelectron spectroscopy, infra-red spectroscopy, ...) the interaction (bond insertion, complexation ...) between the evaporated atoms and the organic molecules in the SAM. When the metal atoms are strongly reactive with the end-groups of the molecules (e.g. Al with COOH or OH groups, Ti with COOCH$_3$, OH or CN groups ....),[35-40] a chemical reaction occurs forming a molecular overlayer on top of the monolayer. This overlayer made of organometallic complexes or metal oxides may perturb the electronic coupling between the metal and the molecule, leading, for instance, to partial or total Fermi-level pinning at the interface.[41] In some cases, if the metal chemically reacts with the end-group of the molecule (e.g. Au on thiol-terminated molecules), this overlayer may further prevent the diffusion of metal atoms into the organic monolayer.[42] The metal/organic interface interactions (e.g. interface dipole, charge transfer, ...) are very critical and they have strong impacts on the electrical properties of the molecular devices. Some reviews are given in

Refs. 43 and 44. If the metal atoms are not too reactive (e.g. Al with $CH_3$ or $OCH_3$ ...),[35-40] they can penetrate into the organic monolayer, diffusing to the bottom interface where they can eventually form an adlayer between this electrode and the monolayer (in addition to metallic filamentary short circuits). In a practical way for device application using organic monolayers, the metal evaporation is generally performed onto a cooled substrate ($\sim$100 K). It is also possible to intercalate blocking baffles on the direct path between the crucible and the sample, or/and to introduce a small residual pressure of inert gas in the vacuum chamber of the evaporator.[10,11,45] These techniques allow reducing the energy of the metal atoms arriving on the monolayer surface, thus reducing the damages.

To avoid these problems, alternative and soft metal deposition techniques were developed. One called nanotransfer printing (nTP), has been described and demonstrated.[46] Nanotransfer printing is based on soft lithographic techniques used to print patterns with nanometric resolution on solid substrates.[47] The principle is briefly described as follows. Gold electrodes are deposited by evaporation onto an elastomeric stamp and then transferred by mechanical contact onto a thiol-functionalized SAM. Transfer of gold is based on the affinity of this metal for thiol function –SH forming a chemical bond Au–S. Loo *et al.*[46] have used the nTP technique to deposit gold electrodes on alkane dithiol molecules self-assembled on gold or GaAs substrates. Nanotransfer printing of gold electrodes was also deposited onto oxidized silicon surface covered by a monolayer of thiol-terminated alkylsilane molecules.[48,49] Soft depositions of pre-formed metal electrodes, e.g. lift-off float-on (LOFO),[50] have also been developped. Recently, another solution has been proposed in which a thin conducting polymer layer has been intercalated as a buffer layer between the organic monolayer and the evaporated metal electrode.[51] It was also reported to use metallic electrode made of a 2D network of carbone nanotubes.[52] Finally, another solution to avoid problems with metal evaporation is to cover a metal wire (about 10 $\mu$m in diameter) with a SAM and then to bring this wire in contact with another wire (crossing each other) using the Laplace force.[53,54] About $10^3$ molecules can be contacted by this way.

At the nanometer-scale, the top electrode can also be a STM tip. The properties of a very small number of molecules (few tens down to a single molecule) can be measured. If one assumes that an intimate contact is provided by the chemical grafting (in case of a SAM) at one end of the molecules on the bottom electrode, the drawback of these STM experiments is the fact that the electrical "contact" at the other end occurs through the air-gap between the SAM surface and the STM tip (or vacuum in case of an UHV-STM). This leads to a difficult estimate of the true conductance of the molecules, while possible through a careful data analysis and choice of experimental conditions.[55,56] Recently, some groups have used a conducting-atomic force microscope (C-AFM) as the upper electrode.[57-59] In that case, the metal-coated tip is gently brought into a mechanical contact with the monolayer surface (this is monitored by the feed-back loop of the AFM apparatus) while an external circuit is used to measure the current-voltage curves. The advantage over the STM is twofold, (i) tip-surface position control and current probing are physically separated (while the same current in the STM is used to control the tip position and to probe the electronic transport properties), (ii) under certain conditions, the molecules may be also chemically bounded to the C-AFM tip at the mechanical contact.[60] The critical point of C-AFM experiments is certainly the very sensitive control of the tip load to avoid excessive pressure on the molecules[61] (which may modify the molecule conformation and thus its electronic transport properties, or even can pierce the monolayer). On the other hand, the capability to apply a controlled mechanical pressure on a molecule to change its conformation is a powerful tool to study the relationship between conformation and electronic transport.[62] A significant improvement has been demonstrated by Xu and Tao[63] to measure the conductance of a single molecule by repeatedly forming few thousands of Au-molecule-Au junctions. This technique is a STM-based break junction, in which molecular junctions are repeatedly formed by moving back and forth the STM tip into and out of contact with a gold surface in a solution containing the molecules of interest. A few molecules, bearing two chemical groups at their ends, can bridge the nano-gap formed when moving back the tip from the surface. Due to the large number of measurements, this technique provides statistical analysis of the conductance data. This technique has been recently used to obtain new insights on the electronic transport through molecular junctions, e.g. on the analysis of the variability of the conductance,[64,65] on the role of the chemical link between the molecule and the metal electrode[65,66] (for instance, it has been shown that the amine group gives a better defined conductance than thiol[65]), on the influence of the atomic configuration of the chemical link.[67] Changes in the electrical conductance of a single molecule as function of a chemical substitution[68] and a conformational change were also evidenced.[69]

The second type of MmM junctions uses a "planar" configuration (two electrodes on the same surface). The advantage over a vertical structure is the possibility to easily add a third gate electrode (3-terminal device) using a bottom gate transistor configuration. The difficulties are (i) to make these electrodes with a nanometer-scale separation; (ii) to deposit molecules into these nano-gaps. Alternatively, if the monolayer is deposited first onto a suitable substrate, it would be very hard to pattern, with a nanometer-scale resolution, the electrodes on top of it. The monolayers have to withstand, without damage, a complete electron-beam patterning process for instance. This has been proved possible for SAMs of alkyl chains[70,71] and alkyl chain functionalized by $\pi$-conjugated oligomers[72] used in nano-scale (15–100 nm) devices. However, recently developed soft-lithographies (micro-imprint contact ...) can be used to pattern organic monolayers or to pattern electrodes on these monolayers.[47] Nowadays, 30 nm width nano-gaps are routinely fabricated by e-beam lithography and 5 nm width nano-gaps are attainable with a lower yield (a few tens %).[73–75] However, these widths are still too large compared to the typical molecule length of 1–3 nm. The smallest nanogaps ever fabricated have a width of about 1 nm. A metal nanowire is e-beam fabricated and a small gap is created by electromigration when a sufficiently high current density is passing through the nanowire.[76] These gold nanogaps where then filled with few molecules (bearing a thiol group at each ends) and Coulomb blockade and Kondo effects were observed in these molecular devices.[77,78] A second approach is to start by making two electrodes spaced by about 50–60 nm, then to gradually fill the gap by electrodeposition until a gap of few nanometers has been reached.[79–81] Recently, carbone nanotubes (CNT) have been used as electrodes separated by a nano-gap (<10 nm).[82] The nano-gap is obtain by a precise oxidation cutting of the CNT, and the two facing CNT ends which are now terminated by carboxylic acids, are covalently bridged by molecules of adapted length derivatized with amine groups at the two ends. It is also possible to functionalize the molecule backbone for further chemical reactions allowing the electrical detection of molecular and biological reactions at the molecule-scale.[82,83] Another approach is to use a breaking junction, bridged by few dithiol-terminated molecules. Reed and coworkers[84] and Kergueris and coworkers[85] have used these breaking junctions to fabricate and to study some MmM junctions based on dithiolbenzene and bisthiolterthiophene, respectively, and this technique was further used with others short oligomers.[86,87] However, these MmM breaking junctions are not stable over a very long period of time (no more than 20–30 min) while the vertical MmM junctions and the "planar" ones based on nanofabricated nano-gaps are stable over months. Weber *et al.* reported some improvements allowing stable MmM breaking junction measurements at low temperature.[88,89] Finally, we mention that Au nanoparticles (NP) can be used to connect a few molecules, these NP (tens of nm in diameter) being themselves deposited between electrodes or contacted with a STM.[60,90,91] Microspheres metallized by Ni/Au can also be magnetically trapped between micro-lithographically patterned electrodes covered by a monolayer of molecules forming two molecular junctions in series.[91] These approaches allow measuring a small number of molecules and avoid the difficult fabrication of few nm size gaps.

To conclude this section, many technological solutions are available to measure the electronic transport properties of molecular monolayers with lateral extension from few molecules to $\sim 10^{10}$ (Fig. 2). A comparison between electrical measurements at the molecular-scale and those on macroscopic devices will be helpful to understand the effect of

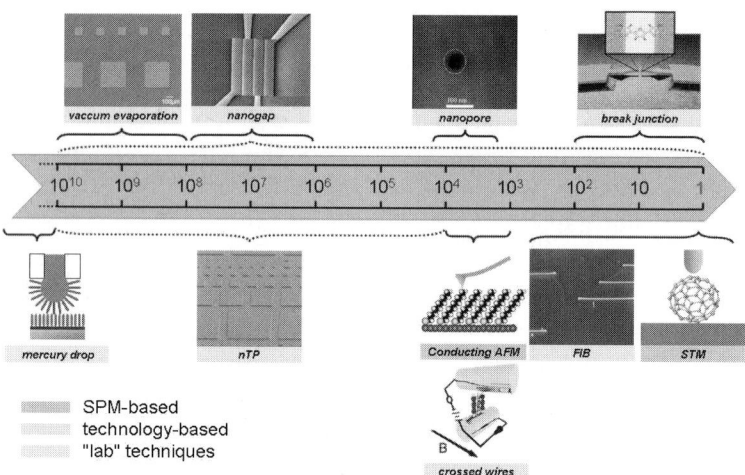

Fig. 2. A schematic overview of the different test-beds used to electrically contact organic molecules. The scale gives the approximate number of molecules contacted from monolayer (left) to single molecule (right). The techniques are (from left to right, upper part of the figure) : micrometer-scale metal evaporation, nano-gap patterned by e-beam lithography, nanopores, break-junction, and (from left to right, lower part of the figure) : mercury drop, nano-transfer printing, conducting AFM, crossed wires, metal deposition by FIB, STM (courtesy of S. Lenfant, IEMN-CNRS).

intermolecular interactions on the transport properties. As a result of these various approaches for making the organic monolayers and the MmM junctions, the nature of the interfaces, and thus the electronic coupling between the molecules and the electrodes are largely depending on the experimental conditions and protocols. This feature requires a multi test-bed approach to assess the intrinsic properties of the molecular devices and not of the contacts.[92] In the following sections, we illustrate and discuss the effects of this molecule/electrode coupling on the electronic transport properties of some molecular devices.

## 3. Molecular Tunneling Barrier

It has long been recognized that a monolayer of alkyl chains sandwiched between two metal electrodes acts as a tunneling barrier. Mann and Kuhn,[1] Polymeropoulos and Sagiv[93,94] have demonstrated that the current through LB monolayers of alkyl chains follows the usual distance-dependant exponential law, $I = I_0 \exp(-\beta d)$, where d is the monolayer thickness and $\beta$ is the distance decay rate. They have found $\beta \sim 1.5$ $\text{Å}^{-1}$. More recently, we found[95] $\beta \sim 0.7$–$0.8$ $\text{Å}^{-1}$ for $n^+$-Si/native $SiO_2$/SAM of alkyl-1-enyl trichlorosilane/metal (Au or Al) junctions and Whitesides's group[29] found $\beta \sim 0.9$ $\text{Å}^{-1}$ for Hg/SAM of alkylthiol/Ag junctions. All these experiments were done with macroscopic-size electrodes. Data taken for alkanethiols in a nanopore junction gave $\sim 0.8$ $\text{Å}^{-1}$.[96] Recently, C-AFM experiments were also done addressing the properties of a small number of molecules. Again, a tunneling law was observed with $\beta \sim 0.9$–$1.4$ $\text{Å}^{-1}$ for Au/SAM of alkylthiols/Au-covered AFM tip junctions.[57,58,97,98] A quite smaller value ($\beta \sim 0.5$ $\text{Å}^{-1}$) was reported for Au/SAM of alkyldithiol/Au-covered AFM tip junctions,[99] but another work reported no significant variation of $\beta$ between alkanethiols and alkanedithiols, but only a contact resistance 1 or 2 decades lower for the alkanedithiols. A more complete review of these data and others is given in Ref. 100. The $\beta$ value is related to the tunneling barrier height ($\Delta$) at the molecule/electrode interface and to the effective mass ($m^*$) of carriers in the monolayer, $\beta = \alpha(m^*/m_0)^{1/2}\Delta^{1/2}$, with $m_0$ the rest mass of the electron and $\alpha = 4\pi(2m_0e)^{1/2}/h = 10.25$ $\text{eV}^{-1/2}$ $\text{nm}^{-1}$ (e is the electron charge and h the Planck constant). The tunneling barrier height may be measured independently by internal photoemission experiment (IPE)[101] where carriers in one of the electrodes are photoexcited over the tunneling barrier and collected at the other electrode (under a small applied dc bias).

Threshold energy of the exciting photons allows the measurement of $\Delta$. We have found an electron tunneling barrier of about 4.3–4.5 eV at the silicon/native $SiO_2$/SAM and aluminum/SAM interfaces in the case of densely packed, well-ordered, SAMs of alkyl chains,[102] a larger value than $\sim$1.4 to 3 eV found in other experiments on LB monolayers and alkylthiol SAM on Au.[1,29,94,96] This high value ($\sim$4.5 eV) is in agreement with theoretical calculations.[106] For the same alkyl chains directly chemisorbed on Si (no native oxide), lower values have been reported from a combination of electrical ($\sim$1–1.5 eV) and UPS/IPES (2.5–3.5 eV) experiments.[104,105] The discrepancy between electrical and spectroscopy data is due to the fact that charge carrier transport is dominated by the presence of interface states localized between the molecular HOMO (highest occupied molecular orbital) and LUMO (lowest unoccupied molecular orbital) and the Si band edges.[105]

These puzzling data may be rationalized if we consider the nature of the molecule/electrode coupling. Figure 3 shows some of these data in a $\beta - \Delta$ plot. The smallest $\beta$ and $\Delta$ values are obtained for a good or "intimate"

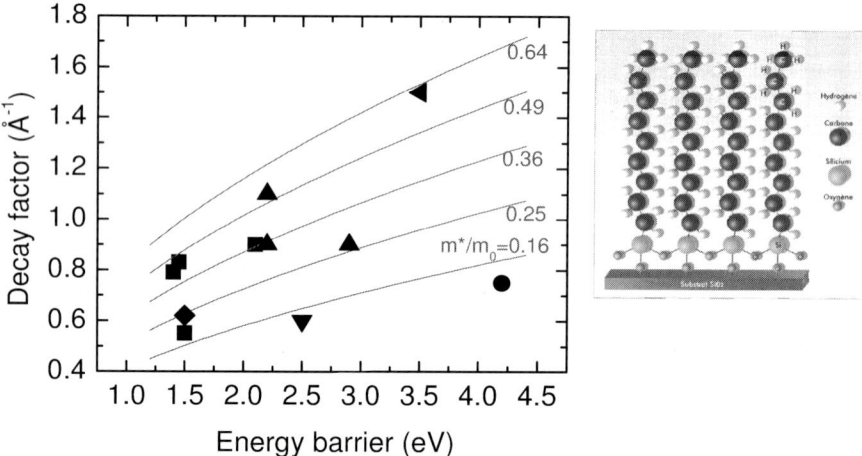

Fig. 3. Left: Tunnel decay factor — energy barrier plot for several molecular tunnel junctions: (■) metal-alkylthiol or dithiol-metal (Au or Hg) junctions,[29,60,96] (▲) Au-alkylthiol or dithiol-Au C-AFM junctions,[58,59,98,103] (◄) LB monolayer,[1] (♦) Si-alkyl-Hg junction,[104,105] (●) Si-native $SiO_2$-alkylsilane-Al junction,[102,106] (▼) Si-native $SiO_2$-mercaptopropyltrimethoxysilane-Au junction.[42] Lines are calculated according to the classical equation (see text) for different values of the effective mass. Right: schematic drawing of alkylsilane monolayer grafted on silicon.

coupling at both the two electrodes. This is the case for SAM of alkyldithiols chemisorbed at the two electrodes,[29,99] and for SAM chemisorbed at one end and contacted at the other one by an evaporated metal.[95] This is also the case for alkyl chains directly attached to Si without native oxide between the substrate and the molecules.[104,105] The largest values are obtained when at least one coupling is weak, as it is the case for physisorbed LB monolayers[1,93,94] and SAM mechanically contacted by C-AFM tip[57,58] or chemisorbed on the native oxide of the Si substrate.[102,106] In this latter case, the top metal electrode (Al or Au) was also weakly coupled with the $CH_3$-terminated molecules. The tunnel barrier height is lowered (2.2–2.5 eV)[42] if Au is used as the top electrode on thiol-terminated SAM of alkyl chains still grafted on naturally oxidized Si, probably due to a better molecule/metal coupling through the S-Au chemical link. This feature reveals that the nature of the molecule/electrode coupling strongly changes the electronic properties of the molecules. The HOMO-LUMO gap of the molecule, and therefore the tunnel barrier height, may be reduced by several eV for a chemisorbed molecule on metal compared to the gas phase molecule.[107] Charge transfer and interface dipole also move the position of the molecular orbitals with respect to Fermi energy of the electrodes. A review on these phenomena is given in Refs. 43 and 44. The molecule/electrode contact is a key parameter in the overall transport properties of the MmM junctions. It was demonstrated that the conductance of a MmM junction is increased when the molecule is chemisorbed at its two ends (via a thiol link on gold for instance) compared to the situation when only one end is chemically connected to one electrode. An increase by a factor $10^3$ was observed for a monolayer of octadecanedithiol molecules as compared to a monolayer of octadecanethiol.[60,103] Another experimental evidence is given by a comparison of two systems (Hg-S-alkyl and Hg/alkyl) where the sulfur linked molecules showed a better electrical conductivity.[31]

Finally, these tunnel junctions are also good prototypal devices to study more detailed phenomena such as: electron — molecular vibration coupling using inelastic electron tunnel spectroscopy (IETS),[108–113] current-induced local heating in a molecular junction,[114] dynamical charge fluctuations using noise measurements[115] and spin-polarized transport.[116,117] Beyond the first results, more of such experiments are now required to achieve a good agreement between a variety of different results, as well as with theoretical predictions. These approaches open very interesting pathway toward a better understanding of electronic transport in molecular junctions.

## 4.  Molecular Semiconducting Wire

Contrary to the case of fully saturated alkyl chains, short oligomers of $\pi$ conjugated molecules are considered as the prototype of molecular semiconducting wires. At low bias, when the LUMO and HOMO of the molecules are not in resonance within the Fermi energy window opened between the two electrodes by the applied bias, the conduction is still dominated by tunneling. However, the decay factor $\beta$ is lower than in the case of alkyl chains (see *supra*), typically $\beta \sim 0.2$ to 0.6 Å$^{-1}$. This is related to the lower HOMO-LUMO gap of the $\pi$-conjugated molecules ($\sim 2$–4 eV, typically, against 8–9 eV for alkyl chains), and therefore to a lower energy barrier for charge injections. A detailed comparison of transport properties between saturated and $\pi$-conjugated molecules is given in Ref. 100 Bumm and coworkers[118] have studied the conductivity of prototypes of molecular wires. A few molecules of di(phenylene-ethynylene)benzenethiolate were inserted in a SAM of dodecanethiols (which are insulating molecules), and the difference in conductivity was investigated using the tip of a STM. With a STM working at a constant current, the tip is retracted when passing over a more conducting molecule than the surrounding matrix of alkyl chains. Thus the apparent amplitude height in the STM image is directly related to the conducting behavior of these molecules. Patrone and coworkers[119,120] have repeated these experiments for thiolterthiophene molecules, another prototype of molecular wires (Fig. 4). However, as explained *supra*, the drawback of these experiments is the fact that the electrical "contact" at the upper end of the molecules occurs through the air-gap between the SAM surface and the STM tip (or vacuum in case of an UHV-STM). This leads to a difficult estimation of the true conductance of the molecules. Reed *et al.*,[84] Kergueris *et al.*,[85] Weber *et al.*[86–88] have used breaking junctions to fabricate and to study some MmM junctions based on short conjugated oligomers. The current-voltage curves are strongly non-linear with steps (peaks in the first derivative) corresponding to resonant charge carrier transfer through the molecular orbitals (MO) of the molecules. The measured conductance corresponds to the conductance through the molecules and the conductance of the molecule/electrode contact. Thus, the influence of the chemical link between the molecules and the electrode is of a prime importance. A change from an asymmetric to a symmetric current-tension (with respect to the bias polarity) curve was observed when comparing MmM junctions of SAMs of monothiolate and dithiolate oligo (phenylene

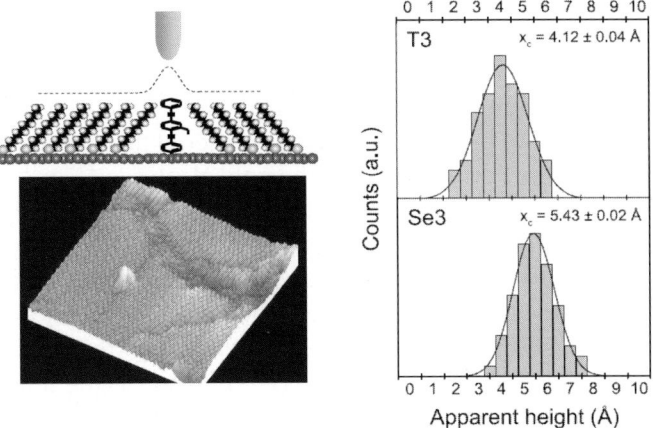

Fig. 4. *Top-left*, schematic view of a mixed monolayer where a few "conducting" molecules (dithiol-terthiophene) are intercalated into "insulating" ones (alkanethiol) used for STM measurements. *Bottom-left*, STM image (28 nm × 28 nm). The bump in the image is due to a higher current when the tip is passing over the more conducting terthiophene molecules. The background corresponds to the tunneling current through the alkanethiols.[119,120] Right, comparison of the apparent height (which is related to the molecular conductance) measured on the STM images for the S- and Se-linked terthiophene molecules — T3 and Se3, respectively (histogram taken from many measurements).

ethynylene) molecules.[54] The current increases by about a factor 10 when a sulfur atom attaches the molecule to the gold electrode compared to a mechanical contact. Today, the thiol group is the most used link to gold. However, theoretical calculations have recently predicted that selenium (Se) and tellurium (Te) are better links than sulfur (S) for the electronic transport through MmM junctions based on phenyl-based molecular wires.[121,123] This was recently demonstrated in a series of experiments using SAMs made of bisthiol- and biselenol-terthiophene molecules inserted in a dodecanethiol matrix.[119,123] Using both STM in ambient air and UHV-STM, the apparent height of the molecular wires above the dodecanethiol matrix (as in the Bumm *et al.* work quoted above[118]) is used to compare the electron transfer through the terthiophene molecule linked to the gold surface by S or Se atoms. Whatever the experimental conditions (air or UHV, tip-substrate bias, tunnel current set-point), the Se-linked molecules always appear higher in the STM images than the ones with a S linker. This feature directly demonstrates that a Se atom provides a better electron coupling between the gold electrode and the molecular wire than a S atom does (at least for the terthiophene molecule used in these experiments). From UPS experiments, this was attributed to a reduction of the energy offset between the highest occupied molecular orbital (HOMO) of the molecules (these molecules are mainly a better hole transport material than an electron transport material) and the Fermi energy of the gold electrode.[119,120] This offset reduction is in agreement with theory.[121,122] Similarly, comparing the electron transport through SAMs of alkylthiols and alkyl-isonitriles (C-AFM measurements), it was established that the contact resistance for the Au/CN link is about 10% lower than for the Au/S interface.[103] Further experiments have shown that : (i) amine group (NH$_2$) give better controlled conductance variability than thiol (SH) and isonitrile (CN)[65] and (ii) the interface contact resistance is lower for amine than for thiol.[66] Further experiments are now required to deeply investigate all possible anchoring atom/electrode couples (S, Se, Te, CN, COOH etc ..., on one side and Au, Ag Pt, Pd, for instance, on the other side) and to determine to which extent the conclusions drawn for a peculiar molecule are valid for any other ones. With all these data on hands, one would optimize the design of future devices for molecular electronics. Electron-molecular vibronic coupling in short semiconducting oligomers has also been recently studied by IETS[108,113] as for alkane molecules, as well as thermoelectricity in these molecular junctions.[124] In this latter case, the Seebeck coefficient of the single molecules has been determined, as well as a clear evidence of hole transport through the junctions. This result allows beginning to explore thermoelectric energy conversion at the molecular-scale.

## 5.    Molecular Rectifying Diode

A basic molecular device is the electrical current rectifier based on suitably engineered molecules. This molecular diode is the organic counterpart of the semiconductor p-n junction. At the origin of this idea, Aviram and Ratner (AR) proposed in 1974 to use D-$\sigma$-A molecules where D and A are respectively electron donor and acceptor, and $\sigma$ is a covalent "sigma" bridge.[2] Several molecular rectifying diodes were synthesized based on this AR paradigm, with donor and acceptor moieties linked by a short $\sigma$ or even $\pi$ bridge.[5,7−11,125,126] This D-b-A (b = bridge) group is also $\omega$-substituted by an alkyl chain to allow a monolayer formation by the Langmuir-Blodgett (LB) method and this

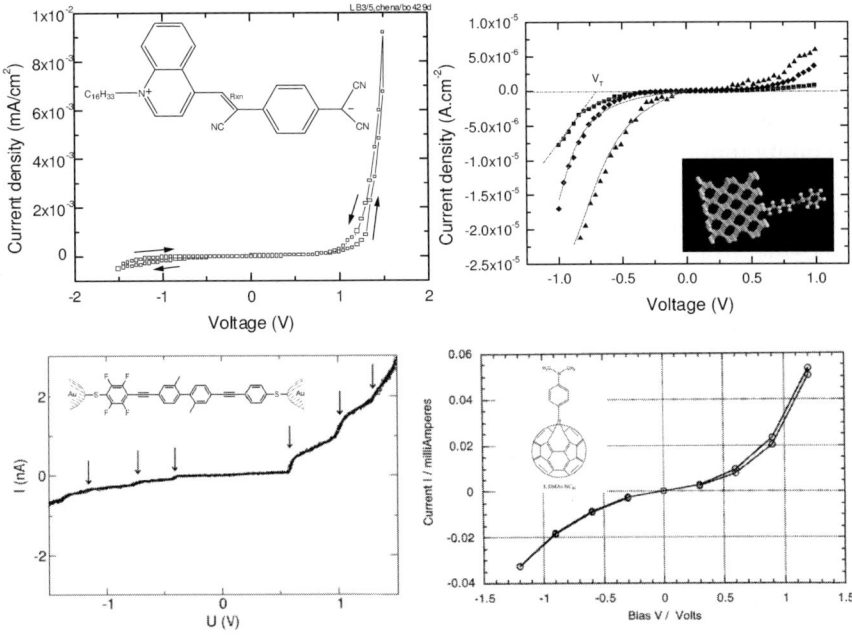

Fig. 5. Typical current-voltage characteristics of some molecular rectifying diodes. From top-left to bottom-right: LB monolayer of D-π-A molecules between metal electrodes, from Refs. 8 and 9, σ − π molecule grafted on Si, from Refs. 41 and 131, D-A molecule inserted in a break-junction (at 30 K in this latter case), from Ref. 88, and D-b-A LB monolayers, from Ref. 12.

LB monolayer is then sandwiched in a metal/monolayer/metal junction. The first experimental results were obtained with the hexadecylquinolinium tricyanoquinodimethanide molecule ($C_{16}H_{33}$-Q-3CNQ for short) — Fig. 5.[5,7−11] However, the chemical synthesis of this molecule was not obvious with several routes leading to erratic and unreliable results. A more reliable synthesis was reported with a yield of 59%.[8] More recently, other D-b-A molecules have been synthesized and tested[127,128] showing rectification with a ratio up to ~2 × $10^4$. We can also mention some other approaches using D-A diblock co-oligomers[129] or CNT asymmetrically functionalized by D and A moieties at their ends[130] with a rectification ratio of ~$10^3$ in this latter case. Even if these results represent an important progress to achieve molecular electronics, the physical mechanism responsible for the rectification is not clear. One critical issue is to know if the AR model can be applied to $C_{16}H_{33}$-Q-3CNQ because it is a D-π-A molecule,[8] and due to the

$\pi$ bridge, the HOMO and LUMO may be more delocalized than expected in the AR model. On the theoretical side, these molecular diodes are complex systems, characterized by large and inhomogeneous electric fields, which result from the molecular dipoles in the monolayer, the applied bias and the screening induced by the molecules themselves and the metallic electrodes. A theoretical treatment of these effects requires a self-consistent resolution of the quantum mechanical problem, including the effect of the applied bias on the electronic structure. Combining *ab initio* and semi-empirical calculations, it was shown[132] that the direction of easy current flow (rectification current) depends not only on the placement of the HOMO and LUMO relative to the Fermi levels of the metal electrodes before bias is applied, but also on the shift induced by the applied bias: this situation is more complex than the AR mechanism, and can provide a rectification current in an opposite direction. The electrical rectification results from the asymmetric profile of the electrostatic potential across the system.[132,133] On other words, this means that the molecule is more strongly coupled with one electrode than with the other one (more closer to one of the electrodes due to the presence of the alkyl chain). The alkyl tail in the $C_{16}H_{33}$-Q-3CNQ molecule plays an important role in this asymmetry, and it was predicted[132] a symmetric current-voltage curve in the case of molecules without the alkyl chain. This asymmetry effect was further theoretically studied more extensively.[134,135] Generally speaking, any asymmetrical coupling of the molecules with the electrodes or any asymmetry in the molecule will result in a rectification effect[88,136] — Fig. 5. This emphasizes the importance of the electrostatic potential profile in a molecular system and suggests that this profile can be chemically engineered to build new devices. For instance, based on these considerations, we have recently reported an experimental demonstration of a simplified and more robust synthesis of a molecular rectifier with only one donor group and an alkyl spacer chain.[41,131] We have used a sequential self-assembly process (chemisorption directly from solution) on silicon substrates. We have analyzed the properties of these molecular devices as a function of the alkyl chain length and for ten different donor groups. We have obtained rectification ratios up to 37 (Fig. 5). We have shown that rectification occurs from resonance through the HOMO of the $\pi$-group in good agreement with our calculations and internal photoemission spectroscopy. However, improvements are still required to suppress Fermi-level pinning at the molecule/metal interface[41] and to allow a clear design and tuning of the electrical behavior of the molecular diode through the right choice of the chemical nature of the molecule. This approach will allows us to fabricate molecular rectifying diodes compatible with silicon nanotechnologies for future hybrid circuitries. Finally, more efforts have been also put forward to design and synthesis new D-b-A molecules not affected by the presence of an asymmetric alkyl chain (see Fig. 5 for one example).[127,128,137]

## 6.    Molecular Switches and Memories

Molecular switches and memories were also suggested at the early stage of the molecular electronics history.[138−140] We generally distinguish three approaches called "conformational memory", "charge-based memory" and 'RTD-based memory' (RTD is resonant tunneling diode). The first one relies on the idea to store a data bit on two bistable conformers of a molecule; the second on different redox states and the third on a negative differential resistance (NDR) due to resonant tunneling through molecular orbitals.

### *6.1.    Conformational memory*

One of the most interesting possibilities for molecular electronics is to take advantage of the soft nature of organic molecules. Upon a given excitation, molecules can undergo conformational changes. If two different conformations are associated with two different conductivity levels of the molecule, this effect can be used to make molecular switches and memories. Such an effect is expected in $\pi$-conjugated oligomers used as molecular wires, if one of the monomer is twisted away from a planar conformation of the molecule.[69] Twisting one monomer breaks the conjugation along the backbone, thus reducing the charge transfer efficiency along the molecule. This has been experimentally observed for a small molecular wire where the central unit was substituted with redox moieties. With the nanopore configuration to fabricate the MmM junction, Chen and coworkers[33,34] have observed that molecules with a nitroamine redox center (2′-amino-4,4′-di(ethynylphenyl)-5′-nitro-1-benzenethiol) exhibit a negative differential resistance behavior. In other words, they have observed that for a certain voltage range (typically between 1.5 and 2.2 V) applied on the MmM junction, the conductivity of the junction increased by a factor $10^3$ (At 60 K, while the on/off ratio dropped to 1 at about 140 K. Other molecules with some changes of the redox moieties have exhibited on/off ratio of about 1.5 at RT[34]). They have also reported the feasibility of molecular random access memory cell using these molecules.[141] The switching behavior of these

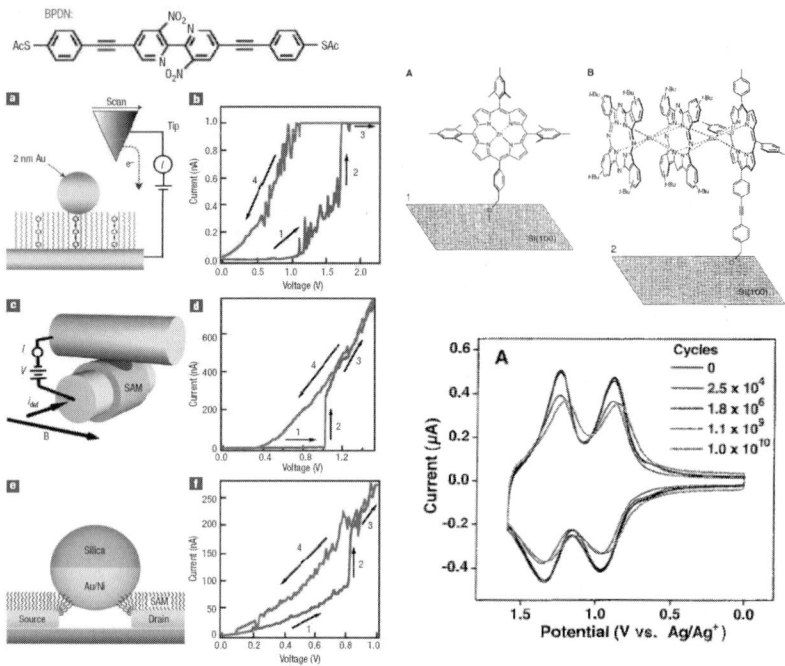

Fig. 6. *Left*: Current-voltage characteristics of bipyridyl-dinitro oligophenylene-ethynylene dithiol connected by Au electrodes using different test-beds, from Ref. 92. Au nanoparticle with STM, crossed-wires put in contact by the Lorentz force and Ni/Au metallized microsphere used as a magnetic bead junction. These experiments demonstrate a clear bias-induced switching behavior, while with a large variability. *Right*: Typical redox molecules (porphyrin derivatives) attached to a silicon substrate used in a charge-based molecular memory device and its electrical response as a function of the number of write/erase cycles. This electrochemical response shows 2 redox states that can be used to implement a multi-level memory, from Ref. 147.

compounds inserted in an alkanethiol SAM was also observed by STM.[142] To separate the intrinsic behavior of the molecules from the molecule/metal interface, the same types of molecules have been measured on various test-beds (Fig. 6).[92] These experiments demonstrated a clear bias-induced switching, while with a large statistical variability. However, it is not firmly established that this switching behavior is solely due to the molecules. Recently, the Lindsey's group showed that another possible mechanism is a random and temporary break in the chemical link between the molecule and the gold surface[143] and this point is still a subject of debate.

Catenane and rotaxane are a class of molecules synthesized to exhibit a bistable bahavior. In brief, these molecules are made of two parts, one allowed to move around or along the other one (e.g. a ring around a rod, two interlocked rings). These molecules adopt two different conformations depending on their redox states, changing the redox state triggers the displacement of the mobile part of the structure to minimize the total energy. This kind of molecules was used to build molecular memories. A MmM junction using a LB monolayer of these molecules mixed with phospholipid acid showed a clear electrical bistable behavior at room temperature.[13,14,144] A voltage pulse of about 1.5 – 2 V was used the switch the device from the "off" state to its "on" state. The state was read at a low bias (typically 0.1– 0.2 V). The on/off ratio was about a few tens. A pulse in reverse bias ($-1.5$ to $-2$ V) returned the device to the "off" state. Using these molecular devices, Chen and coworkers[15,16] have demonstrated a 64 bits non-volatile molecular memory cross-bar with an integration density of 6.4 Gbit/cm$^2$ (a factor $\sim 10$ larger than the state-of-the-art today's silicon memory chip). The fabrication yield of the 64 bits memory is about 85%, the data retention is about 24 h and about 50–100 write/erase cycles are possible before the collapse of the on/off ratio to 1. Recently a 160 kbit based on the same class of molecules has been reported, patterned at a 33 nm pitch ($10^{11}$ bits/cm$^2$).[145] About 25% of the tested memory points passed an on/off ratio larger than 1.5 with an average retention time of $\sim$ 1 h. However, it has also been observed that similar electrical switching behaviors can be obtained without such a class of bistable molecules (i.e. using simple alkyl chains instead of the rotaxanes).[146] The switching behavior is likely due to the formation and breaking of metallic micro-filaments introduced though the monolayer during the top metal evaporation. The presence of such filaments is not systematic (see discussion *supra*), however caution has to be taken before to definitively ascribe the memory effect as entirely due to the presence of the molecules. While having rather poor performances at the moment, these demonstrations allow us to envision the coming era of hybrid-electronics, where molecular cross-bar memories like these ones, will be addressed by multiplexer/demultiplexer and so one fabricated with standard semiconductor CMOS technologies.[15] The advantage of such molecular cross-bar memories are

(i)  a low cost,
(ii)  a very high integration density,
(iii)  a defect-tolerant architecture,

(iv) an easy post-processing onto a CMOS circuitry and

(v) a low power consumption.

For instance, it has been measured that an energy of $\sim$50 zJ (or $\sim$ 0.3 eV) is sufficient to rotate the dibutyl-phenyl side group of a single porphyrin molecule.[148] This is $\sim 10^4$ lower than the energy required to switch a state-of-the art MOSFET, and near the kTLn2 (2.8 zJ at 300 K, or 0.017 eV) thermodynamic limit.

## 6.2.  Charge-based memory

The redox-active molecules, such as mettalocene, porphyrin and triple-decker sandwich coordination compounds attached on a silicon substrate have been found to act as charge storage molecular devices.[147,149−151] The molecular memory works on the principle of charging and discharging of the molecules into different chemically reduced or oxidized (redox) states. It has been demonstrated that porphyrins

(i) offer the possibility of multibit storage at a relatively low potentials (below $\sim$ 1.6 V),

(ii) can undergo trillions of write/read/erase cycles,

(iii) exhibit charge retention times that are long enough (minutes) compared with those of semiconductor DRAM (tens of ms) and

(iv) are extremely stable under harsh conditions (400°C − 30 min) and therefore meet the processing and operating conditions required for use in hybrid molecule/silicon devices.[147]

Moreover, the same principle works with semiconducting nanowires dressed with redox molecules in a transistor configuration.[152−154] Optoelectronic memories have also been demonstrated with polymer-functionalized CNT transistors.[155,156] However, in all cases, further investigations on the search of other molecules and, understanding the factors that control parameters such as, charge transfer rate, which limit write/read times, and charge retention times, which determines refresh rates, are needed.

## 6.3.  RTD-based memory

Memory can also be implemented from RTD devices following cell architecture already used for semiconductor devices. Memory cell based on RTD can be set up with 2 RTD and 2 transistors in a cross-bar architecture.[157] The advantages compared to "resistive" and "capacitive" molecular memories are fast switching times and possible long retention times. RTD devices are characterized by a NDR behavior in their current-voltage curves, however a NDR may be also induced by other physical phenomena such as conformational changes already discussed *supra*. The principle of a RTD molecular device is similar to that of his solid state counter-part (a potential well separated of the electrodes by two tunnel barriers). In the molecular analogue, the barriers should consist of aliphatic chains (of variable length) and the well should be made up of a short conjugated oligomer. Even if NDR behavior has been observed from STM results on single molecule attached to Si[158] and has been ascribed to resonance through the molecular orbitals in agreement with a theoretical result,[159] this interpretation has been ruled out both experimentally[160] and theoretically.[161] The exact origin of the molecular NDR behavior is still an open question, and therefore the RTD molecular device was not yet clearly demonstrated.

## 7.    Molecular Transistor

A true transistor effect (i.e. the current through 2 terminals of the device controlled by the signal applied on a third terminal) embedded in a single three-terminal molecule (e.g. a star-shaped molecule) has not been yet demonstrated. Up to date, only hybrid-transistor devices have been studied. The typical configuration consists of a single molecule or an ensemble of molecules (monolayer) connected between two source and drain electrodes separated by a nanometer-scale gap, separated from an underneath gate electrode by a thin dielectric film — Fig. 7. At a single molecule level (single-molecule transistor), these devices have been used to study Coulomb blockade effects and Kondo effects at very low temperature. For instance, Coulomb blockade (electron flowing one-by-one between source and drain through the molecule due to electron-electron Coulomb repulsion, the molecule acting as a quantum dot) was observed for molecules such as fullerene ($C_{60}$) and oligo-phenyl-vinylene (OPV) weakly coupled to the source-drain electrodes.[162,163] In this latter case, up to eight successive charge states of the molecule have been observed. With organo-metallic molecules bearing a transition metal, such as Cobalt terpiridynil complex and divanadium complex, Kondo resonance (formation of a bound state between a local spin on the molecule, or an island, or a quantum dot,

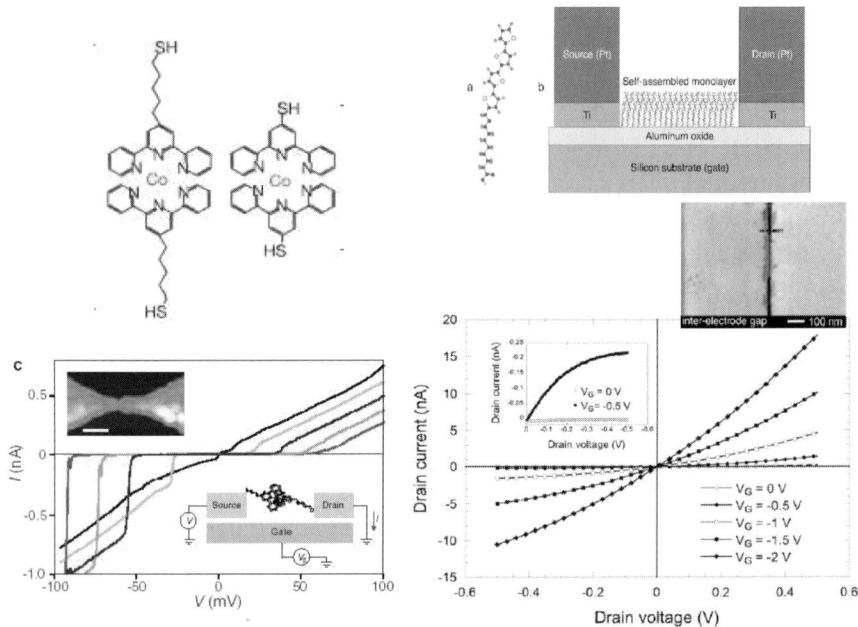

Fig. 7. *Left*: Co-terpirydinyl complex molecules, AFM image of the source-drain nanogaps (∼1–2 nm) made by electromigration, typical I-V with Coulomb blockade gaps measured at 100 mK for various gate voltage, and schematic diagram of the device; from Ref. 78 *Right*: Schematic diagram of the SAMFET and the 4T-octanoic acid molecule, SEM image of the 16 nm source-drain gap, and typical drain current-drain voltage curve for various gate voltage measured at 300 K, from Ref. 72.

and the electrons in the electrodes leading to an increase of the conductance at low bias, around zero volt) has also been observed in addition to Coulomb blockade.[77,78] Kondo resonance is observed when increasing the coupling between the molecule and the electrodes (for instance by changing the length of the insulating tethers between the metal ion and the electrodes). At a monolayer level, self-assembled monolayer field-effect transistors (SAMFET) have been demonstrated at room temperature.[72,164] The transistor effect is observed only if the source and drain length is lower than about 50 nm, that is, more or less matching the size of domains with well organized molecules in the monolayer. This is mandatory to enhance π stacking within the monolayer and to obtain a measurable drain current. SAM of tertacene,[164] terthiophene and quaterthiophene[72] derivatives have been formed in this nano-gap. Under this condition, a field effect mobility of about $3.5 \times 10^{-3}$ cm$^2$V$^{-1}$s$^{-1}$ was measured for a SAMFET made with a quaterthiophene (4T) moiety linked to a short alkyl chain (octanoic acid) grafted on a thin aluminum oxide dielectric (Fig. 7). This value is on a par with those reported for organic transistor made of thicker films of evaporated 4T ($10^{-3}$ to $10^{-2}$ cm$^2$V$^{-1}$s$^{-1}$).[72] The on/off ratio was about $2 \times 10^4$. For some devices, a clear saturation of the drain current vs. drain voltage curve has been observed, but usually, these output characteristics display a super linear behavior. This feature has been explained by a gate-induced lowering of the charge injection energy barrier at the source/organic channel interface.[70]

## 8.    Conclusion

We have described several functions and devices that have been studied at the molecular scale: tunnel barrier, molecular wire, rectifying and NDR diodes, bistable devices and memories. However, a better understanding and further improvements of their electronic properties are mandatory and need to be confirmed. These results suffer from a large dispersion and more efforts are now required to improve reproducibility and repeatability. For viable applications, more efforts are also mandatory to test the integration of molecular devices with silicon-CMOS electronics (hybrid molecular-CMOS nanoelectronics). Moreover most of these devices are 2-terminal, what's about a true/fully molecular 3-terminals device? We have also pointed out that the molecule-electrode coupling and conformation strongly modify the molecular-scale device properties. Molecular engineering (changing ligand atoms for example) may be used to improve or adjust the electrode-molecule coupling. Nevertheless, a better control of the interface (energetics and atomic conformation) is still compulsory. Beyond the study of single or isolated devices, more works towards molecular architectures and circuits are required. Up to now, mainly the ⟨cross-bar⟩ architecture has been studied. Is it sufficient? More new architectures must be explored (e.g. non von Neuman, neuronal, quantum computing, …). Open questions concern the right approaches for inter-molecular device connections and nano-to-micro connections, the interface with the outer-world, hybridation with CMOS and 3D integration.[165−168] Beyond the CMOS probably bets on non-charge based devices. Molecular devices using other state variables (e.g. spin, molecule conformation, …) to code a logic state are still challenging and exciting objectives. Finally, other reviews, current status and challenges on charge transfer on the nanoscale can be found in Refs. 169–172.

## Acknowledgments

The works done at IEMN were financially supported by CNRS, ministry of research, ANR-PNANO, IRCICA, EU-FEDER Region Nord-Pas de Calais and IFCPAR. I thank all the colleagues in the "molecular nanostructures & devices" group at IEMN and many others outside our group for fruitful collaborations.

## References

1.  B. Mann and H. Kuhn, *J. Appl. Phys.* **42**, 4398 (1971).
2.  A. Aviram and M. A. Ratner, *Chem. Phys. Lett.* **29**, 277 (1974).
3.  J.-M. Lehn, *Angew. Chem. Int. Ed. Engl.* **27**, 89 (1988).
4.  A. Ulman, *An Introduction to Ultrathin Organic Films: From Langmuir-Blodgett to Self-Assembly* (Academic Press, Boston, 1991).
5.  G. J. Ashwell, J. R. Sambles, A. S. Martin, W. G. Parker and M. Szablewski, *J. Chem. Soc., Chem. Commun.* **19**, 1374 (1990).
6.  N. J. Geddes, J. R. Sambles and A. S. Martin, *Adv. Materials for Optics and Electronics* **5**, 305 (1995).
7.  A. S. Martin, J. R. Sambles and G. J. Ashwell, *Phys. Rev. Lett.* **70**, 218 (1993).
8.  R. M. Metzger, B. Chen, U. Höpfner, M. V. Lakshmikantham, D. Vuillaume, T. Kawai, X. Wu, H. Tachibana, T. V. Hughes, H. Sakurai, J. W. Baldwin, C. Hosch, M. P. Cava, L. Brehmer and G. J. Ashwell, *J. Am. Chem. Soc.* **119**, 10455 (1997).
9.  D. Vuillaume, B. Chen and R. M. Metzger, *Langmuir* **15**, 4011 (1999).
10. T. Xu, I. R. Peterson, M. V. Lakshmikantham and R. M. Metzger, *Angew. Chem. Int. Ed. Engl.* **40**, 1749 (2001).
11. R. M. Metzger, T. Xu and I. R. Peterson, *J. Phys. Chem.* B **105**, 7280 (2001).
12. C. P. Collier, E. W. Wong, M. Belohradsky, F. M. Raymo, J. F. Stoddart, P. J. Kuekes, R. S. Williams and J. R. Heath, *Science* **285**, 391 (1999).

13. C. P. Collier, G. Mattersteig, E. W. Wong, Y. Luo, K. Beverly, J. Sampaio, F. Raymo, J. F. Stoddart and J. R. Heath, *Science* **289**, 1172 (2000).

14. A. R. Pease, J. O. Jeppesen, J. F. Stoddart, Y. Luo, C. P. Collier and J. R. Heath, *Acc. Chem. Res.* **34**, 433 (2001).

15. Y. Chen, G.-Y. Jung, D. A. A. Ohlberg, X. Li, D. R. Stewart, J. O. Jeppesen, K. A. Nielsen, J. F. Stoddart and R. S. Williams, *Nanotechnology* **14**, 462 (2003).

16. Y. Chen, D. A. A. Ohlberg, X. Li, D. R. Stewart, R. S. Williams, J. O. Jeppesen, K. A. Nielsen, J. F. Stoddart, D. L. Olynick and E. Anderson, *Appl. Phys. Lett.* **82**, 1610 (2003).

17. M. Bruening, E. Moons, D. Yaron-Marcovitch, D. Cahen, J. Libman and A. Shanzer, *J. Am. Chem. Soc.* **116**, 2972 (1994).

18. R. Cohen, S. Bastide, D. Cahen, J. Libman, A. Shanzer and Y. Rosenwaks, *Optical Mat.* **9**, 394 (1998).

19. R. Cohen, N. Zenou, D. Cahen and S. Yitzchaik, *Chem. Phys. Lett.* **279**, 270 (1997).

20. R. Compano, L. Molenkamp and D. J. Paul, (European Commission, IST Proramme, Future and Emerging Technologies, Brussel, 2000).

21. C. Joachim, J. K. Gimzewski and A. Aviram, *Nature* **408**, 541 (2000).

22. F. Schreiber, *Progress in Surf. Sci.* **65**, 151 (2000).

23. W. C. Bigelow, D. L. Pickett and W. A. Zisman, *J. Colloid Sci.* **1**, 513 (1946).

24. R. Maoz and J. Sagiv, *J. Colloid and Interface Sciences* **100**, 465 (1984).

25. J. B. Brzoska, N. Shahidzadeh and F. Rondelez, *Nature* **360**, 719 (1992).

26. J. B. Brzoska, I. Ben Azouz and F. Rondelez, *Langmuir* **10**, 4367 (1994).

27. D. L. Allara, A. N. Parikh and F. Rondelez, *Langmuir* **11**, 2357 (1995).

28. A. N. Parikh, D. L. Allara, I. Ben Azouz and F. Rondelez, *J. Phys. Chem.* **98**, 7577 (1994).

29. R. E. Holmlin, R. Haag, M. L. Chabinyc, R. F. Ismagilov, A. E. Cohen, A. Terfort, M. A. Rampi and G. M. Whitesides, *J. Am. Chem. Soc.* **123**, 5075 (2001).

30. M. A. Rampi, O. J. A. Schueller and G. M. Whitesides, *Appl. Phys. Lett.* **72**, 1781 (1998).

31. Y. Selzer, A. Salomon and D. Cahen, *J. Am. Chem. Soc.* **124**, 2886 (2002).
32. Y. Selzer, A. Salomon and D. Cahen, *J. Phys. Chem.* B **106**, 10432 (2002).
33. J. Chen, M. A. Reed, A. M. Rawlett and J. M. Tour, *Science* **286**, 1550 (1999).
34. J. Chen, W. Wang, M. A. Reed, A. M. Rawlett, D. W. Price and J. M. Tour, *Appl. Phys. Lett.* **77**, 1224 (2000).
35. G. C. Herdt and A. W. Czanderna, *J. Vac. Sci. Technol.* **A13**, 1275 (1995).
36. D. R. Jung and A. W. Czanderna, *Critical Reviews in Solid State and Materials Sciences* **191**, 1 (1994).
37. D. R. Jung, A. W. Czanderna and G. C. Herdt, *J. Vac. Sci. Technol.* A **14**, 1779 (1996).
38. G. L. Fisher, A. E. Hooper, R. L. Opila, D. L. Allara and N. Winograd, *J. Phys. Chem.* B **104**, 3267 (2000).
39. G. L. Fisher, A. V. Walker, A. E. Hooper, T. B. Tighe, K. B. Bahnck, H. T. Skriba, M. D. Reinard, B. C. Haynie, R. L. Opila, N. Winograd and D. L. Allara, *J. Am. Chem. Soc.* **124**, 5528 (2002).
40. K. Konstadinidis, P. Zhang, R. L. Opila and D. L. Allara, *Surf. Sci.* **338**, 300 (1995).
41. S. Lenfant, D. Guerin, F. Tran Van, C. Chevrot, S. Palacin, J.-P. Bourgoin, O. Bouloussa, F. Rondelez and D. Vuillaume, *J. Phys. Chem.* B **110**, 13947 (2006).
42. D. K. Aswal, S. Lenfant, D. Guerin, J. V. Yakhmi and D. Vuillaume, Small **1**, 725 (2005).
43. D. Cahen, A. Kahn and E. Umbach, *Materials Today*, 32 (2005).
44. A. Kahn, N. Koch and W. Gao, *J. Polymer Sci.: Part B: Polymer Phys.* **41**, 2529 (2003).
45. N. Okazaki and J. R. Sambles, in *International Symposium on Organic Molecular Electronics*, Nagoya, Japan), p 66 (2000).
46. Y.-L. Loo, D. V. Lang, J. A. Rogers and J. W. P. Hsu, *Nano Lett.* **3**, 913 (2003).
47. Y. Xia and G. M. Whitesides, *Angew. Chem. Int. Ed. Engl.* **37**, 550 (1998).
48. Y.-L. Loo, R. L. Willet, K. W. Baldwin and J. A. Rogers, *J. Am. Chem. Soc.* **124**, 7654 (2002).
49. D. Guerin, C. Merckling, S. Lenfant, X. Wallart and D. Vuillaume, *J. Phys. Chem.* C **111**, 7947 (2007).

50. A. Vilan and D. Cahen, *Adv. Func. Mater.* **12**, 795 (2002).
51. H. B. Akkerman, P. W. M. Blom, D. M. De Leeuw and B. De Boer, *Nature* **441**, 69 (2006).
52. J. He, B. Chen, A. K. Flatt, J. J. Stephenson, C. D. Doyle and J. M. Tour, *Nature Materials* **5**, 63 (2006).
53. J. G. Kushmerick, D. B. Holt, S. K. Pollack, M. A. Ratner, J. C. Yang, T. L. Schull, J. Naciri, M. H. Moore and R. Shashidhar, *J. Am. Chem. Soc.* **124**, 10654 (2002).
54. J. G. Kushmerick, D. B. Holt, J. C. Yang, J. Naciri, M. H. Moore and R. Shashidhar, *Phys. Rev. Lett.* **89**, 086802 (2002).
55. L. A. Bumm, J. J. Arnold, T. D. Dunbar, D. L. Allara and P. S. Weiss, *J. Phys. Chem. B* **103**, 8122 (1999).
56. A. P. Labonté, S. L. Tripp, R. Reifenberger and A. Wei, *J. Phys. Chem. B* **106**, 8721 (2002).
57. D. J. Wold and C. D. Frisbie, *J. Am. Chem. Soc.* **122**, 2970 (2000).
58. D. J. Wold and C. D. Frisbie, *J. Am. Chem. Soc.* **123**, 5549 (2001).
59. D. J. Wold, R. Haag, M. A. Rampi and C. D. Frisbie, *J. Phys. Chem. B*, 23 (2002).
60. X. D. Cui, A. Primak, X. Zarate, J. Tomfohr, O. F. Sankey, A. L. Moore, T. A. Moore, D. Gust, G. Harris and S. M. Lindsay, *Science* **294**, 571 (2001).
61. K.-A. Son, H. I. Kim and J. E. Houston, *Phys. Rev. Lett.* **86**, 5357 (2001).
62. F. Moresco, G. Meyer, K.-H. Rieder, H. Tang, A. Gourdon and C. Joachim, *Phys. Rev. Lett.* **86**, 672 (2001).
63. B. Xu and N. J. Tao, *Science* **301**, 1221 (2003).
64. J. Ulrich, D. Esrail, W. Pontius, L. Venkataraman, D. Millar and L. H. Doerrer, *J. Phys. Chem. B* **110**, 2462 (2006).
65. L. Venkataraman, J. E. Klare, I. W. Tam, C. Nuckolls, M. S. Hybertsen and M. L. Steigerwald, *Nano Lett.* **6**, 458 (2006).
66. F. Chen, X. Li, J. Hihath, Z. Huang and N. J. Tao, *J. Am. Chem. Soc.* **128**, 15874 (2006).
67. X. Li, J. He, J. Hihath, B. Xu, S. M. Lindsay and N. J. Tao, *J. Am. Chem. Soc.* **128**, 2135 (2006).
68. L. Venkataraman, Y. S. Park, A. C. Whalley, C. Nuckolls, M. S. Hybertsen and M. L. Steigerwald, *Nano Lett.* **7**, 502 (2007).
69. L. Venkataraman, J. E. Klare, C. Nuckolls, M. S. Hybertsen and M. L. Steigerwald, *Nature* **442**, 904 (2006).

70. J. Collet, O. Tharaud, A. Chapoton and D. Vuillaume, *Appl. Phys. Lett.* **76**, 1941 (2000).
71. J. Collet and D. Vuillaume, *Appl. Phys. Lett.* **73**, 2681 (1998).
72. M. Mottaghi, P. Lang, F. Rodriguez, A. Rumyantseva, A. Yassar, G. Horowitz, S. Lenfant, D. Tondelier and D. Vuillaume, *Adv. Func. Mater.* **17**, 597 (2007).
73. S. Cholet, C. Joachim, J. P. Martinez and B. Rousset, Eur. Phys. *J. Appl. Phys.* **8**, 139 (1999).
74. A. Bezryadin and C. Dekker, *J. Vac. Sci. Technol. B* **15**, 793 (1997).
75. M. A. Guillorn, D. W. Carr, R. C. Tiberio, E. Greenbaum and M. L. Simpson, *J. Vac. Sci. Technol. B* **18**, 1177 (2000).
76. H. Park, A. K. L. Lim, P. A. Alivisatos, J. Park and P. L. McEuen, *Appl. Phys. Lett.* **75**, 301 (1999).
77. W. Liang, M. P. Shores, M. Bockrath, J. R. Long and H. Park, *Nature* **417**, 725 (2002).
78. J. Park, A. N. Pasupathy, J. I. Goldsmith, C. Chang, Y. Yaish, J. R. Petta, M. Rinkoski, J. P. Sethna, H. D. Abrunas, P. L. McEuen and D. C. Ralph, *Nature* **417**, 722 (2002).
79. S. Boussaad and N. J. Tao, *Appl. Phys. Lett.* **80**, 2398 (2002).
80. C. Z. Li, H. X. He and N. J. Tao, *Appl. Phys. Lett.* **77**, 3995 (2000).
81. Y. V. Kervennic, H. S. Van der Zant, A. F. Morpurgo, L. Gurevitch and L. P. Kouwenhoven, *Appl. Phys. Lett.* **80**, 321 (2002).
82. X. Guo, J. P. Small, J. E. Klare, Y. Wang, M. S. Purewal, I. W. Tam, B. H. Hong, R. Caldwell, L. Huang, S. O'Brien, J. Yan, R. Breslow, S. J. Wind, J. Hone, P. Kim and C. Nuckolls, *Science* **311**, 356 (2006).
83. X. Guo, A. C. Whalley, J. E. Klare, L. Huang, S. O'Brien, M. L. Steigerwald and C. Nuckolls, *Nano Lett.* **7**, 1119 (2007).
84. M. A. Reed, C. Zhou, C. J. Muller, T. P. Burgin and J. M. Tour, *Science* **278**, 252 (1997).
85. C. Kergueris, J. P. Bourgoin, S. Palacin, D. Esteve, C. Urbina, M. Magoga and C. Joachim, *Phys. Rev. B* **59**, 12505 (1999).
86. J. Reichert, R. Ochs, D. Beckmann, H. B. Weber, M. Mayor and H. v. Löhneysen, *Phys. Rev. Lett.* **88**, 176804 (2002).
87. H. B. Weber, J. Reichert, F. Weigend, R. Ochs, D. Beckmann, M. Mayor, R. Ahlrichs and H. v. Löhneysen, *Chemical Physics* **281**, 113 (2002).
88. M. Elbing, R. Ochs, M. Koentopp, M. Fischer, C. von Hänisch, F. Weigend, F. Evers, H. B. Weber and M. Mayor, *Proc. Natl. Acad. Sci. USA* **102**, 8815 (2005).

89.    J. Reichert, H. B. Weber, M. Mayor and H. v. Lölneysen, *Appl. Phys. Lett.* **82**, 4137 (2003).

90.    T. Dadosh, Y. Gordin, R. Krahne, I. Khrivrich, D. Mahalu, V. Frydman, J. Sperling, A. Yacoby and I. Bar-Joseph, *Nature* **436**, 677 (2005).

91.    D. P. Long, C. H. Patterson, M. H. Moore, D. S. Seferos, G. Bazan and J. G. Kushmerick, *Appl. Phys. Lett.* **86**, 153105 (2005).

92.    A. Szuchmacher Blum, J. G. Kushmerick, D. P. Long, C. H. Patterson, J. C. Yang, J. C. Henderson, Y. Yao, J. M. Tour, R. Shashidhar and B. R. Ratna, *Nature Materials* **4**, 167 (2005).

93.    E. E. Polymeropoulos, *J. Appl. Phys.* **48**, 2404 (1977).

94.    E. E. Polymeropoulos and J. Sagiv, *J. Chem. Phys.* **69**, 1836 (1978).

95.    S. Lenfant, (PhD, Univ. of Lille, 2001).

96.    W. Wang, T. Lee and M. A. Reed, *Phys. Rev. B* **68**, 035416 (2003).

97.    H. Sakaguchi, A. Hirai, F. Iwata, A. Sasaki, T. Nagamura, E. Kawata and S. Nakabayashi, *Appl. Phys. Lett.* **79**, 3708 (2001).

98.    V. B. Engelkes, J. M. Beebe and C. D. Frisbie, *J. Am. Chem. Soc.* **126**, 14287 (2004).

99.    X. D. Cui, A. Primak, X. Zarate, J. Tomfohr, O. F. Sankey, A. L. Moore, T. A. Moore, D. Gust, L. A. Nagahara and S. M. Lindsay, *J. Phys. Chem. B* **106**, 8609 (2002).

100.   A. Salomon, D. Cahen, S. M. Lindsay, J. Tomfohr, V. B. Engelkes and C. D. Frisbie, *Adv. Mat.* **15**, 1881 (2003).

101.   R. J. Powell, *J. Appl. Phys.* **41**, 2424 (1970).

102.   C. Boulas, J. V. Davidovits, F. Rondelez and D. Vuillaume, *Phys. Rev. Lett.* **76**, 4797 (1996).

103.   J. M. Beebe, V. B. Engelkes, L. L. Miller and C. D. Frisbie, *J. Am. Chem. Soc.* **124**, 11268 (2002).

104.   A. Salomon, M. Boecking, C. K. Chan, F. Amy, O. Girshevitz, D. Cahen and A. Kahn, Phys. *Rev. Lett.* **95**, 266897 (2005).

105.   A. Salomon, M. Boecking, O. Seitz, T. Markus, F. Amy, C. K. Chan, W. Zhao, D. Cahen and A. Kahn, *Adv. Mat.* **19**, 445 (2007).

106.   D. Vuillaume, C. Boulas, J. Collet, G. Allan and C. Delerue, Phys. *Rev. B* **58**, 16491 (1998).

107.   T. Vondrak, C. J. Cramer and X.-Y. Zhu, *J. Phys. Chem. B* **103**, 8915 (1999).

108.   J. G. Kushmerick, J. Lazorcik, C. H. Patterson, R. Shashidhar, D. S. Seferos and G. Bazan, *Nano Lett.* **4**, 643 (2004).

109.   W. Wang, T. Lee, I. Krestchmar and M. A. Reed, *Nano Lett.* **4**, 643 (2004).
110.   C. Petit, G. Salace, S. Lenfant and D. Vuillaume, *Microelectronic Engineering* **80**, 398 (2005).
111.   D. K. Aswal, C. Petit, G. Salace, D. Guérin, S. Lenfant, J. V. Yakhmi and D. Vuillaume, *Phys. Stat. Sol.* (a) **203**, 1464 (2006).
112.   J. M. Beebe, H. J. Moore, T. R. Lee and J. G. Kushmerick, *Nano Lett.* **7**, 1364 (2007).
113.   D. P. Long, J. L. Lazorcik, B. A. Mantooth, M. H. Moore, M. A. Ratner, A. Troisi, Y. Yao, J. W. Ciszek, J. M. Tour and R. Shashidhar, *Nature Materials* **5**, 901 (2006).
114.   Z. Huang, B. Xu, Y. Chen, M. Di Ventra and N. J. Tao, *Nano Lett.* **6**, 1240 (2006).
115.   N. Clement, S. Pleutin, O. Seitz, S. Lenfant and D. Vuillaume, *Phys. Rev. B* **76**, 205407 (2007).
116.   J. R. Petta, S. K. Slater and D. C. Ralph, *Phys. Rev. Lett.* **93**, 136601 (2004).
117.   W. Wang and C. A. Richter, *Appl. Phys. Lett.* **89**, 153105 (2006).
118.   L. A. Bumm, J. J. Arnold, M. T. Cygan, T. D. Dunbar, T. P. Burgin, L. Jones II, D. L. Allara, J. M. Tour and P. S. Weiss, *Science* **271**, 1705 (1996).
119.   L. Patrone, S. Palacin, J.-P. Bourgoin, J. Lagoute, T. Zambelli and S. Gauthier, *Chemical Physics* **281**, 325 (2002).
120.   L. Patrone, S. Palacin, J. Charlier, F. Armand, J.-P. Bourgoin, H. Tang and S. Gauthier, *Phys. Rev. Lett.* **91**, 096802 (2003).
121.   M. Di Ventra and N. D. Lang, *Phys. Rev. B* **65**, 045402 (2001).
122.   S. N. Yaliraki, M. Kemp and M. A. Ratner, *J. Am. Chem. Soc.* **121**, 3428 (1999).
123.   L. Patrone, S. Palacin and J.-P. Bourgoin, *Appl. Surf. Sci.* **212**, 446 (2003).
124.   P. Reddy, S.-Y. Jang, R. A. Segalman and A. Majumdar, *Science* **315**, 1568 (2007).
125.   R. M. Metzger, *J. Mater. Chem.* **9**, 2027 (1999).
126.   R. M. Metzger and C. A. Panetta, *New J. Chem.* **15**, 209 (1991).
127.   R. M. Metzger, J. W. Baldwin, W. J. Shumate, I. R. Peterson, P. Mani, G. J. Mankey, T. Morris, G. Szulczewski, S. Bosi, M. Prato, A. Comito and Y. Rubin, *J. Phys. Chem. B* **107**, 1021 (2003).
128.   J. W. Baldwin, R. R. Amaresh, I. R. Peterson, W. J. Shumate, M. P. Cava, M. A. Amiri, R. Hamilton, G. J. Ashwell and R. M. Metzger, *J. Phys. Chem. B* **106**, 12158 (2002).

129. M. Ng, -K., D.-C. Lee and L. Yu, *JACS* **124**, 11862 (2002).
130. Z. Wei, M. Kondratenko, L. H. Dao and D. F. Perepichka, *JACS* **128**, 3134 (2006).
131. S. Lenfant, C. Krzeminski, C. Delerue, G. Allan and D. Vuillaume, *Nano Lett.* **3**, 741 (2003).
132. C. Krzeminski, G. Allan, C. Delerue, D. Vuillaume and R. M. Metzger, *Phys. Rev. B* **64**, 085405 (2001).
133. K. Stokbro, J. Taylor and M. Brandbyge, *J. Am. Chem. Soc.* **125**, 3674 (2003).
134. P. E. Kornilovitch, A. M. Bratkovsky and R. S. Williams, *Phys. Rev. B* **66**, 165436 (2002).
135. J. Taylor, M. Brandbyge and K. Stokbro, *Phys. Rev. Lett.* **89**, 138301 (2002).
136. S. Datta, W. Tian, S. Hong, R. Reifenberger, J. I. Henderson and C. P. Kubiak, *Phys. Rev. Lett.* **79**, 2530 (1997).
137. A. M. Honciuc, R.M., A. Gong and C. W. Spangler, *J. Am. Chem. Soc.* **129**, 8310 (2007).
138. A. Aviram, C. Joachim and M. Pomerantz, *Chem. Phys. Lett.* **146**, 490 (1988).
139. A. Aviram, C. Joachim and M. Pomerantz, *Chem. Phys. Lett.* **162**, 416 (1989).
140. A. Aviram, *J. Am. Chem. Soc.* **110**, 5687 (1988).
141. M. A. Reed, J. Chen, A. M. Rawlett, D. W. Price and J. M. Tour, *Appl. Phys. Lett.* **78**, 3735 (2001).
142. Z. J. Donhauser, B. A. Mantooth, K. F. Kelly, L. A. Bumm, J. D. Monnell, J. J. Stapleton, D. W. Price, A. M. Rawlett, D. L. Allara, J. M. Tour and P. S. Weiss, *Science* **292**, 2303 (2001).
143. G. K. Ramachandran, T. J. Hopson, A. M. Rawlett, L. A. Nagahara, A. Primak and S. M. Lindsay, *Science* **300**, 1413 (2003).
144. C. P. Collier, J. O. Jeppesen, Y. Luo, J. Perkins, E. W. Wong, J. R. Heath and J. F. Stoddart, *J. Am. Chem. Soc.* **123**, 12632 (2001).
145. J. E. Green, J. W. Choi, A. Boukai, Y. Bunimovich, E. Johnston-Halperin, E. Delonno, Y. Luo, B. A. Sherrif, K. Xu, Y. S. Shin, H.-R. Tseng, J. F. Stoddart and J. R. Heath, *Nature* **445**, 414 (2007).
146. D. R. Stewart, D. A. A. Ohlberg, P. A. Beck, Y. Chen, R. S. Williams, J. O. Jeppesen, K. A. Nielsen and J. F. Stoddart, *Nano Lett.* **4**, 133 (2004).
147. Z. Liu, A. A. Yasseri, J. S. Lindsey and D. F. Bocian, *Science* **302**, 1543 (2003).

148. C. Loppacher, M. Guggiesberg, O. Pfeiffer, E. Meyer, M. Bammerlin, R. Lüthi, R. G. Schlitter, J.K., H. Tang and C. Joachim, *Phys. Rev. Lett.* **90**, 066107 (2003).

149. Q. Li, G. Mathur, M. Homsi, S. Surthi, V. Misra, V. Malinovskii, K.-H. Schweikart, L. Yu, J. S. Lindsey, Z. Liu, R. B. Dabke, A. A. Yasseri, D. F. Bocian and W. G. Kuhr, *Appl. Phys. Lett.* **81**, 1494 (2002).

150. K. M. Roth, J. S. Lindsey, D. F. Bocian and W. G. Kuhr, *Langmuir* **18**, 4030 (2002).

151. K. M. Roth, A. A. Yasseri, Z. Liu, R. B. Dabke, V. Malinovskii, K.-H. Schweikart, L. Yu, H. Tiznado, F. Zaera, J. S. Lindsey, W. G. Kuhr and D. F. Bocian, *J. Am. Chem. Soc.* **125**, 505 (2003).

152. X. Duan, Y. Huang and C. M. Lieber, *Nano Lett.* **2**, 487 (2002).

153. C. Li, W. Fan, B. Lei, D. Zhang, S. Han, T. Tang, X. Liu, Z. Liu, S. Asano, M. Meyyappan, J. Han and C. Zhou, *Appl. Phys. Lett.* **64**, 1949 (2004).

154. C. Li, J. Ly, B. Lei, W. Fan, D. Zhang, J. Han, M. Meyyappan, M. Thompson and C. Zhou, *J. Phys. Chem. B* **108**, 9646 (2004).

155. J. Borghetti, V. Derycke, S. Lenfant, P. Chenevier, A. Filoramo, M. Goffman, D. Vuillaume and J.-P. Bourgoin, *Adv. Mat.* **18**, 2535 (2006).

156. A. Star, Y. Lu, K. Bradley and G. Grüner, *Nano Lett.* **4**, 1587 (2004).

157. J. P. A. van der Wagt, A. C. Seabaugh and E. A. Beam, *IEEE Electron Dev. Lett.* **19**, 7 (1998).

158. N. P. Guisinger, M. E. Greene, R. Basu, A. S. Baluch and M. C. Hersam, *Nano Lett.* **4**, 55 (2004).

159. T. Rakshit, G.-C. Liang, A. W. Ghosh and S. Datta, *Nano Lett.* **4**, 1803 (2004).

160. J. L. Pitters and R. A. Wolkow, *Nano Lett.* **6**, 390 (2006).

161. S. Y. Quek, J. B. Neaton, M. S. Hybertsen, E. Kaxiras and S. G. Louie, *Phys. Rev. Lett.* **98**, 066807 (2007).

162. H. Park, J. Park, A. K. L. Lim, E. H. Anderson, P. A. Alivisatos and P. L. McEuen, *Nature* **407**, 57 (2000).

163. S. Kubatkin, A. Danilov, M. Hjort, J. Cornil, J. L. Brédas, N. Stuhr-Hansen, P. Hedegard and T. Bjornholm, *Nature* **425**, 698 (2003).

164. G. S. Tulevski, Q. Miao, M. Fukuto, R. Abram, B. M. Ocko, R. Pindak, M. L. Steigerwald, C. R. Kagan and C. Nuckolls, *J. Am. Chem. Soc.* **126**, 15048 (2004).

165. A. Dehon, P. Lincoln and J. E. Savage, *IEEE Trans. Nanotechnol.* **2**, 165 (2003).

166. J. M. Tour, W. L. van Zandt, C. P. Husband, S. M. Husband, L. S. Wilson, P. D. Franzon and D. P. Nackashi, *IEEE Trans. Nanotechnol.* **1**, 100 (2002).

167. K. K. Likharev and D. B. Strukov, in *Introduction to Molecular Electronics*, edited by G. Cuniberti Springer, p 447 (2005).

168. S. C. Goldstein and M. Budiu, in *Int. Symp. on Computer Architecture*, p 178 (2001).

169. D. M. Adams, L. Brus, C. E. D. Chidsey, S. Creager, C. Creutz, C. R. Kagan, P. V. Kamat, M. Lieberman, S. M. Lindsay, R. A. Marcus, R. M. Metzger, M. E. Michel-Beyerle, J. R. Miller, M. D. Newton, D. R. Rolison, O. Sankey, K. S. Schanze, J. Yardley and X. Zhu, *J. Phys. Chem. B* **107**, 6668 (2003).

170. A. Nitzan, *Annu. Rev. Phys. Chem.* **52**, 681 (2001).

171. N. J. Tao, *Nature Nanotechnol.* **1**, 173 (2006).

172. A. Nitzan and M. A. Ratner, *Science* **300**, 1384 (2003).

# 12

# Carbon Nanotube Electronics

Vincent Derycke, Arianna Filoramo and Jean-Philippe Bourgoin*

Service de Physique de l'Etat Condensé (CNRS URA 2464),
DSM/IRAMIS/SPEC, CEA Saclay
91191 Gif sur Yvette Cedex, France.

*jean-philippe.bourgoin@cea.fr

·······························

Carbon nanotubes have excellent electrical properties that make
them one of the most promising building blocks for future nanotechnologies. The performances of individual devices, in particular field-effect transistors, already compete favorably with
standard CMOS devices. However, there are still some serious
issues that remain to be solved before a viable technology could
be developed. In particular, the main concern regards the controlled synthesis and positioning of nanotubes. The combination of their electrical properties with their chemical, mechanical
and/or thermal properties has already opened very promising
routes toward new type of applications in electronics.

## 1.    Introduction

After 40 years of scaling, the silicon technology still follows an exponential
growth law. But, since no exponential grow can last forever, it is clear that
the scaling of silicon based transistors will stop at some point. For many
years, a famous "brick-wall" was predicted to be reached within the next
ten years. But up to now, the technology always succeeded in meeting the
ever more difficult challenges of the scaling. Even though physical and
economical limits will finally prevail, the CMOS technology still has good
years ahead. Nevertheless, a very interesting consequence of these predicted

difficulties was to stimulate huge research efforts in many fields related to information processing. So far, none of these "emerging technologies" has demonstrated its economical viability. However, in the last years, great advances were made and there is no doubt that eventually some of these discoveries will find their way to the market, though maybe not in the originally predicted way. One of the field that benefits the most from these research efforts and associated funding is the so called "molecular electronics" field. In this field, molecular scale objects are used to build devices in the "bottom up" approach (in contrast with CMOS that relies on the "top-down" approach). Among the considered molecules and nano-objects, carbon nanotubes (CNTs) occupy a central place. Due to their exceptional physical properties they attracted a lot of attention and their relative compatibility with standard technologies allowed fast progress in the study of nanotube based electronic devices. In this chapter, we present the most important results concerning nanotube electronics by emphasizing the state of the art and remaining challenges.

## 2. Definition and Structure

Carbon nanotubes and fullerenes are different allotropic forms of carbon. Fullerenes are close-cages molecules containing only carbon atoms disposed in a hexagonal and pentagonal interatomic bonding network. Nanotubes are like large, cylindrical fullerenes with aspect ratio as large as $10^3$ to $10^5$ (see Fig. 1). More precisely, a single-wall carbon nanotube (SWNT) is a cylinder that one obtains by rolling-up a graphene sheet of hexagonal carbon rings (with half-fullerenes potentially capping the shell ends). Similarly, multi-wall nanotubes (MWNTs) can be schematized like a rolled-up stack of graphene sheets in concentric shells (like Russian dolls).

Each SWNT can be unambiguously identified by two integer numbers $n$ and $m$. The nomenclature $(n, m)$, with $n > m$ defines a bidimensional vector, the so called chiral vector (C in Fig. 1(c)), on the graphene lattice plane. The direction of the chiral vector in the graphene plane determines the direction along which the graphene sheet is rolled to form the nanotube. In addition, its modulus is related to the nanotube diameter.

A SWNT can be either metallic or semiconductor, depending on its chiral vector $(n, m)$. This remarkable property, which relates a basic physical feature (here the conductive character) of a system to its geometrical characteristics, is actually ultimately linked to the particular band structure

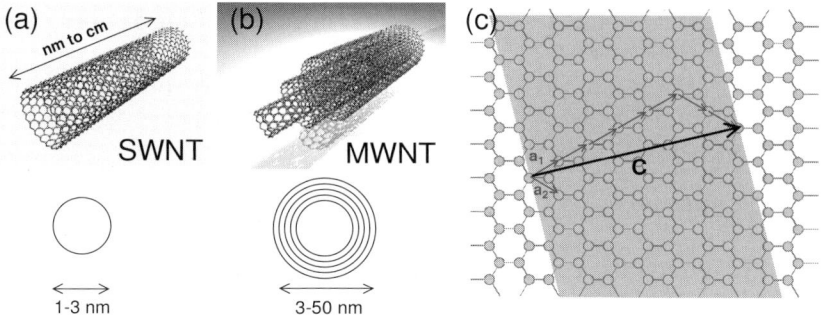

Fig. 1. Schematic view and typical size of (a) a SWNT and (b) a MWNT. (c) Example of the $(n, m)$ definition in the case of a (5,2) nanotube. The chiral vector C (joining two atoms of a graphene stripe (gray), which become equivalent once the stripe has been rolled into a cylinder) is defined as $C = 5a_1 + 2a_2$.

of the graphene sheet and to the existence of cyclic boundary conditions imposed by the wrapping in the cylindrical shape. Indeed, the graphene is an ideal semi-metal, with zero gap value in only six points of its Brillouin zone. As a consequence, whenever the states of these particular zero-gap points of the graphene fulfill the circunferencial boundary condition, the resulting SWNT will ideally (i.e., at zero temperature) display metallic behavior, otherwise it will display a semiconductor behavior. In practice, at room temperature, one has as a general rule that a SWNT is metallic if the difference n-m is an integer multiple of 3, while in all the other cases a semiconductor nanotube is obtained. It follows from this rule that, if all $(n, m)$ configurations are equally probable, one has a semiconductor-to-metallic abundance ratio of 2/3-1/3. Concerning MWNTs, the stacking of nanotubes with different chiralities can lead to more unexpected physical properties, due to the interlayer coupling.[1−6]

Since their discovery in the early 90's by Sumio Ijima,[7] carbon nanotubes have been a privileged subject of research due to their extraordinary physical properties in terms of transport, superlative resilience, tensile strength and thermal stability.[8−20] This large panel of interesting properties is reflected in the large number of studies on carbon nanotubeapplications reported in the literature.[21] They span from field emission electron sources,[22−27] supercapacitors,[28−30] artificial muscles,[31−33] nanoelectromechanical systems,[34−39] photoactuators,[40] controlled drug delivery/release,[41,42] reinforcement of materials,[16,43−47] composite printable conductors,[48] optical components,[49,50] nanoelectronic devices,[51] scanning probe tips,[52−54] etc.

## 3.  Synthesis and Positioning

### 3.1.  Synthesis

There are mainly three methods of synthesis of CNTs: arcdischarge,[55−61] laser ablation[62−64] and chemical vapor deposition (CVD).[65−69]

The first two approaches are evaporation methods that employ solid-state carbon precursors as carbon sources for the nanotube growth and involve carbon vaporization at high temperatures assisted respectively by an arc-discharge or by laser ablation. In order to achieve the SWNTs growth some metal catalysts are added in the solid graphite source while this is not the case for MWNTs. The most commonly used catalysts are transition metals or rare earth metals or a mixture of them. Historically, the first vaporization process to be developed was the arc-discharge one. [55,56] In this approach an electric arc is set between two graphite electrodes and while consuming the anode it forms a high temperature plasma (up to 6000°C). Then, the plasma condenses carbon nanotube soot on the cathode. The laser approach was originally developed by Smalley.[70] It consists in the evaporation by laser ablation of a graphite target placed into a background gas (typically ∼500 Torr of Ar) which is gently flowing through a quartz tube inside a high temperature oven (1100–1200°C). The hot evaporation cloud (plume) is carried by the gas flow onto a cool copper collector where the soot condenses. Since then, various configurations of laser-ablation experiments have been reported, from pulsed laser systems to continuous lasers ones. In both evaporation methods the nanotubes are not the only component of the soot and an additional purification step is often necessary to remove by-products, like amorphous carbon, catalyst particles (if present in the target), graphitic particles, fullerenes, etc.

On the contrary, the chemical vapor deposition method utilizes hydrocarbon gases as sources for carbon atoms. Also in this method metal catalyst particles are needed to act as "seeds" for the nanotube (SWNTs and MWNTs) growth but the process takes place at relatively lower temperatures (500–1000°C). The first step is the energy activated decomposition of the hydrocarbon gas. The energy source can be either a plasma or a resistively heated coil, and its function is to "crack" the gaseous molecules to provide reactive carbon atoms. Such carbon atoms diffuse towards the substrate, which is heated and coated with a transition metal catalyst. Then, the carbon atoms are fixed on the substrate and, if the appropriate conditions are fulfilled, the carbon nanotube growth takes place. The most commonly used gaseous carbon sources are methane,[71-79] ethylene,[67,80-82] propylene,[82-84] carbon monoxide[64,71,80,85,86] and acetylene.[87-91] Acetylene is widely used as carbon precursor for the growth of MWNTs, which occurs at temperatures typically in the 600–800°C range. Carbon monoxide or methane have proven to be more effective for the growth of SWNTs, since the temperature required is usually higher (800–1000°C) and acetylene is not stable at these temperatures. Hydrogen, nitrogen or argon are often used as diluent gas. In the aerosol CVD the catalyst is in spray form mixed with the hydrocarbon gas precursor.[92,93] Recent developments of catalytic CVD generation of SWNTs using alcohol as the carbon source have been also reported.[94] In these works the chirality distribution and purity of SWNTs is quite promising for the used low-temperature CVD conditions.[95-97]

The arc-discharge, laser ablation and CVD methods have not been equally explored in the literature. The CVD route is by far the most used one, and nowadays a large variety of production approaches based on this method are explored with success. This is in particular explained by the fact that the CVD technique is expected to be the solution for mass production of SWNTs or MWNTs. As a consequence of this important effort, one observes a rapid evolution of the state-of-the-art in the synthesis of CNTs by CVD methods. Presently, one has currently: the production of cm-long CNTs,[98,99] the synthesis of CNTs along the direction of an applied electric field[100] and a very regular ordering of CNTs grown on templates.[101,102] Despite the increasing success of the CVD-related approaches, some major problems in the synthesis of SWNTs still remain, notably: (i) the difficulty to produce nanotubes with narrow diameter distribution, (ii) the tubes produced at lower temperature are generally more defective. It is nevertheless worth to note that, even if the best quality SWNTs are so far those produced by evaporation-related methods, the differences with high temperature CVD ones are nowadays less significant.

Two important issues in the growth of CNTs are the control of the tube diameter ($d$) and the control of the chirality of the distributions. This results from the tight link between physical property and geometry of the nanotube, as mentioned above. The control of the diameter distribution is of particular importance. For instance, the nanotube diameter is strictly related with the bandgap of a semiconducting nanotube ($E_G \sim 1/d$), and plays consequently a major role in the performances of CNT-based electronic devices such as transistors. So far, none of the three synthesis methods has yielded bulk materials with homogeneous diameters and chiralities. Evaporation techniques still remain the best ones for the selective synthesis of SWNTs with narrow diameter distribution.[103,104] It is nevertheless worth to point out the dramatic and important developments on the CVD method these past years. Indeed, a better control of the growth parameters has allowed the optimization of both the yield and average diameter of SWNTs[105] (even if the diameter distribution is still not as narrow as for laser ablation synthesis). Finally, two interesting kind of reports should be noted: the first achievement concerns a preferential growth of semiconducting SWNTs (with a yield of 90%),[106] while the second class concerns some post synthesis method to separate metallic from semiconducting nanotubes.[107–109]

## 3.2. Positioning

Carbon nanotubes can be used to fabricate nanodevices, like field effect transistors, with very interesting performances. However, the possible use of carbon nanotubes as active elements in future nanoelectronics is closely related with a question of legacy/compatibility with the present information technology. To fully take advantage of the unique electrical properties of SWNTs in device/circuit applications, it is very desirable to be able to selectively place them -for connection- at specific locations on a substrate with a low cost and high yield, self-assembly based technique. Nowadays, the state-of-the-art on this issue can be divided in two different classes of self-assembly methods: (i) the *in situ* CVD growth where the localization arises from the catalyst controlled positioning and (ii) a post-growth deposition on a substrate. In the latter case, the nanotubes are first grown, handled in solution, and subsequently positioned on the substrate. Obviously, the technique chosen for this selective placement of the nanotubes must not degrade the electrical characteristics of the devices.

Concerning the fabrication of carbon nanotube devices by CVD, the basic idea is to achieve the *in situ* localized growth of nanotubes by controlling the localization of the metal catalyst. Indeed, the CVD carbon nanotube synthesis is essentially a two-step process consisting in an initial catalyst preparation step followed by the actual growth of the nanotubes, which starts at the places where the catalysts are present. Following this strategy, examples of localized growth of SWNTs have been realized since 1998.[72,101,102,110] However, there is an important issue to be solved before integrating such CVD method with nowadays CMOS technology. Indeed, for the direct growth by CVD of CNTs on silicon, the temperature range is at the moment incompatible with the CMOS integration. In this sense, a substantial progress has been achieved by the use of a plasma enhanced CVD (PECVD) method.[106] In this work, the nanotubes growth was carried out at 600°C on $SiO_2$/Si wafers on which some discrete ferritin particles were randomly adsorbed to act as catalysts. It should be added that another advantage of lowering the CVD growth temperature is related to the diameter distribution and chirality issue. Indeed, it is likely that the size and shape of the catalytic nanoparticles should be more stable at lower temperatures, leading to a better control of the size and potential chirality of nanotubes. More recently, lower temperature (down to 350°C) have been demonstrated effective for the growth of SWNTs.[111] This work may open new possibilities for full integration of CVD method into present complementary metal-oxide semiconductor (CMOS) technology, provided that the quality of such low temperature nanotube is improved.

In order to fully overcome the growth temperature issue and ensure the compatibility with CMOS technology, a very interesting (but fundamentally different) solution can be envisioned. Indeed, this kind of limitations can be avoided by preparing the nanotubes *ex situ*, functionalizing them, and then selectively depositing the nanotubes into the CMOS circuit. This is the philosophy of the post-growth strategies, as discussed in the following.

The advantage of any post-growth deposition method is that, before deposition, CNTs can be purified and chemically treated in order to separate them by diameter,[112–114] lengths [115] or chirality.[107,108,116,117] Moreover, in this pre-deposition step the nanotubes can also be chemically functionalized to add to their exceptional features other interesting chemical or physical properties.[118,119] As discussed previously, the drawback to overcome in this case is mainly related to the deposition issue since if no strategy is employed it is generally random on the substrate. To solve this SWNTs random deposition issue, three post-synthesis approaches can be drafted: (i) by surface treatements, (ii) by electric field and (iii) by a bio-directed assembly. These three methods are discussed separately in the following.

The first one is to achieve a selective placement of SWNTs on regions of the substrate that are predefined by surface treatements. This post-growth selective placement method is based on the use of self-assembled monolayers (SAMs) to modify the surface properties of certain regions of a substrate. This in-turn affects the interactions between the sidewalls of a CNT and the surface, and the CNTs are preferentially attracted there. This kind

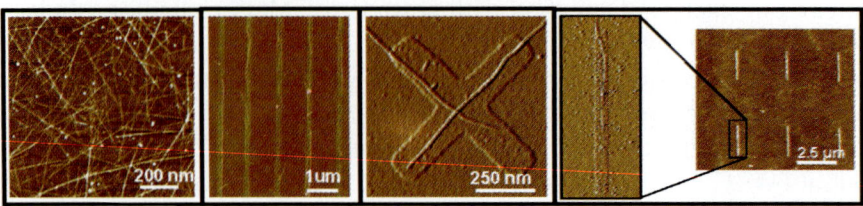

Fig. 2. AFM images showing the efficiency of APTS for the selective deposition of SWNTs. From left to right: deposition on a completely covered surface and on various patterns (continuous stripes, cross, and finite size stripes).

of studies started with the pioneering works of Liu *et al.*[120] or of Muster *et al.*[121] and Choi *et al.*[122] The approach relies either on a local chemical functionalisation of the surface[120] or on an electrostatic anchoring of surfactant covered SWNTs on amino-silane functionalised surfaces.[121,122] The basic idea beyond these processes is the same, but the use of amino-silane surfaces has allowed achieving for isolated SWNTs, the control of both the deposition density and their selective placement in predefined area of the substrate[123−125] as shown in Fig. 2. It should be noted that in some reports the surface properties have been combined with the molecular combing technique.[126,127] More recently, SWNTs selective deposition has been reported where SAMs on gold are patterned by dip pen nanolithography (DPN).[128] In this work SWNTs are positioned thanks to their specific attraction to the boundary between hydrophilic and hydrophobic surfaces made of 16-mercaptohexadecanoic acid (MHA) and 1-octadecanethiol (ODT) SAMs, respectively.

The second method is based on the dielectrophoresis (DEP) to position nanotubes on a set of predefined microelectrodes.[129−134] Dielectrophoresis is based on the appearance of a force on a dielectric object when it is placed in a non-uniform electric field. In the case of an object with a high aspect ratio, such as carbon nanotubes, the dielectrophoretic force aligns the nanotubes along the electric field lines. In this approach, a droplet of solution containing nanotubes is deposited onto a substrate patterned with a set of microelectrodes. The alternating (AC) electric field is applied and traps the nanotubes in the high field region between the microelectrodes. This deposition process depends on various parameters: the electrode geometry, the dielectric characteristic of both the nanotubes and the solvent, the concentration of nanotubes in the solution, the amplitude and frequency of the AC signal and the duration of application of the field. Recently, this method has also been used by Krupke *et al.*[135] to attract predominantly metallic SWNTs to a set of electrodes, exploiting the fact that the magnitude of the DEP force depends on the dielectric and conducting properties of a particle. Moreover, this technique presents another potential advantage, since it can be envisioned, by tuning electrodes geometry and by sequential application of electric fields, to fabricate more complex device structures, such as multiterminal transistors and branching interconnects.[133]

Finally, the third approach could solve the deposition challenge using biological scaffolds, as DNA molecules, to realize a site-controlled implementation of nanocomponents. Indeed, the unique intra- and intermolecular recognition properties of DNA have already been used to build-up scaffold structures and position nanoparticles.[136–140] First demonstrations of carbon nanotube field effect transistor using DNA-directed assembly have already been reported[141–143] even if the realization of a structured circuit hosting more than one nanotube device is, at the present time, still to be done.

## 4. Electronics Devices

While most of the properties of CNTs were theoretically predicted in the couple of years following their discovery, the device oriented studies really took off in 1996 when high quality SWNTs became more largely available for the research community.[61,62] From that year on, proofs of concepts for most conventional electronic devices were demonstrated using CNTs: Single Electron Transistors in 1997,[144] Field Effect Transistors the next year,[145,146] followed by Diodes (intra-tube,[147] inter-tubes[148] and p-n junctions[149]), Memory Devices,[150–152] elementary Logic Gates,[153–155] etc. Most of these realizations were more than small size replica of conventional devices. The truly nanometer size of the active element and its one-dimensional (1D) character often gave rise to original physical behaviors. Still, they also reflected mostly experimental skills in the sense that the associated challenges were often related to the handling and connection of these individual small size objects. During the period 1996–2001, the performances were not an issue. Once it was clear that any device that can be built with usual semiconductors could be reproduced with nanotubes, the field entered in a new period where performances became central.

In the following, we will focus our attention on carbon nanotube field effect transistors. While they are not the only nanotube-based electronic

devices they are the most studied and most advanced in terms of performances. They will serve to illustrate the potential capabilities and remaining
blocking issues of the field. Toward the end of the chapter, other nanotube
devices will be presented briefly to illustrate the diversity and versatility of
nanotube electronic devices. This description will include nanotube chemical and bio-sensors, flexible electronic devices, opto-electronic devices and
interconnects.

## 4.1.    Carbon nanotube field-effect transistors (CNTFETs)

First CNTFETs were demonstrated in 1998 by two groups from Delft
University[145] and IBM.[146] In these early versions, a single semiconducting
nanotube was deposited on top of gold (or platinum) electrodes prefabri-
cated on an oxidized silicon wafer, which served as a global back-gate.
Figures 3(a) and 3(b) present one of these early CNTFETs and the corresponding p-type transistor characteristics. These elegant proofs of concept
showed that a single molecular object can serve as the channel of a field
effect transistor with remarkably good separation between the conducting

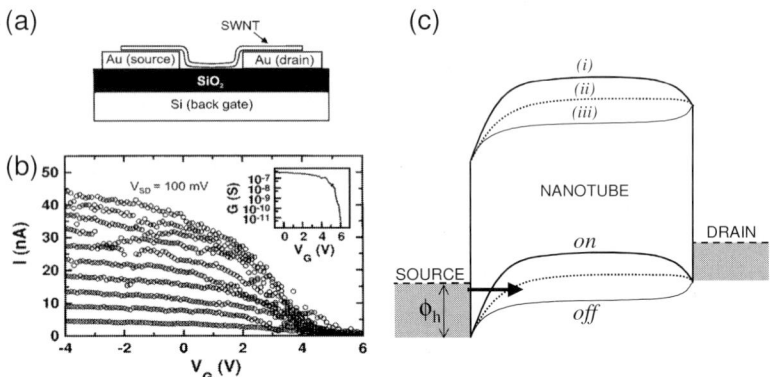

Fig. 3.   (a) and (b) schematic representation and electrical characteristics of the 1998 CNT-
FETs from the IBM group. Reprinted with permission from Ref. 146. Copyright (1998)
American Institute of Physics. (c) Schematic representation of the band bending conditions
in a typical SB-CNTFET at three different gate biases corresponding to (i) and (ii) ON-
states and (iii) OFF-state. The thick arrow illustrates tunneling injection of holes through
the barrier at the source contact, the thickness of which depends on $V_{GS}$. As an example, the
nanotube band. gap is $E_G = 650$ meV, $V_{DS} = -150$ mV and the Schottky barrier height
for holes is $\Phi_h = 200$ meV.

and insulating states (see the on-off ratio of $\sim 10^6$ in the inset of Fig. 3b). Still, they were limited to low current drives and transconductances.

There are great similarities between the electrical characteristics of these CNTFETs and those of regular silicon MOSFET. Therefore, it was first assumed that a similarity exists also in their operation mode. However, very important differences exist. In a conventional MOSFET, the source and drain metal electrodes do not contact directly the channel but are separated from the latter by highly doped regions so as to insure ohmic contacts. The performances of this kind of silicon devices are essentially limited by the quality of the transport within the channel (the carrier mobility). In a CNTFET, a semiconducting nanotube is directly connected to metal electrodes. As in most metal-semicondutor junctions, a Schottky barrier (SB) is formed. Injecting charges in the channel of a CNTFET thus requires overcoming, or tunneling through, the energy barrier at the source contact. Once a charge is injected in the nanotube it is very efficiently transported through the channel. Indeed, in semiconducting SWNTs, the carrier mean free path is of several hundreds of nanometers (at room temperature and moderate electric field), longer than the typical channel length in CNTFETs.[156,157] Thus, the performances of CNTFETs are mostly limited by the efficiency of the carriers injection rather than by the carriers' mobility.

The importance of the Schottky barriers at the metal-nanotube interface was early realized[158,159] and extensively studied, especially by the IBM group.[158,160–162] It was made clear that the switching in these SB-CNTFETs was due to the modulation of the thickness of the injection barrier by the gate potential (see Fig. 3c) and that the currents in both the ON- and OFF-states were mainly tunneling currents through this barrier of adjustable transparency.[163] This mode of operation has important consequences on the scaling as it was shown both theoretically and experimentally.[164–166]

## 4.2. Performances of CNTFETs

### 4.2.1. DC performances

Starting in 2001, lots of efforts were made to improve the DC performances of CNTFETs. Progress was very fast, in particular because most of the problems were similar to those faced many years before by the traditional semiconductor industry. In particular, trying to improve the gate efficiency using very thin and high permittivity dielectrics was a natural move. But with respect to that issue, nanotubes have a very important advantage in comparison with silicon: the natural chemical inertness of their surface made them directly compatible with high-k dielectrics and no reduction in carrier mobility is observed when changing the chemical nature of the dielectric. Several studies concerning the gate oxide scaling were done using for example: $SiO_2$,[167] $Al_2O_3$,[154] $HfO_2$,[162] $ZrO_2$,[168] $TiO_2$[169] or $SrTiO_3$[170] with drastic consequences in terms of performances.

At the same time, the issue of carrier injection was also tackled. The most significant improvement came from the discovery by the Stanford group that palladium can form ohmic contacts for the injection of holes into carbon nanotubes.[156] The reasons for the specificity of the nanotube-Pd contact quality are still largely unknown and ohmic contacts are only obtained with nanotubes of relatively large diameters (corresponding to reduced bandgaps).[171] Still, the combination of Pd with scaled dielectrics lead to the fabrication of the best CNTFETs to date[172−174] as illustrated in Fig. 4. With such a simple design, scaling of the channel length is only limited by lithography capabilities and CNTFETs in the 10–50 nm range were demonstrated.[174−176]

According to simulations, direct metal-nanotube contacts, even when ohmic, cannot give the ultimate performances.[166,177] In particular, the tunneling processes in CNTFETs are so efficient that in a p-type FET, avoiding electrons injection from the drain electrode (through a very high

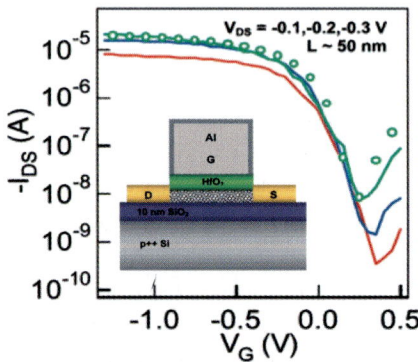

Fig. 4. Schematic view and transfer characteristics of one of the best CNTFET to date reprinted with permission from Ref. 173. Copyright (2004) American Chemical Society. The channel length is 50 nm, the $HfO_2$ dielectric is ∼7 nm and Pd source and drain electrodes are used. Max. transconductance ∼30 $\mu S$, max. linear ON-state conductance ∼$0.5 \times 4.e^2/h$, saturation current ∼25 $\mu A$ and sub-threshold slope S ∼ 110 mV/dec.

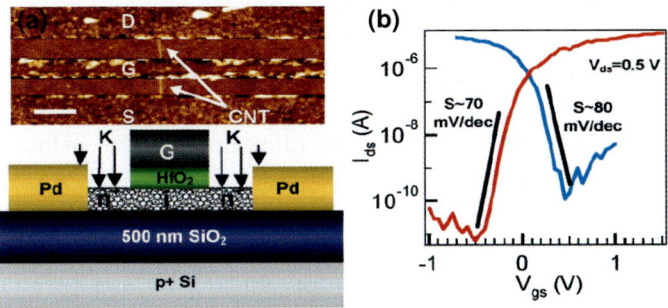

Fig. 5. Principle of access doping using potassium (in vacuum) and corresponding transfer characteristics reprinted with permission from Ref. 181. Copyright (2005) American Chemical Society. Note that without potassium doping, the global back-gate can also be used to improve holes injection by electrostatically doping the nanotube.

but potentially very thin barrier) imposes very strong constraints on the $V_{DS}$ scaling, with high risks of degradation of the off-state.[166,177] An alternative route to high performances is to try to mimic a more conventional situation where injection occurs through highly doped semiconducting sections as in Si-MOSFETs.[178] But the absence of substitutional doping techniques for carbon nanotubes prevents the easy fabrication of such structures. This can however be done, as illustrated in Fig. 5 (for a n-type FET), using two techniques: multiple gate configurations[179] or chemical doping of the accesses.[180,181] In the first case, a global back gate is used to electrostatically "dope" the nanotube sections close to the source and drain contacts (or the full channel) and another local-gate is used for the switching. In the second case, a top-gate is protecting a central section of the nanotube and chemical treatments are used to dope the open sections close to the contacts. Note however, that while these techniques can improve device performances, they do not allow a very aggressive scaling of the channel length.

A large part of the above mentioned high performances CNTFETs are p-type transistors, because usual (high work function) metal electrodes favors holes injection into carbon nanotubes. But n-type CNTFETs have also been demonstrated using different methods such as tuning the metal-nanotube interface,[160] doping the channel in vacuum using potassium[153,160,181,182] or in air using polymers[183,184] and multiple gates that allow the electrostatic control of the type of injected carriers.[179] Most interestingly, it was shown that because tunneling processes through the contact barriers are very efficient, ambipolar transistors can also be produced with carbon nanotubes.[158]

An important consequence of the implementation of multiple gate configurations, combined with the efficiency of tunneling processes in CNTs, is the possibility to built new types of tunneling devices, which could, in principle, outperform traditional CNTFETs. In particular, it was shown that using the phenomena called band-to-band tunneling[185] (carriers injected in the valence band tunnel through the band-gap into the conduction band and back into the valence band), leads to a sub-threshold slope steeper than the well-known 60 mV/dec — the limit for traditional switching at room temperature. In specially designed p-i-n structures (see Fig. 6)[186] optimal performances of nanotube-based FETs were predicted. Such performances were recently demonstrated by the Stanford group with a sub-threshold slope of 25 mV/dec at room temperature.[187]

Comparisons between CNTFETs and conventional Si-MOSFETs have been attempted by several authors.[167,172,178,188−191] The most difficult issue is a proper comparison of the maximum on-current. Indeed, while the

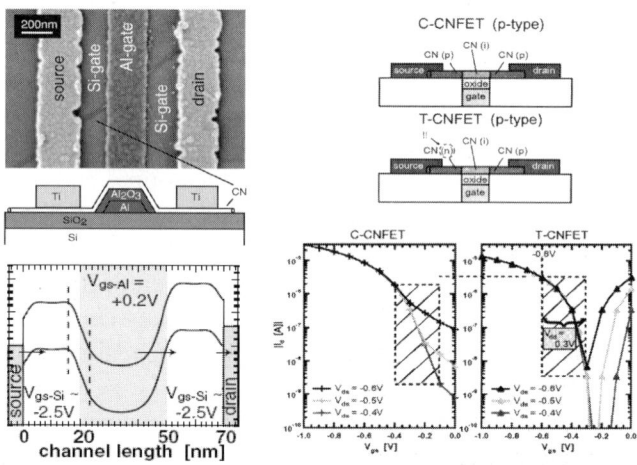

Fig. 6. (*Left*) double back-gate configuration and band bending conditions to observe band-to-band tunneling, reprinted with permission from Ref. 185. Copyright (2004) by the American Physical Society. (*Right*) structure and simulated characteristics of two types of CNT-FETs, a conventional and a tunneling (p-i-n) device, reprinted with permission from Ref. 186. Copyright (2005) IEEE. In the shaded area of the characteristics, the tunneling device does not suffer from degradation of the off-state at high $V_{DS}$.

current density in the channel of a CNTFET is extraordinary (considering its very small width of 1–3 nm), the total current barely exceeds 20 $\mu$A due to increased phonon generation at high bias. To compare performances with silicon-based technologies, it is convenient to scale the figures of merit per unit of channel width ($I_{on}$ in $\mu$A/$\mu$m, $g_m$ in $\mu$S/$\mu$m etc.). It is equivalent to consider the channel of the CNTFET as a 2D network of parallel CNTs (with a typical spacing between CNTs equal to twice the CNTs diameter). While this scaling is convenient, such perfect and dense network has not been experimentally realized yet. With this in mind, it comes out of a simple comparison that CNTFETs are indeed excellent transistors ($g_m$ up to 17650 $\mu$S/$\mu$m, $I_{on}$ up to 11600 $\mu$A/$\mu$m at $V_{ds} = 0.4$ V with $t_{ox} \sim 7$ nm and $\varepsilon_r \sim$ 15,[173] channel length as low as 10-20 nm,[174,175] subthreshold slope as low as 70–80 mV/dec[172,181]) that outperform present Si-MOSFETs. The projected performances of CNTFETs also outperform those of future silicon MOSFETs in particular due to the high carrier velocity and to the very long carrier mean free path in CNTs[178] that allow near perfect ballistic transport for any realistic channel length ($<300$ nm). Nevertheless, these projected performances only reflect the ultimate capabilities of individual devices. The dense integration and device-to-device dispersion issues have barely been addressed at present and the expected advantages will likely not be sufficient to justify the large R&D effort to develop CNTs into a technology for replacing Si.

### 4.2.2. HF performances

Due to their very high carrier mobility and very high current density capability, carbon nanotube are considered very promising for high frequency (HF) applications.[192–197] A way to roughly estimate their potential is to look at the current gain cut-off frequency $f_t \sim g_m/2\pi C_g$ where $g_m$ is the transconductance and $C_g$ is the total gate capacitance. According to several theoretical studies the $f_t$ of short CNTFETs would be in the THz range. Figure 7 compares CNTs with other semiconductors with respect to $f_t$. Mainly due to the higher carrier velocity, CNTFETs are predicted to be faster than transistors made with any other semiconductor,[192] in particular silicon, including ultra-thin body double gate Si-MOSFETs.[193]

As an example, Guo *et al.* considered the HF performances of one of the best reported CNTFET,[173] for which a DC transconductance of $\sim 26$ $\mu$S was achieved using a geometry with $C_g \sim 2.3$ aF. This would yield an intrinsic $f_t$ of $\sim 1.8$ THz (channel length 50 nm).[193] In fact, fully optimized

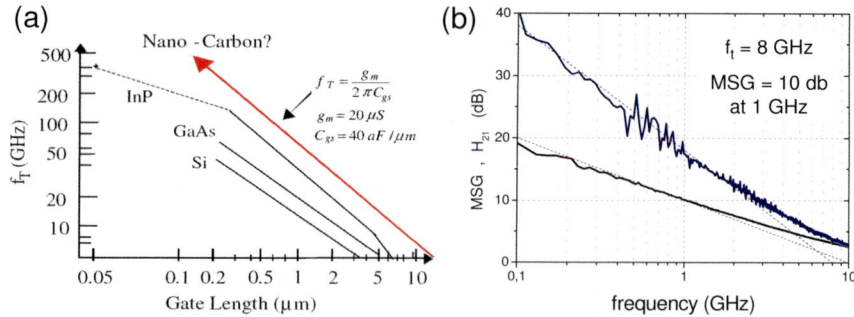

Fig. 7. (a) Comparison of maximum predicted current gain cut-off frequencies (f_t) for different semiconductors. Reprinted with permission from Ref. 192. Copyright (2004) Elsevier. (b) Measured f_t and maximum stable gain (MSG) on a multiple nanotube device. Reprinted with permission from Ref. 202. Copyright (2006) IEEE.

ballistic devices will be ultimately limited only by the carrier velocity, which corresponds to $f_t = v_F/2\pi L \sim 130$ GHz/$\mu$m,[194] with $v_F \sim 8.10^5$ m.s$^{-1}$ the Fermi velocity for nanotubes and L the channel length.

These optimistic views have to be moderated by looking at the extrinsic performances, i.e. by including the parasitic capacitances. Because the channel width of a CNTFET is very small (1–2 nm) compared to typical connecting electrodes, parasitic capacitances dominate the HF behavior of actual devices. The projected $f_t \sim 1.8$ THz of the device discussed above reduces to $\sim 1.7$ GHz if the parasitic capacitances are included.[193] However, it is worth to note that this device was optimized for DC measurements only.

When it comes to actually measuring the HF performances of nanotube devices, researchers face two main challenges: the limited drive current (or high impedance) of CNT devices and the predominance of parasitic capacitances. The first issue implies a very poor matching of single tube devices ($R_{ON} > 10$ k $\Omega$) with conventional 50 $\Omega$ equipment and the second one is hiding the true potential of CNTs at HF. Mostly due to these problems, direct measurements of HF performances of CNTFETs are still sparse but their number is increasing very fast.

To circumvent these problems, two categories of experiments were carried out. The first one is based on *indirect* assessments of the HF capability. Appenzeller *et al.* used the non linearity of CNTFET characteristics to obtain indication of HF behavior up to 580 MHz.[198] Other groups used mixing techniques to measure at low frequencies the impact of HF excitation up to 23 GHz[199] and 50 GHz.[200] These studies are very interesting but to really evaluate the potential of CNTs for HF applications, more information is needed such as HF current and power gains and HF equivalent circuit models. The second category of experiments is based on *direct* S-parameters measurements of multiple nanotube devices. By increasing the number of nanotubes forming the channel, the total impedance of the device can be very much decreased so that conventional 50 Ω equipment can be used. At the same time, it decreases the parasitic capacitance per nanotube. This technique was employed by different groups and $f_t$ in the GHz range were directly measured.[201−203] Figure 7(b) presents an example of such a measurement showing an $f_t$ of 8 GHz and a maximum stable gain (MSG) of 10 dB at 1 GHz (after de-embedding).[202] Noticeably, using very dense and mostly aligned SWNTs networks, Le Louarn *et al.* reported an $f_t$ of 30 GHz for 200 nm long channel devices.[203] While this value is the highest reported to date, it is still far from taking full advantage of the high potential of CNTs. Indeed, this measurement, as all the other published to date, is still limited mostly by parasitic capacitances rather than the intrinsic properties of the CNTs. Nevertheless it shows that high frequency devices based on nanotubes are already feasible and that there is still plenty of room for significant improvement.

## 4.3.    *Beyond conventional field effect transistors*

CNTFETs are just one example of CNT-based devices. If they are very well suited for benchmarking CNTs again other materials, they may not be — in the present form — the most promising devices in terms of applications, in particular because they don't take full advantage of all the specific properties of CNTs. Reviewing all the possible applications of CNTs in electronics is beyond the scope of this chapter. In the following, we present briefly some potentially interesting routes, which could bring CNTs to the market and propose some relevant references where the reader would find the full device descriptions.

### 4.3.1.   *Gas and bio-sensors*

One important property of CNTs is that they can be chemically functionalized. Combining the high charge sensitivity of CNTFETs with the molecular recognition capabilities of certain classes of molecules allows the fabrication of highly sensitive and highly specific gas and bio-sensors.

The potential of CNTs as part of sensitive sensors was realized very soon after the first demonstration of CNTFETs.[110,204] It is now clear that CNTFETs can detect molecules adsorbed on the nanotube but also at the nanotube-electrode interface[160,205,206] and at the nanotube-dielectric interface.[207] To prevent unselective detection, appropriate functionalization of CNT sensors must be performed. It was in particular shown that CNT-based bio-sensors can detect with great selectivity enzyme-protein binding with a sensitivity approaching the single molecule level.[208,209]

In the past few years, the field of nanotube-based sensors has grown very fast and has now reached a certain degree of maturity with, for example, the recent commercialization of specific gas sensors by Nanomix, Inc.[210]

## 4.3.2. Flexible electronics

Another important property of CNTs, when compared with usual semiconductors, is that they are naturally flexible. Combined with their compatibility with most substrates, this property makes them particularly promising for flexible electronic applications. In this field, they are direct competitors for the usual materials of organic electronics: polymers (PPV, P3HT ...) and small molecules (pentacene, rubrene ...). But they can count on there very high carrier mobility as a fundamental advantage over other organic materials.

While the early demonstrations of flexible CNTFETs were based on very simple technological processes, they showed that the same level of performances as the one obtained with other organic materials (typical carrier mobility in the 1–10 $cm^2$/V.s range) could be easily reached[211,212] and that stability was not an issue as it can be in conventional organic electronic devices. More recently, large progresses were made in the use of CNTs for flexible electronics. In particular, Rogers *et al.* showed that CNTs could be used both as the channel and as electrodes of thin film transistors by tuning the density of nanotube networks.[213] The same group then showed that oriented growth of nanotubes followed by transfer on flexible substrates could yield very high performance flexible devices.[102] The achieved mobility of $\sim$500 $cm^2$/V.s is the highest reported for a *p*-type device on a plastic substrate showing that nanotubes are probably the material of choice for most applications requiring flexible electronics, as for example, "electronic paper". Noticeably, we were able to recently demonstrate GHz operation frequencies for flexible transistor based on CNTs deposited by dielectrophoresis onto a plastic film.[214]

Note that when used as part of flexible devices, CNTs are never used as individual objects but as part of large networks. This presents several advantages: first it allows the development of low cost and large area processes, a requirement to compete with other organic materials. Next, it limits the effects of device-to-device dispersions by averaging the performances over large numbers of nanotubes. And finally, it limits the impact of metallic nanotubes. Indeed, it was shown that when the size and density of a 2D percolating networks of nanotubes are adjusted, the effect of metallic nanotubes can be minimized, and good off-states can be systematically obtained. On the other hand, the mobility in these networks cannot approach the one of individual nanotubes ($\sim 10^5$ cm$^2$/V·s).[215]

### 4.3.3. Nanotube opto-electronics

Since CNTs have a direct band-gap, they can, in principle, be used for opto-electronic applications. The development of this activity really took off in 2003 when the IBM group showed that CNTFETs can be used to emit[216] or detect[217] infra-red photons. In the latter case, photons absorbed in the channel generate electron-hole pairs separated by the source-drain electric field giving rise to a photo-current. In the first one, an ambipolar transistor was used, in which electrons and holes injected at opposite contacts could recombine within the channel and emit light at wavelengths set by the nanotube band structure. It was later shown that light can be emitted by CNT devices following other mechanisms such as impact excitation[218,219] or phonon-assisted activation of excited charges in quasi-metallic nanotubes.[220]

In these experiments, the optical properties of the devices come from the nanotube itself. Another strategy is now increasingly followed, which combines the electrical properties of CNTs with the optical properties of molecules or polymers. It was shown in particular that chemically functionalized CNTFETs can form very interesting photo-transistors[221,222] or optical memory devices.[223,224] As it was the case with bio-sensors, these studies set the basis for new classes of applications for which the excellent properties of CNTs are completed by additional properties coming from their chemical functionalization.

## 4.3.4. *Interconnects*

One important class of applications that can be targeted by CNTs is interconnects, in particular the vertical ones (VIA). Indeed, the ITRS roadmap predicts a dramatic increase of the current density that vertical interconnects will have to sustain in a near future. While conventional copper interconnects may have difficulties to meet the requirements, CNTs may prove a better choice. Indeed, because of their natural immunity against electromigration coming from the very strong C-C bonds, CNTs can sustain current density as high as $10^9$ A.cm$^2$ [225] at least an order of magnitude higher than any metallic nanowire.

Still the localized growth of high quality nanotubes within small holes remains an open issue, as well as the optimization of the contact resistance. Up to now, it is not yet fully clear whether CNTs can indeed outperform traditional interconnect technologies or exactly at which technology node CNTs would be a realistic alternative. Nevertheless, very interesting preliminary studies have been performed, lead in particular by Infineon,[190,191] Fujitsu[226-228] and the NASA[229] in an attempt to clarify the real potential of CNTs for interconnects. A key issue is to reach the predicting optimal performances of CNTs as VIA but within the technological constrains of a CMOS process, in particular in terms of materials and temperature compatibility.

## 5. Conclusions

Through the different examples of devices presented, it appears clearly that carbon nanotubes combine a set of physical properties that make them very promising for applications in electronics. Nevertheless, the example of germanium and III-V semiconductors showed in the past that superior intrinsic physical properties are generally not enough to impose a material as a standard. The quality of the Si/SiO$_2$ interface was to a large extend responsible for the prevalence of silicon. On the other hand, nanotubes can count on very strong advantages, in particular their compatibility with other materials (including high-k dielectrics), their high carrier velocity and their long carrier mean free path, to cite just a few. Before important economical breakthroughs can be made with nanotubes in electronics, the critical processing issues have to be tackled at a large scale. In the broad context of electronic development toward nanosize dimensions, new type of functions, beyond conventional transistors, have to be invented that would really take full advantage of the original properties of nanotubes specific to their 1D character and nanoscale.

# References

1.  R. Saito, G. Dresselhaus and M. S. Dresselhaus, *J. Appl. Phys.* **73**, 494 (1993).
2.  S. Roche, F. Triozon, A. Rubio and D. Mayou, *Phys. Rev. B* **64**, 121401 (2001).
3.  Y. A. Yoon, P. Delaney and S. G. Louie, *Phys. Rev. B* **66**, 073407 (2002).
4.  K.-H. Ahn, Y.-H. Kim, J. Wiersig and K. J. Chang, *Phys. Rev. Lett.* **90**, 026601 (2003).
5.  M. A. Tunney and N. R. Cooper, *Phys. Rev. B* **74**, 075406 (2006).
6.  S. Uryu and T. Ando, *Phys. Rev. B* **72**, 245403 (2005).
7.  S. Iijima, *Nature* **354**, 56 (1991).
8.  R. Saito, G. Dresselhaus and M. S. Dresselhaus, *Physical Properties of Carbon Nanotubes*, London, Imperial College Press, 1998.
9.  J.-P. Salvetat, J.-M. Bonard, N. H. Thomson, A. J. Kulik, L. Forro, W. Benoit and L. Zuppiroli, *Appl. Phys. A* **69**, 255 (1999).
10. S. Reich, C. Thomsen and P. Ordejon, *Phys. Rev. B* **65**, 153407 (2002).
11. T. Natsuki, K. Tantrakarn and M. Endo, *Carbon* **42**, 39 (2004).
12. J. P. Lu, *Phys. Rev. Lett.* **79**, 1297 (1997).
13. M. M. J. Treacy, T. W. Ebbesen and J. M. Gibson, *Nature* **381**, 678 (1996).
14. A. Krishnan, E. Dujardin, T. W. Ebbesen, P. N. Yianilos and M. M. J. Treacy, *Phys. Rev. B* **58**, 14013 (1998).
15. B. G. Demczyk, Y. M. Wang, J. Cumings, M. Hetman, W. Han, A. Zettl and R. O. Ritchie, *Mater. Sci. Eng. A* **334**, 173 (2002).
16. M. Cadek, J. N. Coleman, V. Barron, K. Hedicke and W. J. Blau, *Appl. Phys. Lett.* **81**, 5123 (2002).
17. J. Hone, M. Whitney and A. Zettl, *Synthetic Metals* **103**, 2498 (1999).
18. C. Qin, X. Shi, S. Q. Bai, L. D. Chen and L. J. Wang, *Mater. Sci. Eng. A* **420**, 208 (2006).
19. K. Bi, Y. Chen, J. Yang, Y. Wang and M. Chen, *Physics Letters A* **350**, 150 (2006).

20. Q. Gong, Z. Li, X. Bai, D. Li, Y. Zhao and J. Liang, *Mater. Sci. Eng. A* **384**, 209 (2004).

21. R. H. Baughman, A. A. Zakhidov and W. A. de Heer, *Science* **297**, 787 (2002).

22. W. A. Deheer, A. Chatelain and D. Ugarte, *Science* **270**, 1179 (1995).

23. N. de Jonge and J. M. Bonard, *Phylos. Trans. Royal Soc. London A* **362**, 2239 (2004).

24. J. M. Bonard, M. Croci, C. Klinke, R. Kurt, O. Noury and N. Weiss, *Carbon* **40**, 1715 (2002).

25. W. I. Milne, K. B. K. Teo, G. A. J. Amaratunga, P. Legagneux, L. Gangloff, J. P. Schnell, V. Semet, V. T. Binh and O. Groening, *Jour. Mat. Chem.* **14**, 933 (2004)

26. J. M. Bonard, T. Stockli, O. Noury and A. Chatelain, *Appl. Phys. Lett.* **78**, 2775 (2001).

27. Y. S. Choi, Y. S. Cho, J. H. Kang, Y. J. Kim, I. H. Kim, S. H. Park, H. W. Lee, S. Y. Hwang, S. J. Lee, C. G. Lee, T. S. Oh, J. S. Choi, S. K. Kang and J. M. Kim, *Appl. Phys. Lett.* **82**, 3565 (2003).

28. I. H. Kim, J. H. Kim and K. B. Kim, *Electrochemical Solid State Lett.* **8**, A369 (2005).

29. C. Y. Lee, H. M. Tsai, H. J. Chuang, S. Y. Li, P. Lin and T. Y. Tseng, *J. Electrochem. Soc.* **152**, A716 (2005).

30. E. S. Snow, F. K. Perkins, E. J. Houser, S. C. Badescu and T. L. Reinecke, *Science* **307**, 1942 (2005).

31. R. H. Baughman, C. Cui, A. A. Zakhidov, Z. Iqbal, J. N. Barisci, G. M. Spinks, G. G. Wallace, A. Mazzoldi, D. De Rossi, A. G. Rinzler, O. Jaschinski, S. Roth and M. Kertesz, *Science* **284**, 1340 (1999).

32. U. Vohrer, I. Kolaric, M. H. Haque, S. Roth and U. Detlaff-Weglikowska, *Carbon* **42**, 1159 (2004).

33. B. J. Landi, R. P. Raffaelle, M. J. Heben, J. L. Alleman, W. VanDerveer and T. Gennett, *Nanoletters* **2**, 1329 (2002).

34. J. Cao, Q. Wang and H. Dai, *Phys. Rev. Lett.* **90**, 157601 (2003).

35. S. Sapmaz, Y. M. Blanter, L. Gurevich and H. S. J. van der Zant, *Phys. Rev. B* **67**, 235414 (2003).

36. B. Bourlon, D. C. Glattli, C. Miko, L. Forro and A. Bachtold, *Nano Lett.* **4**, 709 (2004).

37. C. Li and T. W. Chou, *Phys. Rev. B* **68**, 073405 (2003).

38. S. W. Lee, D. S. Lee, R. E. Morjan, S. H. Jhang, M. Sveningsson, O. A. Nerushev, Y. W. Park and E. E. B. Campbell, *Nano Lett.* **4**, 2027 (2004).

76. J.-F. Colomer, C. Stephan, S. Lefrant, G. van Tendeloo, I. Willems, Z. Konya, A. Fonseca, C. Laurent and J. B. Nagy, *Chem. Phys. Lett.* **317**, 83 (2000).
77. Q. Liang, L. Gao, Q. Li, S. Tang, B. Liu and Z. Yu, *Carbon* **39**, 897 (2001).
78. Y. Ning, X. Zhang, Y. Wang, Y. Sun, L. Shen, X. Yang and G. van Tendeloo, *Chem. Phys. Lett.* **366**, 555 (2002).
79. J. Cumings, W. Mickelson and A. Zettl, *Solid State Commun.* **126**, 359 (2003).
80. J. Hafner, M. Bronikowski, B. Azamian, P. Nikolaev, A. Rinzler, D. Colbert, K. Smith and R. Smalley, *Chem. Phys. Lett.* **296**, 195 (1998).
81. J.-F. Colomer, G. Bister, I. Willems, Z. Konya, A. Fonseca, G. van Tendeloo and J. B. Nagy, *Chem. Comm.* 1343 (1999).
82. A. Cassell, J. Raymakers, J. Kong and H. Dai, *J. Phys. Chem. B* **103**, 6484 (1999).
83. K. Hernadi, A. Fonseca, J. B. Nagy, A. Siska and I. Kiricsi, *Appl. Catal. A* **General 199**, 245 (2000).
84. M. Mabudafhasi, R. Bodkin, C. Nicolaides, X.-Y. Liu, M. Witcomb and N. Coville, *Carbon* **40**, 2737 (2002).
85. H. Dai, A. Rinzler, P. Nikolaev, A. Thess, D. Colbert and R. Smalley, *Chem. Phys. Lett.* **260**, 471 (1996).
86. B. Kitiyanan, W. Alvarez, J. Harwell and D. Resasco, *Chem. Phys. Lett.* **317**, 497 (2000).
87. Y. Soneda, L. Duclaux and F. Beguin, *Carbon* **40**, 965 (2002).
88. H. Ago, T. Komatsu, S. Ohshima, Y. Kuriki and M. Yumura, *Appl. Phys. Lett.* **77**, 79 (2000).
89. L. Ci, S. Xie, D. Tang, X. Yan, Y. Li, Z. Liu, X. Zou, W. Zhou and G. Wang, *Chem. Phys. Lett.* **349**, 191 (2001).
90. Z. Zhou, L. Ci, L. Song, X. Yan, D. Liu, H. Yuan, Y. Gao, J. Wang, L. Liu, W. Zhou, G. Wang and S. Xie, *Carbon* **41**, 2607 (2003).
91. D. Geohegan, A. Puretzky, I. Ivanov, S. Jesse and G. Eres, *Appl. Phys. Lett.* **83**, 1851 (2003).
92. M. Mayne, N. Grobert, M. Terrones, R. Kamalakaran, M. Rühle, H. W. Kroto and D. R. M. Walton, *Chem. Phys. Lett.* **338**, 101 (2001).
93. M. Pinault, M. Mayne-L'Hermite, C. Reynaud, O. Beyssac, J. N. Rouzaud and C. Clinard, *Diamond and Related Materials* **13**, 1266 (2004).

94.   S. Maruyama, E. Einarsson, Y. Murakami and T. Edamura, *Chem. Phys. Lett.* **403**, 320 (2005).
95.   S. Maruyama, R. Kojima, Y. Miyauchi, S. Chiashi and M. Kohno, *Chem. Phys. Lett.* **360**, 229 (2002).
96.   Y. Murakami, Y. Miyauchi, S. Chiashi and S. Maruyama, *Chem. Phys. Lett.* **374**, 53 (2003).
97.   Y. Miyauchi, S. Chiashi, Y. Murakami, Y. Hayashida and S. Maruyama, *Chem. Phys. Lett.* **387**, 198 (2004).
98.   H. W. Zhu, C. L. Xu, D. H. Wu, B. Q. Wei, R. Vajtai and P. M. Ajayan, *Science* **296**, 884 (2002).
99.   L. Zheng, M. O'Connel, S. Doorn, X. Liao, Y. Zhao, E. Akhadov, M. Hoffbauer, B. Roop, Q. Jia, R. Dye, D. Peterson, S. Huang, J. Liu and Y. Zhu, *Nat. Mater.* **3**, 673 (2004).
100.  A. Ural, Y. Li and H. Dai, *Appl. Phys. Lett.* **81**, 3464 (2002).
101.  A. Javey, Q. Wang, A. Ural, Y. Li and H. Dai, *Nanoletters* **2**, 929 (2002).
102.  S. G. Kang, C. Kocabas, T. Ozel, M. Shim, N. Pimparkar, M. A. Alam, S. V. Rotkin and J. A. Rogers, *Nature Nanotechnology* **2**, 230 (2007).
103.  O. Jost, A. A. Gorbunov, J. Möller, W. Pompe, X. Liu, P. Georgi, L. Dunsch, M. S. Golden and J. Fink, *J. Phys. Chem.* **B 106**, 2875 (2002).
104.  O. Jost, A. A. Gorbunov, X. J. Liu, W. Pompe and J. Fink, *J. Nanoscience Nanotechnology* **4**, 433 (2004).
105.  P. Nikolaev, M. J. Bronikoswski, K. Bradeley, F. Rohmund, D. T. Colbert, K. A. Smith and R. E. Smalley, *Chem. Phys. Lett.* **313**, 91 (1999).
106.  Y. Li, D. Mann, M. Rolandi, W. Kim, A. Ural, S. Hung, A. Javey, J. Cao, D. Wang, E. Yenilmez, Q. Wang, J. F. Gibbons, Y. Nishi and H. Dai, *Nano Lett.* **4**, 317 (2004).
107.  M. Zheng, A. Jagota, E. D. Semke, B. A. Diner, R. S. Mclean, S. R. Lustig, R. E. Richardson andN. G. Tassi, *Nature Materials* **2**, 338 (2003).
108.  M. S. Arnold, A. A. Green, J. F. Hulvat, S. I. Stupp and M. C. Hersam, *Nature Nanotechnology* **1**, 60 (2006).
109.  G. Y. Zhang, P. F. Qi, X. R. Wang, Y. R. Lu, X. L. Li, R. Tu, S. Bangsaruntip, D. Mann, L. Zhang and H. J. Dai, *Science* **314**, 974 (2006).
110.  J. Kong, N. R. Franklin, C. W. Zhou, M. G. Chapline, S. Peng, K. J. Cho and H. J. Dai, *Science* **287**, 622 (2000).

111. M. Cantoro, S. Hofmann, S. Pisana, V. Scardaci, A. Parvez, C. Ducati, A. C. Ferrari, A. M. Blackburn, K.-Y. Wang and J. Robertson, *Nano. Lett.* **6**,1107 (2006).
112. E. Menna, F. Della Negra, M. Dalla Fontana and M. Meneghetti, *Phys. Rev. B* **68**, 193412 (2003).
113. L. Kavan and L. Dunsch, *Nanoletters* **3**, 969 (2003).
114. J. G. Wiltshire, A. N. Khlobystov, L. J. Li, S. G. Lyapin, G. A. D. Briggs and R. J. Nicholas, *Chem. Phys. Lett.*386, 239 (2004).
115. G. S. Duesberg, J. Muster, H. J. Byrne, S. Roth and M. Burghard, *Appl. Phys. A* **69**, 269 (1999).
116. M. S. Strano, *J. Am. Chem. Soc.* **125**, 16148 (2003).
117. Z. Chen, X. Du, M.-H. Du, C. D. Rancken, H.-P. Cheng and A. G. Rinzler, *Nano. Lett.* **3**, 1245 (2003).
118. K. Balasubramanian and M. Burghard, *Small* **1**, 180 (2005).
119. Y. P. Sun, K. Fu, Y. Lin and W. Huang, *Acc. Chem. Res.* **35**, 1096 (2002).
120. J. Liu, M. J. Casavant, M. Cox, D. A. Walters, P. Boul, W. Lu, A. J. Rimberg, K. A. Smith, D. T. Colbert and R. E. Smalley, *Chem. Phys. Lett.* **303**, 125 (1999).
121. J. Muster, M. Burghard, S. Roth, G. S. Duesberg, E. Hernandez and A. Rubio, *J. Vac. Sci. Tech. B* **16**, 2796 (1998).
122. K. H. Choi, J. P. Bourgoin, S. Auvray, D. Esteve, G. S. Duesberg, S. Roth and M. Burghard, *Surf. Sci.* **462**, 195 (2000).
123. E. Valentin, S. Auvray, J. Goethals, J. Lewenstein, L. Capes, A. Filoramo, A. Ribayrol, R. Tsui, J. P. Bourgoin and J. N. Patillon, *Microelect. Eng.* **61**, 491 (2002).
124. E. Valentin, S. Auvray, A. Filoramo, A. Ribayrol, M. F. Goffman, L. Capes, J. P. Bourgoin and J. N. Patillon, *Mat. Res. Soc. Symp. Proc.* **772**, 201 (2003).
125. E. Valentin, S. Auvray, A. Filoramo, A. Ribayrol, M. Goffman, J. Goethals, L. Capes, J. P. Bourgoin and J. N. Patillon, NATO SCIENCE SERIES: Molecular Nanowires and other Quantum Objects, Eds. S. Alexandrov, J. Demsar and I. Yanson (2004).
126. T. Ondarcuhu, C. Joachim and S. Gerdes, *Europhys. Lett.* **52**, 178 (2000).
127. V. V Tsukruk and V. N. Bliznyuk, *Langmuir* **14**, 446 (1998).
128. Y. Wang, D. Maspoch, S. Zou, G. C. Schatz, R. E. Smalley and C. A. Mirkin, *Proc. Natl. Acad. Sci,* **103**,2026 (2006).
129. X. Chen, T. Saito, H. Yamada and K. Matsushige, *Appl. Phys. Lett.* **78**, 3714 (2001).

130. L. A. Nagahara, I. Amlani, J. Lewenstein and R. K. Tsui, *Appl. Phys. Lett.* **80**, 3826 (2002).
131. Z. Chen, Y. Yang, F. Chen, Q. Qing, Z. Wu and Z. Liu, *J. Phys. Chem. B* **109**, 11420 (2005).
132. H. W. Seo, C. S. Han, D. G. Choi, K. S. Kim and Y. H. Lee, *Microelec. Eng.* **81**, 83 (2005).
133. S. Banerjee, B. E. White, L. M. Huang, B. J. Rego, S. O'Brien and I. P. Herman, *J. Vac. Sci. Technol. B* **24**, 3173 (2006) and *Appl. Phys. A* **86**, 415 (2007).
134. F. Hennrich, R. Krupke, M. M. Kappes and H. V. Lohneysen, *J. of Nanoscience and Nanotechnology* **5**, 1166 (2005).
135. R. Krupke, S. Linden, M. Rapp and F. Hennrich, *Adv. Mater.* **18**, 1468 (2006 ).
136. N. C. Seeman, *Nature* **421**, 427 (2003).
137. A. P. Alivisatos, K. P. Johnsson, X. G. Peng, T. E. Wilson, C. J. Loweth, M. P. Bruchez and P. G. Schultz, *Nature* **382**, 609 (1996).
138. C. Niemeyer and B. Ceyhan, *Angew. Chem. Int. Ed.* **40**, 3685 (2001).
139. C. A. Mirkin, *Inorg. Chem.* **39**, 2258 (2000).
140. H. Li, S. A. Park, J. H. Reif, T. H. LaBean and H. Yan, *J. Am. Chem. Soc.* **126**, 418 (2004).
141. K. Keren, R. S. Berman, E. Buchstab, U. Sivan and E. Braun, *Science* **302**, 1380 (2003).
142. M. Hazani, D. Shvarts, D. Peled, V. Sidorov and R. Naaman, *Appl. Phys. Lett.* **85**, 5025 (2004).
143. M. Hazani, F. Hennrich, M. Kappes, R. Naaman, D. Peled, V. Sidorov and D. Shvarts, *Chem. Phys. Lett.* **391**, 389 (2004).
144. S. J. Tans, M. H. Devoret, H. J. Dai, A. Thess, R. E. Smalley, L. J. Geerligs and C. Dekker, *Nature* **386**, 474 (1997).
145. S. J. Tans, A. R. M. Verschueren and C. Dekker, *Nature* **393**, 49 (1998).
146. R. Martel, T. Schmidt, H. R. Shea, T. Hertel and P. Avouris, *Appl. Phys. Lett.* **73**, 2447 (1998).
147. Z. Yao, H. W. Ch. Postma, L. Balents and C. Dekker, *Nature* **402**, 273 (1999).
148. M. S. Fuhrer, J. Nygard, L. Shih, M. Forero, Y. G. Yoon, M. S. C. Mazzoni, H. J. Choi, J. Ihm, S. G. Louie, A. Zettl and P. L. McEuen, *Science* **288**, 494 (2000).
149. C. W. Zhou, J. Kong, E. Yenilmez and H. Dai, *Science* **290**, 1552 (2000).

150. T. Rueckes, K. Kim, E. Joselevich, G. Y. Tseng, C. L. Cheung and C. M. Lieber, *Science* **289**, 94 (2000).

151. M. S. Fuhrer, B. M. Kim, T. Durkop and T. Brintlinger, *Nano Lett.* **2**, 755 (2002).

152. M. Radosavljevic, M. Freitag, K. V. Thadani and A. T. Johnson, *Nano Lett.* **2**, 761 (2002).

153. V. Derycke, R. Martel, J. Appenzeller and P. Avouris. *Nano Lett.* **1**, 453 (2001).

154. A. Bachtold, P. Hadley, T. Nakanishi and C. Dekker, *Science* **294**, 1317 (2001).

155. Z. H. Chen, J. Appenzeller, Y. M. Lin, J. Sippel-Oakley, A. G. Rinzler, J. Y. Tang, S. J. Wind, P. M. Solomon and P. Avouris, *Science* **311**, 1735 (2006).

156. A. Javey, J. Guo, Q. Wang, M. Lundstrom and H. Dai, *Nature* **424**, 6949 (2003).

157. S. Wind, J. Appenzeller and P. Avouris, *Phys. Rev. Lett.* **91**, 058301 (2003).

158. R. Martel, V. Derycke, C. Lavoie, J. Appenzeller, K. Chen, J. Tersoff and Ph. Avouris, *Phys. Rev. Lett.* **87**, 256805 (2001).

159. M. Freitag, M. Radosavljevic, Y. X. Zhou, A. T. Johnson and W. F. Smith, *Appl. Phys. Lett.* **79**, 3326 (2001).

160. V. Derycke, R. Martel, J. Appenzeller and P. Avouris, *Appl. Phys. Lett.*, **80**, 2773 (2002).

161. S. Heinze, J. Tersoff, R. Martel, V. Derycke, J. Appenzeller and P. Avouris, *Phys. Rev. Lett.* **89**, 106801 (2002).

162. J. Appenzeller, J. Knoch, V. Derycke, R. Martel, S. Wind and P. Avouris, *Phys. Rev. Lett.* **89**, 126801 (2002).

163. J. Appenzeller, M. Radosavljevic, J. Knoch and P. Avouris, *Phys. Rev. Lett.* **92**, 048301 (2004).

164. S. Heinze, M. Radosavljevic, J. Tersoff and P. Avouris, *Phys. Rev. B* **68**, 235418 (2003).

165. M. Radosavljevic, S. Heinze, J. Tersoff and P. Avouris, *Appl. Phys. Lett.* **83**, 2435 (2003).

166. J. Guo, S. Datta and M. Lundstrom, *IEEE Trans. Elec. Dev.* **51**, 172 (2004).

167. S. J. Wind, J. Appenzeller, R. Martel, V. Derycke and P. Avouris, *Appl. Phys. Lett.* **80**, 3817 (2002).

168. A. Javey, H. Kim, M. Brink, Q. Wang, A. Ural, J. Guo, P. McIntyre, P. McEuen, M. Lundstrom and H. Dai, *Nat. Mater.* **1**, 241 (2002).

169. F. Nihey, H. Hongo, M. Yudasaka and S. Iijima, *Jpn. J. Appl. Phys.* **41**, 1049 (2002).

170. B. M. Kim, T. Brintlinger, E. Cobas, M. S. Fuhrer, H. M. Zheng, Z. Yu, R. Droopad, J. Ramdani and K. Eisenbeiser, *Appl. Phys. Lett.* **84**, 1946 (2004).

171. Z. H. Chen, J. Appenzeller, J. Knoch, Y. M. Lin and P. Avouris, *Nano Lett.* **5**, 1497 (2005).

172. A. Javey, J. Guo, D. B. Farmer, Q. Wang, D. W. Wang, R. G. Gordon, M. Lundstrom and H. Dai, *Nano Lett.* **4**, 447 (2004).

173. A. Javey, J. Guo, D. B. Farmer, Q. Wang, E. Yenilmez, R. G. Gordon, M. Lundstrom and H. Dai, *Nano Lett.* **4**, 1319 (2004).

174. R. V. Seidel, A. P. Graham, J. Kretz, B. Rajasekharan, G. S. Duesberg, M. Liebau, E. Unger, F. Kreupl and W. Hoenlein, *Nano Lett.* **5**, 147 (2005).

175. A. Javey, P. Qi, Q. Wang and H. Dai, *Proc. Natl. Acad. Sci. USA* **101**, 13408 (2004).

176. Y. M. Lin, J. Appenzeller, Z. H. Chen, Z. G. Chen, H. M. Cheng and P. Avouris, *IEEE Elec. Dev. Lett.* **26**, 823 (2005).

177. J. Knoch, S. Mantl and J. Appenzeller, *Sol. Stat. Elec.* **49**, 73 (2005).

178. J. Guo, M. Lundstrom and S. Datta, *Appl. Phys. Lett.* **80**, 3192 (2002).

179. Y. M. Lin, J. Appenzeller, J. Knoch and P. Avouris, *IEEE Trans. on Nanotechnol.* **4**, 481 (2005).

180. J. Chen, C. Klinke, A. Afzali and P. Avouris, *Appl. Phys. Lett.* **86**, 123108 (2005).

181. A. Javey, R. Tu, D. B. Farmer, J. Guo, R. G. Gordon and H. Dai, *Nano Lett.* **5**, 345 (2005).

182. M. Radosavljevic, J. Appenzeller, P. Avouris and J. Knoch, *Appl. Phys. Lett.* **84**, 3693 (2004).

183. J. Kong and H. Dai, *J. Phys. Chem. B* **105**, 2890 (2001).

184. M. Shim, A. Javey, N. W. S. Kam and H. Dai, *J. Am. Chem. Soc.* **123**, 11512 (2001).

185. J. Appenzeller, Y. M. Lin, J. Knoch and P. Avouris, *Phys. Rev. Lett.* **93**, 196805 (2004).

186. J. Appenzeller, Y. M. Lin, J. Knoch, Z. H. Chen and P. Avouris, *IEEE Trans on Elec. Dev.* **52**, 2568 (2005).

187. G. Zhang, X. Wang, X. Li, Y. Lu, A. Javey and H. Dai, *IEDM Tech. Dig.* 431 (2006).

188. H. S. P. Wong, *IBM J. of Res. and Dev.* **46**, 133 (2002).

189. R. Chau, S. Datta, M. Doczy, B. Doyle, B. Jin, J. Kavalieros, A. Majumdar, M. Metz and M. Radosavljevic, *IEEE Trans. on Nanotechnol.* **4**, 153 (2005).
190. W. Hoenlein, F. Kreupl, G. S. Duesberg, A. P. Graham, M. Liebau, R. V. Seidel and E. Unger, *IEEE Trans. on Comp. and Pack. Techn.* **27**, 629 (2004).
191. A. P. Graham, G. S. Duesberg, W. Hoenlein, F. Kreupl, M. Liebau, R. Martin, B. Rajasekharan, W. Pamler, R. Seidel, W. Steinhoegl and E. Unger, *Appl. Phys. A* **80**, 1141 (2005).
192. P. J. Burke, *Sol. Stat. Elec.* **48**, 1981 (2004).
193. J. Guo, S. Hasan, A. Javey, G. Bosman and M. Lundstrom, *IEEE Trans. Nanotechnol.* **4**, 715 (2005).
194. S. Hasan, S. Salahuddin, M. Vaidyanathan and A. A. Alam, *IEEE Trans. Nanotechnol.* **5**, 14 (2006).
195. D. Akinwande, G. E. Close and H. S. P. Wong, *IEEE Trans. Nanotechnol.* **5**, 599 (2006).
196. K. Alam and R. Lake, *Appl. Phys. Lett.* **87**, 073104 (2005).
197. L. C. Castro and D. L. Pulfrey, *Nanotechnol.* **17**, 300 (2006).
198. J. Appenzeller and D. J. Frank, *Appl. Phys. Lett.* **84**, 1771 (2004).
199. A. A. Pesetski, J. E. Baumgardner, E. Folk, J. X. Przybysz, J. D. Adam and H. Zhang, *Appl. Phys. Lett.* **88**, 113103 (2006).
200. S. Rosenblatt, H. Lin, V. Sazonova, S. Tiwari and P. L. McEuen, *Appl. Phys. Lett.* **87**, 153111 (2005).
201. X. Huo, M. Zhang, P. C. H. Chan, Q. Liang and Z. K. Tang, *Proc. IEEE IEDM Tech. Dig.* p 691 (2004).
202. J. M. Bethoux, H. Happy, G. Dambrine, V. Derycke, M. F. Goffman and J. P. Bourgoin, *IEEE Elec. Dev. Lett.* **27**, 681 (2006).
203. A. Le Louarn, F. Kapche, J.-M. Bethoux, H. Happy, G. Dambrine, V. Derycke, P. Chenevier, N. Izard, M. F. Goffman and J.-P. Bourgoin, *Appl. Phys. Lett.* **90**, 233108(2007)
204. P. G. Collins, K. Bradley, M. Ishigami and A. Zettl, *Science* **287**, 5459 (2000).
205. X. D. Cui, M. Freitag, R. Martel, L. Brus and P. Avouris, *Nano Lett.* **3**, 783 (2003).
206. S. Auvray, J. Borghetti, M. F. Goffman, A. Filoramo, V. Derycke, J. P. Bourgoin and O. Jost, *Appl. Phys. Lett.* **84**, 5106 (2004).
207. S. Auvray, V. Derycke, M. Goffman, A. Filoramo, O. Jost and J. P. Bourgoin, *Nano Lett.* **5**, 451 (2005).
208. K. Besteman, J. O. Lee, F. G. M. Wiertz, H. A. Heering and C. Dekker, *Nano Lett.* **3**, 727 (2003).

209. R. J. Chen, S. Bangsaruntip, K. A. Drouvalakis, N. W. S. Kam, M. Shim, Y. M. Li, W. Kim, P. J. Utz and H. Dai, *Proc. Natl. Acad. Sci. USA* **100**, 4984 (2003).
210. www. nano. com
211. K. Bradley, J. C. P. Gabriel and G. Gruner, *Nano Lett.* **3**, 1353 (2003).
212. E. Artukovic, M. Kaempgen, D. S. Hecht, S. Roth and G. Gruner, *Nano Lett.* **5**, 757 (2005).
213. Q. Cao, S. H. Hur, Z. T. Zhu, Y. Sun, C. J. Wang, M. A. Meitl, M. Shim and J. A. Rogers, *Adv. Mater.* **18**, 304 (2006).
214. N. Chimot, V. Derycke, M. F. Goffman, J. P. Bourgoin, H. Happy and G. Dambrine, *Appl. Phys. Lett.* **91**, 153111 (2007).
215. T. Durkop, S. A. Getty, E. Cobas and M. S. Fuhrer, *Nano Lett.* **4**, 35 (2004).
216. J. A. Misewich, R. Martel, P. Avouris, J. C. Tsang, S. Heinze and J. Tersoff, *Science* **300**, 783 (2003).
217. M. Freitag, Y. Martin, J. A. Misewich, R. Martel and P. Avouris, *Nano Lett.* **3**, 1067 (2003).
218. J. Chen, V. Perebeinos, M. Freitag, J. Tsang, Q. Fu, J. Liu and P. Avouris *Science* **310**, 5751 (2005).
219. L. Marty, E. Adam, L. Albert, R. Doyon, D. Menard and R. Martel, *Phys. Rev. Lett.* **96**, 136803 (2006).
220. D. Mann, Y. K. Kato, A. Kinkhabwala, E. Pop, J. Cao, X. Wang, L. Zhang, Q. Wang, J. Guo and H. Dai, *Nat. Nanotechnol.* **2**, 33 (2007).
221. X. F. Guo, L. M. Huang, S. O'Brien, P. Kim and C. Nuckolls, *J. Am. Chem. Soc.* **127**, 15045 (2005).
222. J. M. Simmons, I. In, V. E. Campbell, T. J. Mark, F. Leonard, P. Gopalan and M. A. Eriksson, *Phys. Rev. Lett.* **98**, 086802 (2007).
223. A. Star, Y. Lu, K. Bradley and G. Gruner, *Nano Lett.* **4**, 1587 (2004).
224. J. Borghetti, V. Derycke, S. Lenfant, P. Chenevier, A. Filoramo, M. Goffman, D. Vuillaume and J. P. Bourgoin, *Adv. Mater.* **18**, 2535 (2006).
225. Z. Yao, C. L. Kane and C. Dekker, *Phys. Rev. Lett.* **84**, 2941 (2000).
226. M. Nihei, M. Horibe, A. Kawabata and Y. Awano, *Jap. J. of Appl. Phys.* **43**, 1856 (2004).
227. M. Nihei, A. Kawabata, D. Kondo, M. Horibe, S. Sato and Y. Awano, *Jap. J. of Appl. Phys.* **44**, 1626 (2005).
228. Y. Awano, *IEICE Trans. on Elec.* **E89C**, 1499 (2006).
229. Q. Ngo, A. M. Cassell, A. J. Austin, J. Li, S. Krishnan, M. Meyyappan and C. Y. Yang, *IEEE Elec. Dev. Lett.* **27**, 221 (2006) and references therein.

# 13

# Spin Electronics

Kyung-Jin Lee and Sang Ho Lim

Department of Materials Science and Engineering, Korea University, Anam-dong, Seongbuk-gu, Seoul 136-713, Korea

........................................

Spintronics, which utilizes additional spin degree of freedom in electronic systems, is an emerging research subject in the field of electronics due to its distinct advantages of non-volatility, higher speed, lower power consumption, more functionality, and higher density. Some important spintronic devices are introduced in this chapter and a particular emphasis is placed on spin-transfer torque magnetic random access memory and spin field-effect transistor which are considered to have a great potential in applications. Recent progress and remaining challenges on these devices are described.

## 1.    Introduction

Spintronics is an acronym for "SPIN TRansport electrONICS" which was coined by S. A. Wolf[1] initially to name a DARPA (Defense Advanced Research Projects Agency) project to develop magnetoresistive memory and sensors. Now, spintronics (or "spin-based electronics") refers to a multidisciplinary research field, the central theme of which is the active manipulation of spin degree of freedom in solid state systems.[2] The technology makes it possible to develop novel sensor, memory and logic device, since the addition of the spin degree of freedom to conventional charge-based electronics will improve capability and performance of present electronic products. The potential advantages of these new devices over conventional charge-based ones would provide non-volatility, higher speed in data processing, lower power consumption, more functionality, and higher density.

The research field of spintronics is quite wide and can be classified into three categories.[3] The first category is metal based devices such as read sensor in hard disc drive and magnetic random access memory (MRAM) where the magnetization configuration is used to control the movement of charge, a phenomenon known as magnetoresistance (MR). Many of the devices in this category were commercialized and even some of them are technologically in a mature stage. Read sensor in hard disc drive based on giant MR (GMR), which was commercially available in 1997, is a good example. Now, more advanced sensors based on tunneling magnetoresistance (TMR), a large value of which at room temperature was discovered in 1995,[4,5] are being actively developed and will soon be implemented in commercial products. TMR is also promising in MRAM. MRAM has an important attribute of non-volatility, and furthermore, possesses excellent intrinsic properties such as high density comparable to dynamic random access memory and high speed comparable to static random access memory, and unlimited read/write operation. In spite of its great potential, the progress is rather slow. Currently, only a low density MRAM (4 Mb) is commercially available[6] and, due to its low density, the price per bit is very high prohibiting wide spread applications. Therefore, the current focus is to increase the density, hopefully to a level comparable to existing memories such as flash memory. There have been several critical issues on the road to high density MRAM: most notably, high switching field and narrow write margin. In recent years, however, there were a couple of breakthroughs, reviving a new ray of hope. One is the discovery of the spin-transfer torque (STT), which was theoretically predicted in 1996[7,8] and was experimentally demonstrated in 2000.[9] The other breakthrough is the realization of a very large TMR (also called giant TMR) in MgO based magnetic tunnel junctions (MTJs). Butler *et al.*[10] and Mathon and Umerski[11] independently predicted giant TMR (about 1000%) in an epitaxial [001] Fe/ MgO/ Fe tunnel junctions. This prediction was quickly realized in 2004 by Parkin *et al.*[12] in sputtered textured junctions and Yuasa *et al.*[13] in epitaxial thin films by MBE. However, these techniques used are rather restrictive in real applications, in particular MBE. Initially, this kind of restriction appeared inevitable because, according to theories, the perfect band matching through the epitaxial FM electrode/ MgO/ FM electrode (FM stands for ferromagnet) is required to achieve the giant TMR. To a pleasant surprise, however, this restriction was removed by Djayaprawira *et al.*[14] who observed 230%

TMR at RT in CoFeB/ MgO/ CoFeB MTJs fabricated by using a Cannon Anelva commercial 200-mm production sputter system. More careful and systematic studies involving high temperature annealing were soon followed, eventually leading to a huge TMR of 472% at RT and of 804% at 5 K by the Hideo Ohno group at Tohoku University in Sendai, Japan.[15] These values are the highest ones so far reported in the literature. Even higher values (500% at RT and 1010% at 5 K) were presented by the same group at a meeting in February 2007.[16] Armed with these two recent breakthroughs, the development of high density MRAM is expected to move at a faster pace. A recent demonstration of 2 Mb chip by Hitachi Central Research Laboratory and Tohoku University (led by Hideo Ohno group), which was made public on February 14, 2007, is simply a beginning towards the commercial development of high density STT-MRAM.[17]

The second category of spintronic devices is semiconductor-based devices such as spin-FET (field-effect transistor) and spin-LED (light emitting diode). In these second category devices, FM electrodes are usually used as a spin source. In this sense, the second category devices are often called hybrid devices consisting of semiconductors and metals. The key to semiconductor-based spin devices is the creation of the spin-polarized currents in semiconductors (often referred to as spins in semiconductors), usually through the injection of spins from FM electrode. Spins in semiconductors were successfully realized by optical means. Optical pumping with circularly polarized light was found to be very effective in generating spin currents in direct-bandgap semiconductors.[18] Through this technique, long spin lifetimes[19] and diffusion lengths[20] in semiconductors were demonstrated. Furthermore, it was shown that the electron spin can traverse the interfaces of two different semiconductors without losing its coherence,[21] showing the realistic possibility of semiconductor-based spintronic devices. However, this optical technique is not practical at all, because it is very hard to miniaturize optical devices at a low price. In this sense, electrical injection from FM into semiconductor is more practical. Initially, ohmic contacts formed by FM and semiconductor were used to inject spins into semiconductors, but no clear spin signal was detected.[22,23] The reason for this was identified to be the conductivity mismatch,[24] which, according to Rashba,[25] could be solved by forming a tunnel contact between FM and semiconductor. Subsequently, some encouraging results were reported by several groups using various tunnel contacts, such as Fe/GaAs Schottky barrier[26,27] AlOx barrier,[28,29] and MgO barrier.[30,31] In these works, the spin polarization in semiconductors was usually measured in a spin-LED geometry. A high spin polarization value of 47% was achieved near room temperature (290 K) with the use of MgO barrier.[30] In spite of this success in spin-LEDs, no one realized a fully operated spin-FET, even though intensive work was carried out by numerous groups around the world after the first proposal in 1990 by Datta and Das.[32] Considering that spin-FETs are more important than spin-LEDs from application point of view, it is important to know the reason behind this. In spin-LEDs, spins are injected from FM or diluted magnetic semiconductor (DMS) into semiconductor, usually a quantum well, where the spins (either electrons or holes) are combined to respective carriers to generate circularly polarized light, which is detected optically usually in a spin-LED geometry. In the view point of spin process, there is only a single process of spin injection involved for the working of the spin-LED. This is not the case for a spin-FET where a series of processes involving the spin injection, spin transport, and spin detection should be done successfully for a workable spin-FET, making it much harder to realize a workable device. In addition to spin-FETs and spin-LEDs, there are other types of spintronic devices in this second category: magnetic logic devices such as magnetic quantum dot cellular automata. In these devices, the role of the spin is to process data without any need to move charge at all. Results reported so far on this type of devices are impressive. Cowburn and Welland demonstrated room temperature magnetic quantum dot cellular automata in 2000[33] and, several years later in 2006, Imre *et al.* demonstrated similar but more complicated devices such as a majority logic gate.[34]

The third category is related to the devices using the spins as quantum bits (qubits) which are the essential ingredient of quantum computing. Although this field has a great technological potential, it is in a very preliminary stage of technological development and, therefore, it will not be treated in this article. From this very brief introduction, it is considered that STT-MRAM and spin-FET have a great potential in applications, but they are not fully developed at this stage. It is therefore natural to pay more attention on these devices. The following sections are devoted to STT-MRAM and spin-FET.

## 2. STT-MRAM

The spin-transfer torque is a quantum mechanical effect. The spins of conduction electrons are filtered when an electrical current passes through a structure consisting of normal metal (NM)/ FM/ NM because the reflection probabilities at the interface of NM/ FM are spin-dependent. As a consequence of the spin-filtering, the spin direction of conduction electrons is oriented toward the magnetization of FM (see Fig. 1(a)). In a spin valve structure consisting of NM/ FM1/ NM/ FM2/ NM, the filtered spin-flow by FM1, a spin-polarized current, is again filtered by FM2 which has a non-collinear magnetization to FM1. In the second spin-filtering process, the spin direction of conduction electrons is reoriented along the magnetization of FM2. Because the spin angular momentum must be conserved, the changed amount of the spin angular momentum of conduction electrons is transferred and exerts a torque to the magnetization of FM2, i.e. the spin-transfer torque (STT) (see Fig. 1(b)).[7,8] The STT enables various types of current-induced magnetic excitations such as magnetization switching, magnetization precession and domain wall motion.[9,35−46]

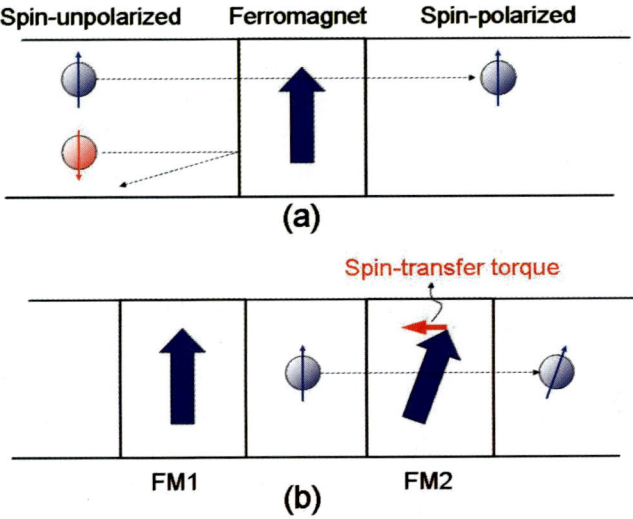

**(a)**

**(b)**

Fig. 1. A schematic illustration of spin-filtering assuming a ferromagnet as a perfect spin-filter. An electrical current in the left side of ferromagnet is unpolarized. When it passes through the ferromagnet, all up spins transmit whereas all down spins reflect at the left interface of ferromagnet. As a consequence, only up spins are in the right side of ferromagnet and, therefore, the electrical current is spin-polarized along the magnetization of ferromagnet. (b) A schematic illustration of spin-transfer torque. Dotted lines in (a) and (b) indicate the movement of conduction electrons.

The STT-MRAM uses the same read scheme with the conventional MRAM technology whereas it adopts the current-induced magnetization switching (CIMS) as a new write scheme instead of the field-induced magnetization switching. The STT theories based on the macrospin concept predicted that the switching current ($I_C$) for the current-induced magnetic excitation is given by,[7,47]

$$I_C = \frac{2e}{\hbar} \frac{\alpha}{g} M_S V (H_K + 2\pi M_S) \tag{1}$$

where $\alpha$ is the intrinsic damping constant, $g$ is the spin polarization factor, $M_S$ is the saturation magnetization, $V$ is the volume of magnetic cell, and $H_K$ is the anisotropy field.

The CIMS provides a scalable write scheme for MRAM since the switching current is proportional to the volume of magnetic cell. The CIMS also solves the issue of write selectivity which is foreseen as an increasingly challenging difficulty in the conventional MRAM technology based on the field-induced magnetization switching. In the field-induced magnetization switching using two orthogonal current lines, the cell located at the cross-point of the two lines is selected to switch by applying currents in both lines. Cells except for the selected one at the cross-point in each current line are inevitably half-selected. Undesired switching of the half-selected cell may happen when the distribution of write current of each cell is not well-controlled. In the CIMS, however, there is no half-selected cell and therefore no write error related to the selectivity issue since the current only flows through the selected cell.

For the application of the CIMS in MRAM, the switching current density must be comparable to that supplied by a typical CMOS circuit of comparable density.[48] There have been a lot of efforts to reduce the current density for magnetization switching. Referring to Eq. 1, the switching current density is approximately proportional to $M_S^2$ since $H_K << 2\pi M_S$. Therefore, an order of magnitude of $I_c$ can be reduced by decreasing $M_S$ by the factor of 3.[49] Another way of reducing the switching current density is to increase the spin polarization factor, $g$. The spin polarization factor can be enhanced by introducing double spin-filters,[50,51] or nano-oxide layer for specular scattering.[52] Replacing a normal metal spacer between two FMs by an MgO insulating barrier provides the enhancement of $g$ by the factor of 3~5.[53−57] Therefore, adopting MgO as the insulating barrier is inevitable not only for increasing the reading signal but also for reducing the switching current density. Many precessions before the magnetization switching

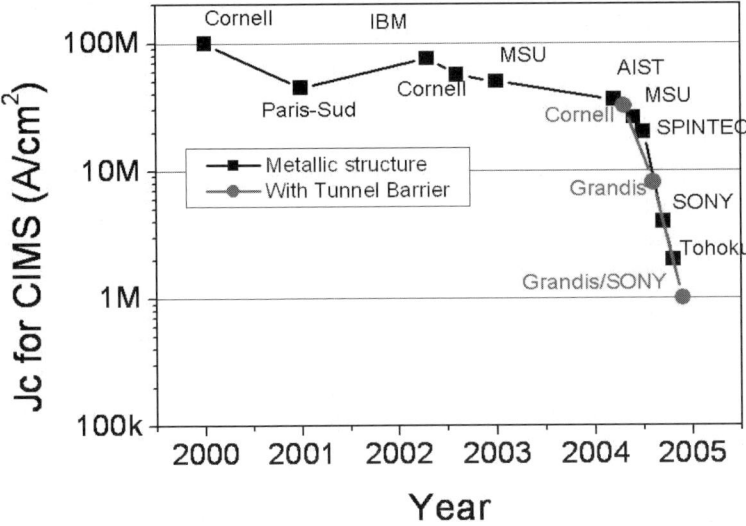

Fig. 2. Yearly change of the switching current density for the current-induced magnetization switching.

is a distinguishing feature of the magnetization dynamics induced by STT. Pre-charging,[58] or applying a hard-axis magnetic field[59] also reduces the switching current density by effectively increasing the precession susceptibility of magnetization.

Figure 2 shows the yearly change of the switching current density reported in the literature. The initial current density was approximately $10^8$ A/cm$^2$ in a Co/ Cu/ Co spin valve.[9] Up to now, the lowest current density is $10^6$ A/cm$^2$ using double MgO barriers with different resistances (Ta/ PtMn/ CoFe/ Ru/ CoFeB/ MgO/ CoFeB/ MgO/ CoFeB/ CoFe/ PtMn/ Ta).[57] However, considering the thermal stability ($KV/k_BT = 60$, where $K$ is the total anisotropy energy density including crystalline and shape anisotropies, $k_B$ is the Boltzmann constant, and $T$ is the temperature in Kelvin.), the desired current density is of the order of $10^5$ A/cm$^2$ for a cell with the lateral dimensions of about 50 nm. Therefore, another order of magnitude reduction would be necessary. Another constraint on the current density is related to the breakdown voltage ($V_B$) and $RA$ (resistance × area of junction) of tunnel barrier. The writing voltage must be smaller than 0.8 $V_B$. The maximum allowed write current density is given by 0.8 $V_B/RA$ and is also of the order of $10^5 \sim 10^6$ A/cm$^2$. Therefore, the development of a tunnel barrier with a low $RA$ and a high $V_B$ is desired for the application of STT-MRAM.

Besides the efforts to reduce the switching current density, the study on the magnetization dynamics induced by STT is essential for the application of STT-MRAM. Initial STT theory predicted a single-domain behavior of magnetization. However, micromagnetic studies[60–63] and direct X-ray imaging[33] revealed that the STT induces incoherent magnetization dynamics. The incoherence results in broadening distribution of write current and incomplete magnetization switching even at a high current density.[63] The circular magnetic field (viz., Oersted field) due to the charge-flow perpendicular to the film plane is responsible for the incoherence. A high spin polarization factor is needed to suppress the incoherence since the competition between the Oersted field and the STT is crucial to determine the degree of the incoherence.[63]

In 2005, SONY demonstrated CMOS integrated 4 kb STT-MRAM using 130 nm technology and magnetic tunnel junction with $100 \times 150$ nm$^2$.[65] The switching current density at 10 ns current pulse was about $3 \times 10^6$ A/cm$^2$. Reproducible spin-torque switching up to $10^{12}$ cycles was experimentally confirmed. Recently, 2 Mb STT-MRAM with bit-by-bit bidirectional current write and parallelizing-direction current read was demonstrated by Hitachi-Tohoku University.[17] Reminding its rapid progress, the STT-MRAM is a very promising candidate for the next generation non-volatile memory.

## 3.    Spin-FET

As was mentioned in the introduction section, a workable spin-FET was not realized yet, although the first proposal of the device was made in 1990.[32] A schematic illustration of a spin-FET is shown in Fig. 3. The basic concept of the device is similar to that of a conventional transistor such as an MOS-FET, except that both the source and drain are FMs or DMSs. Spin carriers are injected from the source and these injected carriers are detected at the drain after the spin transport along the channel. If the spin direction at the drain is identical to that at the source, then the conductance is high. The opposite is true when the spin directions are antiparallel. The spin direction can be modulated during the spin transport along the channel. This can be done by applying a gate voltage, which modulates the spin precession through the Rashba effect.[66] This kind of spin control by an applied

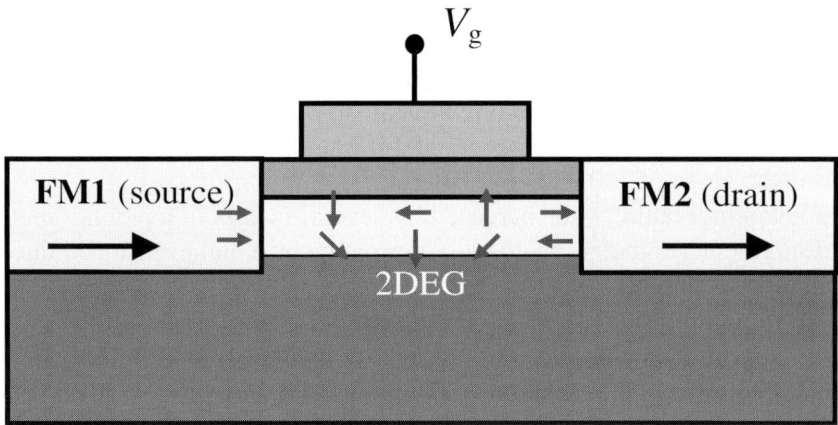

Fig. 3. A schematic illustration of a spin-FET. The overall structure is similar to that of an MOSFET, except that both source and drain are ferromagnetic materials. The spin direction is modulated by a gate voltage, as the spin is transported along the channel such as 2 dimensional electron gas (2DEG).

electric field is the most novel feature of a spin-FET. Several advantages expected, among many others, are a fast operation and small power consumption. Spin-FETs, like MRAM, are considered to be more robust being less susceptible to environment than conventional charge-based devices. Furthermore, spin-FETs can handle more complicated functions with relative ease by manipulating the magnetization direction of FM electrodes, for example, during the device operation.

Active research has been carried out by many research groups. Instead of introducing various research activities and related results, only selected results from limited groups are introduced here simply for a better and clear understanding of the research direction. The Johnson group at NRL[67−69] fabricated spin-FETs based on an InAs 2DEG (two-dimensional electron gas) structure. The group was able to observe a spin-valve type signal from the device (without the gate control), but the detected signal was very weak. In order to improve the spin signal, a group at KIST (Korea Institute of Science and Technology) fabricated similar devices but with much reduced dimensions. The dimensions of the original devices by Johnson *et al.* were all micron-sized, but those of the new devices by the KIST group were nanoscaled. It is well-accepted that the number of spin-polarized carriers decays exponentially with increasing channel length. So, a short channel length is essential for a large spin signal. Also, there are many negative effects related to a wide channel width and some of them include a local Hall and a fringe field effect as well as the effect due to the DP mechanism.[70,71] Several fabrication processes were developed to fabricate InAs based spin-FETs with nanochannels and one typical structure is shown in Fig. 4. The channel length was in the range of 150–1200 nm, while the channel width was in the range of 200–800 nm. Expectedly, the spin signal, only observed at a low temperature, was increased by several factors in a potentiometric measurement geometry[72] and this improvement is believed to be due to smaller spin de-phasing resulting from the size reduction. There are still many problems to be solved on the road to a fully workable spin-FET and some of them may be the further improvement of spin signal, the observation of spin signal at room temperature, and the gate control of the spin procession.

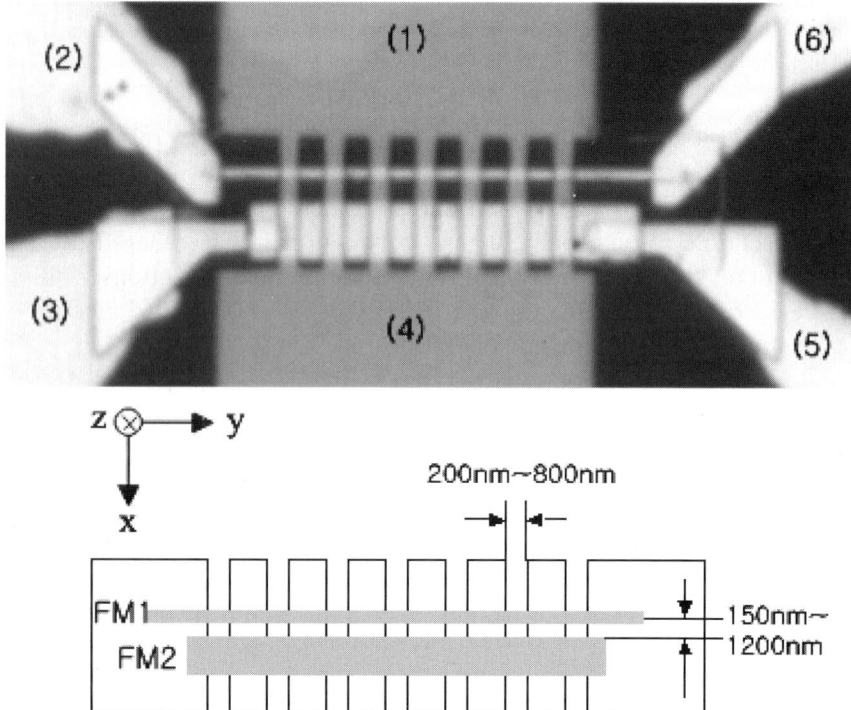

Fig. 4. A scanning electron microscopy image of an InAs based spin-FET with nanochannels (upper panel), together with the schematic diagram showing the channel dimensions (lower panel).

In spite of the improvement in the spin signal through a quasi 1D channel with a nanoscale channel length, the observed spin signal is considered to be still not high enough for a workable spin-FET. There may be several reasons and, among these, the spin scattering at the interfaces can be an important factor. It was demonstrated in a spin-LED that spin injection efficiency is reduced significantly by interface defect spin scattering.[73] In the spin-FETs with nanochannels described earlier, there are several interfaces involving FM/ SC (semiconductor) and SC/ SC interfaces. That is the reason why the spin injection and detection across various interfaces are currently real hot issues. A novel idea was proposed to overcome this problem.[74] The main idea is to use the well-known compound semiconductor, HgCdTe (MCT), with a large Zeeman effect (5.7 meV at 1 T)[75] and a large Rashba coupling ($\sim$11.5 meV).[76] Also the compound is known to have a long spin life time (356 ps at T=150 K).[77] The device structure is simple, as can be expected from no interfaces along the spin transport. An MCT channel was fabricated by using conventional lithography and source and drain contacts were formed at both ends. Then, in the middle of the channel, a gate electrode with a length of 5 $\mu$m was positioned. MCT is non-magnetic, but, due to a large Zeeman effect, spin imbalance can be generated with an applied magnetic field. Surprisingly, a clear resistance modulation was observed at low temperatures by modulating the gate voltage.[74] In Fig. 5, some of the results for the magnetoconductance ($\tau$) versus applied field curves are shown at various gate voltages. The variation of the Rashba coefficient ($\alpha$) and spin-orbit scattering time ($\tau_{SO}$) with the gate voltage, extracted from the experimental data, is summarized in the inset. It is clear from the figure that the magnetoconductance varies appreciably with the gate voltage. The obtained Rashba coefficients are comparable to those reported in the literature,[78] confirming that the observed resistance modulation is due to the gate effect through the Rashba coupling. It is of interest to consider the reason for the large resistance modulation. Quite likely, the key to the success is the absence of any interface along the spin transport that can prevent efficient spin injection from ferromagnetic material to semiconductor. This new device has several disadvantages, some of which include the application of a large magnetic field during the device operation and low temperature operation.

FM metals such as Fe, FeCo or FeNi are commonly used as a spin aligner in spintronic devices. This is a natural choice because ordinary FM metals are good spin aligners and possess high Curie temperatures. Furthermore, thin film deposition of these metals can be done with ease. However,

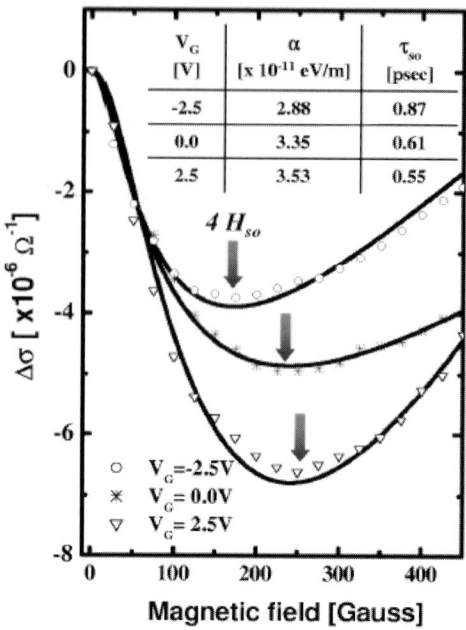

| $V_G$ [V] | $\alpha$ [x $10^{-11}$ eV/m] | $\tau_{so}$ [psec] |
|---|---|---|
| -2.5 | 2.88 | 0.87 |
| 0.0 | 3.35 | 0.61 |
| 2.5 | 3.53 | 0.55 |

Fig. 5. The results for the magnetoconductance versus applied magnetic field curve in a HgCdTe based spin-FET. The results were obtained at three different gate voltages of $-2.5$ V (circles), 0 V (asterisks) and $+2.5$ V (inverse triangles). The minimum points are indicated by the arrows. The variation of the Rashba coefficient ($\alpha$) spin-orbit scattering time ($\tau_{SO}$) with the gate voltage, obtained from the results, is summarized in the inset.[74] [Reprinted with permission from J. Hong, J. Lee, S. Joo, K. Rhie, B. C. Lee, J. Lee, S.-Y. An, J. Kim, and K.-H. Shin, *J. Kor. Phys. Soc.* **45**, 197 (2004). Copyright (2004), The Korean Physical Society.]

it was observed experimentally[22,23] and also predicted theoretically[24,79−81] that the spin injection from FM metals into SCs occurs very inefficiently, if it does not occur at all, when there is an Ohmic contact at the FM/ SC interface due to the conductivity or energy band mismatch. One way to overcome this problem is to form a tunnel contact between FM and semiconductor,[25] as was mentioned in the introduction. If the conductivity or energy band mismatch is a source of the problem of spin injection and also spin detection, then the natural extension is to use a magnetic material with a SC energy band and resistivity, known as magnetic SCs or DMSs. Great interest in DMSs were revived with the discovery of ferromagnetic III–V based materials such as InMnAs in 1992[82] and GaMnAs in 1996,[83] although Eu-based chalcogenides were reported in the 1960s and early 1970s[84] and II–VI based alloys in the 1980s.[85] In both II–VI and III–V based systems, the semiconductors become ferromagnetic with the incorporation of magnetic transition metals (TM) such as Mn. The incorporation of TM into II–VI semiconductors is rather easy, because TM typically exhibits the valence state of $+2$, being the same as one of the constituents in II–V. However, this is not the case for III–V systems; actually, TM is thermodynamically not stable in III–V, resulting in TM segregation. So, low temperature molecular beam epitaxy (LTMBE) was usually used to fabricate III–V based magnetic semiconductors. Even in this case, the amount of TM incorporated in III–V is limited up to 10 at.%. So far, InMnAs and GaMnAs are the most extensively studied systems and the origin of its ferromagnetism is reasonably well-established. Recent progress on these materials was described in a review article by MacDonald *et al.*[86] The highest Curie temperature of as-grown GaMnAs sample is ~110 K for a wide range of Mn compositions. After careful annealing, the Curie temperature can reach as high as 170 K.[87] Active work is currently undergoing to further increase the Curie temperature, for example, by co-doping with other materials (mainly to increase the carrier density and hence carrier-mediated exchange coupling) and by wavefunction engineering (to increase the effectiveness of the exchange coupling).[86,87] Recently, group IV based magnetic semiconductors also received much attention, due to the theoretical prediction of a high Curie temperature based on the Zener model.[88] Among many group IV semiconductors, Ge has been studied most extensively.[89–91] Ge has an important advantage in that it is lattice-matched to the AlGaAs/GaAs family, thus facilitating incorporation into III–V heterostructures. Furthermore, Ge has higher intrinsic hole mobility than GaAs and Si. Results reported so far on group IV based magnetic semiconductors are encouraging and the most significant results were observed by Park *et al.* in epitaxial Ge–Mn thin films by MBE.[89] Ferromagnetic ordering with reasonably high Curie temperatures was reported. Also, voltage controlled ferromagnetic order was demonstrated with a low gate voltage of 0.5 V, opening up the possibility of new spintronic devices. However, the highest Curie temperature achieved so far is 116 K, being still far lower than room temperature. The main reason for the low Curie temperature may be the limited Mn content incorporated into Ge. Recent first principles calculations predict that the Mn–Mn exchange interactions and hence Curie temperature can be increased by increasing Mn content.[92] However, the introduction of Mn into Ge was found to be very much limited; the highest amount of Mn incorporated without segregation was reported to be 3.3 at.% even with a very low temperature (70°C) MBE process.[89] The brief discussion on DMSs indicates that DMSs can play an important role in realizing a workable spin-FET, but the main issue is to increase the Curie temperature exceeding room temperature. Another issue is related with the carrier type. Electron spins are better than hole spins, because the spin diffusion length of the former is much longer than that of the latter. Unfortunately, however, the main carriers in most DMSs are holes.

Theoretical calculations in both diffusive[24,79−81] and ballistic transports[93] predict that the injected electron current is completely spin polarized, if the FM or DMS contacts are 100% spin polarized, namely half metals. Half metals, discovered in the early 1980s by de Groot *et al.*,[94] have

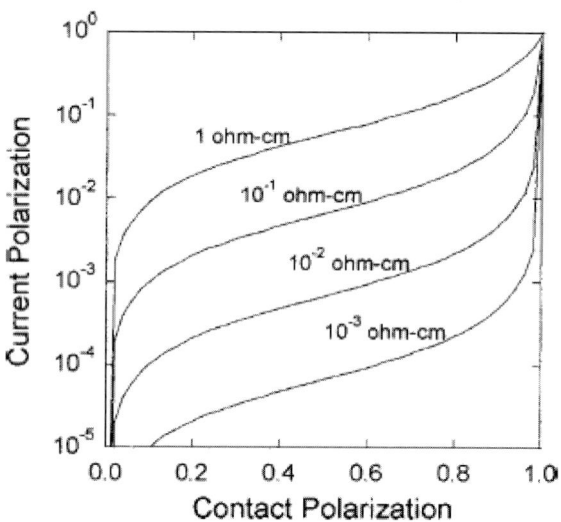

Fig. 6. Calculated results for the injected current spin polarization as a function of the contact spin polarization for various values of the contact resistivity.[79] The interface resistance was assumed to be zero. Irrespective of the contact resistivity, the injected current is completely spin-polarized as the contact polarization approaches 100% (half-metallicity). Note that the injected current spin polarization drops off extremely rapidly as the contact polarization is deviated from the half metallicity. [Reprinted (figure) with permission from D. L. Smith and R. N. Silver, *Phys. Rev. B* **64**, 045323 (2001). Copyright (2001) by the American Physical Society.]

a unique band structure in that, at the Fermi level, one spin band is completely empty (insulator) while the other spin band is occupied (metallic), leading to 100% spin polarized conduction electrons at the Fermi surface. An important point of theoretical calculations is that the complete spin polarization occurs whenever half metallic materials are used as the contact, irrespective of the conductivity mismatch. This is really exciting for a spin-FET, because complete spin injection and detection can be possible with half metallic electrodes. However, no experimental demonstration of half metals has been made so far, mainly because, among many others, real samples contain defects and the electronic state of surface (which dominantly affects the transport properties) is different from that of bulk where half metallicity was predicted theoretically. Furthermore, the transport theories predict that the spin injection efficiency drops off extremely rapidly as the spin polarization of the electrodes is deviated from the half metallicity,[24,79] as shown in Fig. 6. Initial results on half metals were very encouraging; for example, more than 90% spin polarization was observed

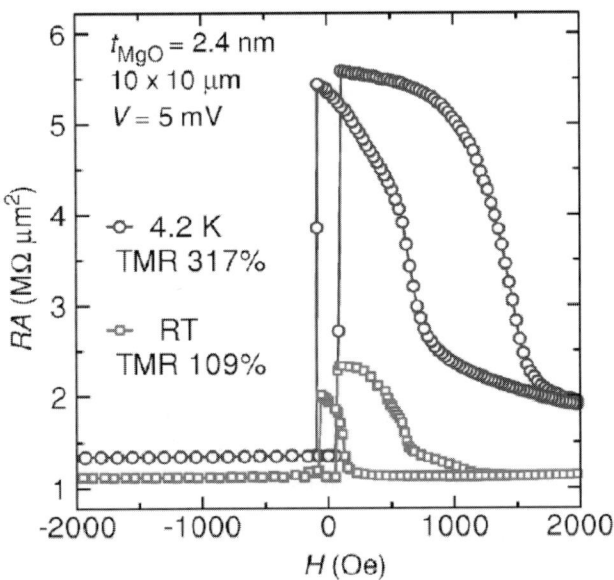

Fig. 7. RA versus applied field curves at 4.2 K and RT in fully epitaxial MTJs, $Co_2 Cr_{0.6} Fe_{0.4} Al/ MgO/ Co_{0.5} Fe_{0.5}$, showing giant TMR values.[97] [Reprinted with permission from T. Marukame, T. Ishikawa, S. Hakamata, K. Matsuda, T. Uemura, and M. Yamamoto, *Appl. Phys. Lett.* **90**, 012508 (2007). Copyright (2007), American Institute of Physics.]

by Park *et al.* in a $La_{0.7}Sr_{0.3}MnO_3$ compound by using photoelectron emission experiments.[95] However, no encouraging results on the spin polarization in a junction structure such as magnetic tunnel junctions were reported, until recently. Sakuraba *et al.* obtained a high TMR ratio of 570% at 2 K (67% at room temperature) in a $Co_2MnSi/AlOx/Co_2MnSi$ junction[96] The insulating barrier was amorphous and the Co-based Heusler alloy $Co_2MnSi$ was highly oriented in both electrodes. Although the low temperature TMR value is very high, the temperature dependence is very strong, resulting in a moderate TMR value at room temperature. Very recently, Marukame *et al.* achieved a TMR ratio of 317% at 4.2 K (109% at room temperature) in fully epitaxial MTJs, $Co_2Cr_{0.6}Fe_{0.4}Al/MgO/Co_{0.5}Fe_{0.5}$.[97] The main results are shown in Fig. 7. The (tunneling) spin polarization at 4.2 K is estimated to be 88% from the Julliere model.[97] A weak temperature dependence of TMR and hence spin polarization is also noted, although the observed temperature dependence is still higher than the conventional FM/ MgO/ FM tunnel junctions. The latest development in half metals is really exciting, although we still need some more breakthroughs on the road towards the complete spin polarization and hence 100% spin injection or detection efficiency.

## 4.    Conclusions and Outlook

Spin electronics (or spintronics) is a new and emerging technology which exploits the spin as well as the charge of electrons. With the additional spin degree of freedom, novel devices with much enhanced performance and functionality can be realized. Some spintronic devices such as read sensors in hard disc drives were commercialized in a decade ago and their impact on the information technology is already enormous and wide-spread. Many more are still to be implemented in real products. A brief overview on spintronic technology is given in this article, followed by some detailed description on STT-MRAM and spin-FET, which are not fully developed at this moment but are considered to be the most important spintronic devices from the application point of view. STT-MRAM has an important advantage of excellent scalability over conventional MRAM, hopefully leading to a density level comparable to existing memories such as Flash memory. Two important recent breakthroughs, current induced magnetization reversal in nanopillars and the realization of giant TMR in MgO based magnetic tunnel junctions (reaching 500% at room temperature), are expected to speed up the development of high density STT-MRAM. This expectation was quickly met by a recent demonstration of 2 Mb STT-MRAM. Considering very active research efforts all over the world, some viable high density MRAM will hit the market in 3~4 years. The development of a spin-FET has been rather slow, in spite of great research efforts, no workable devices being demonstrated so far after the first theoretical proposal in 1990. Nowadays, the research efforts appear to run out of steam, but it is hard to neglect the advantages which spin-FETs may have, such as low power consumption and fast operation. Fortunately, most of the fundamentals for the device operation are already developed some of which include the long spin life times and diffusion lengths, and an efficient spin injection into semiconductors (though realized in a simpler spin-LED geometry). The observation of a strong spin signal after a series of spin processes involving spin injection, spin transport, and spin detection may be the key factor for a workable spin-FET. Accumulated knowledge in the field, together with focused efforts, may eventually lead to the development of a workable spin-FET, hopefully at room temperature, in several years from now.

## 5.    Acknowledgments

This work was supported by the Korea Science and Engineering Foundation (KOSEF) through the National Research Laboratory program funded by the Korean Ministry of Science and Technology (Project No. M10600000198-06J0000-19810).

## References

1.    S. A. Wolf, A. Y. Chtchelkanova and D. M. Treger, *IBM J. Res. & Dev.* **50**, 101 (2006).
2.    I. Zutic, J. Fabian and S. Das Sarma, *Rev. Mod. Phys.* **76**, 323 (2004).
3.    D. D. Awschalom, M. E. Flatte and N. Samarth, *Scientific American* **286**, 52 (2002).
4.    J. S. Moodera, L. R. Kinder, T. M. Wong and R. Meservey, *Phys. Rev. Lett.* **74**, 3273 (1995).
5.    T. Miyazaki and N. Tezuka, *J. Magn. Magn. Mater.* **139**, L21 (1995).
6.    http://www.freescale.com
7.    J. C. Slonczewski, *J. Magn. Magn. Mater.* **159**, L1 (1996).
8.    L. Berger, *Phys. Rev. B* **54**, 9353 (1996).

9.  J. A. Katine, F. J. Albert, R. A. Buhrman, E. B. Myers and D. C. Ralph, *Phys. Rev. Lett.* **84**, 3149 (2000).
10. W. H. Butler, X. G. Zhang, T. C. Schulthess and J. M. MacLaren, *Phys. Rev. B* **63**, 054416 (2001).
11. J. Mathon and A. Umerski, *Phys. Rev. B* **63**, 220403 (2001).
12. S. S. P. Parkin, C. Kaiser, A. Panchula, P. M. Rice, B. Hughes, M. Samant and S.-H. Yang, *Nature Mater.* **3**, 862 (2004).
13. S. Yuasa, T. Nagahama, A. Fukushima, Y. Suzuki and K. Ando, *Nature Mater.* **3**, 868 (2004).
14. D. D. Djayaprawira, K. Tsunekawa, M. Nagai, H. Maekawa, S. Yamagata, N. Watanabe, S. Yuasa, Y. Suzuki and K. Ando, *Appl. Phys. Lett.* **86**, 092502 (2005).
15. S. Ikeda, J. Hayakawa, Y. M. Lee, T. Tanikawa, F. Matsukura and H. Ohno, *J. Appl. Phys.* **99**, 08A907 (2006).
16. Y. M. Lee, S. Ikeda, J. Hayakawa, F. Matsukura and H. Ohno, *The 2nd RIEC International Workshop on Spintronics – MgO-based Magnetic Tunnel Junctions*, Feb. 15–16, 2007, Tohoku University, Sendai, Japan.
17. T. Kawahara, R. Takemura, K. Miura, J. Hayakawa, S. Ikeda, Y. Lee, R. Sasaki, Y. Goto, K. Itoh, T. Meguro, F. Matsukura, H. Takahashi, H. Matsuoka and H. Ohno, ISSCC 2007, San Franciso (2007).
18. M. I. D'ykonov and V. I. Perel, in *Optical Orientation*, edited by F. Meier and B. P. Zakharchenya, North Holland, Amsterdam, pp 11–50 (1984).
19. J. M. Kikkawa and D. D. Awschalom, *Phys. Rev. Lett.* **80**, 4313 (1998).
20. J. M. Kikkawa and D. D. Awschalom, *Nature (London)* **397**, 139 (1999).
21. I. Malajovich, J. J. Berry, N. Samarth and D. D. Awschalom, *Nature (London)* **411**, 770 (2001).
22. F. G. Monzon and M. L. Roukes, *J. Magn. Magn. Mater.* **198**, 632 (1999).
23. A. T. Filip, B. H. Hoving, F. J. Jedema, B. J. van Wees, B. Dutta and S. Borghs, *Phys. Rev. B* **62**, 9996 (2000).
24. G. Schmidt, D. Ferrand, L. W. Molenkamp, A. T. Filip and B. J. van Wees, *Phys. Rev. B* **62**, R4790 (2000).
25. E. I. Rashba, *Phys. Rev. B* **62**, R16267 (2000).
26. A. T. Hanbicki, B. T. Jonker, G. Itskos, G. Kioseoglou and A. Petrou, *Appl. Phys. Lett.* **80**, 1240 (2002).

27. A. T. Hanbicki, O. M. J. van't Erve, R. Magno, G. Kioseoglou, C. H. Li, B. T. Jonker, G. Itskos, R. Mallory, M. Yasar and A. Petrou, *Appl. Phys. Lett.* **82**, 4092 (2003).

28. V. F. Motsnyi, J. de Boeck, J. Das, W. van Roy, G. Borghs, E. Goovaerts and V. I. Safarov, *Appl. Phys. Lett.* **81**, 265 (2002).

29. V. F. Motsnyi, P. van Dorpe, W. van Roy, E. Goovaerts, V. I. Safarov, G. Borghs and J. de Boeck, *Phys. Rev. B* **68**, 245319 (2003).

30. X. Jiang, R. Wang, R. M. Shelby, R. M. Macfarlane, S. R. Bank, J. S. Harris and S. S. P. Parkin, *Phys. Rev. Lett.* **94**, 056601 (2005).

31. R. Wang , X. Jiang, R. M. Shelby, R. M. Macfarlane, S. S. P. Parkin, S. R. Bank and J. S. Harris, *Appl. Phys. Lett.* **86**, 052901 (2005).

32. S. Datta and B. Das, *Appl. Phys. Lett.* **56**, 665 (1990).

33. R. P. Cowburn and M. E. Welland, *Science* **287**, 1466 (2000).

34. A. Imre, G. Csaba, L. Ji, A. Orlov, G. H. Bernstein and W. Porod, *Science* **311**, 205 (2006).

35. M. Tsoi, A. G. M. Jansen, J. Bass, W. C. Chiang, M. Seck, V. Tsoi and P. Wyder, *Phys. Rev. Lett.* **80**, 4281 (1998).

36. E. B. Myers, D. C. Ralph, J. A. Katine, R. N. Louie and R. A. Buhrman, *Science* **285**, 867 (1999).

37. J. Z. Sun, *J. Magn. Magn. Mater.* **202**, 157 (1999).

38. S. I. Kiselev, J. C. Sankey, I. N. Krivorotov, N. C. Emley, R. J. Schoelkopf, R. A. Buhrman and D. C. Ralph, *Nature* (*London*) **425**, 308 (2003).

39. W. H. Rippard, M. R. Pufall, S. Kaka, S. E. Russek and T. J. Silva, *Phys. Rev. Lett.* **92**, 027201 (2004).

40. M. Yamanouchi, D. Chiba, F. Matsukura and H. Ohno, *Nature* (*London*) **428**, 539 (2004).

41. E. Saitoh, H. Miyajima, T. Yamaoka and G. Tatara, *Nature* (*London*) **432**, 203 (2004).

42. A. Yamaguchi, T. Ono, S. Nasu, K. Miyake, K. Mibu and T. Shinjo, *Phys. Rev. Lett.* **92**, 077205 (2004).

43. I. N. Krivorotov, N. C. Emley, J. C. Sankey, S. I. Kiselev, D. C. Ralph and R. A. Buhrman, *Science* **307**, 228 (2005).

44. S. Kaka, M. R. Pufall, W. H. Rippard, T. J. Silva, S. E. Russek and J. A. Katine, *Nature* (*London*) **437**, 389 (2005).

45. F. B. Mancoff, N. D. Rizzo, B. N. Engel and S. Tehrani, *Nature* (*London*) **437**, 393 (2005).

46. L. Thomas, M. Hayashi, X. Jiang, R. Moriya, C. Rettner and S. S. P. Parkin, *Nature* (*London*) **443**, 197 (2006).

47. J. Z. Sun, *Phys. Rev. B* **62**, 570 (2000).
48. J. Z. Sun, *IBM J. Res. & Dev.* **50**, 95 (2006).
49. K. Yagami, A. A. Tulapurkar, A. Fukushima and Y. Suzuki, *Appl. Phys. Lett.* **85**, 5634 (2004).
50. L. Berger, *J. Appl. Phys.* **93**, 7693 (2003).
51. Y. Jiang, T. Nozaki, S. Abe, T. Ochiai, A. Hirohata, N. Tezuka and K. Inomata, *Nature Mater.* **3**, 361 (2004).
52. H. Y. T. Nguyen, H. Yi, S. J. Joo, K. H. Shin, K. J. Lee and B. Dieny, *Appl. Phys. Lett.* **89**, 094103 (2006).
53. Y. Huai, M. Pakala, Z. Diao and Y. Ding, *Appl. Phys. Lett.* **87**, 222510 (2005).
54. Z. Diao, D. Apalkov, M. Pakala, Y. Ding, A. Panchula and Y. Huai, *Appl. Phys. Lett.* **87**, 232502 (2005).
55. G. D. Fuchs, I. N. Krivorotov, P. M. Braganca, N. C. Emley, A. G. F. Garcia, D. C. Ralph and R. A. Buhrman, *Appl. Phys. Lett.* **86**, 152509 (2005).
56. G. D. Fuchs, J. A. Katine, S. I. Kiselev, D. Mauri, K. S. Wooley, D. C. Ralph and R. A. Buhrman, *Phys. Rev. Lett.* **96**, 186603 (2006).
57. Z. Diao, A. Panchula, Y. Ding, M. Pakala, S. Wang, Z. Li, D. Apalkov, H. Nagai, A. Driskill-Smith, L.-C. Wang, E. Chen and Y. Huai, *Appl. Phys. Lett.* **90**, 132508 (2007).
58. T. Devolder, C. Chappert, P. Crozat, A. Tulapurkar, Y. Suzuki, J. Miltat and K. Yagami, *Appl. Phys. Lett.* **86**, 062505 (2005).
59. T. Devolder, P. Crozat, J.-V. Kim, C. Chappert, K. Ito, J. A. Katine and M. J. Carey, *Appl. Phys. Lett.* **88**, 152502 (2006).
60. J. Miltat, G. Albuquerque, A. Thiaville and C. Vouille, *J. Appl. Phys.* **89**, 6982 (2001).
61. J. G. Zhu and X. Zhu, *IEEE Trans. on Magn.* **40**, 182 (2004).
62. K. J. Lee, A. Deac, O. Redon, J. P. Nozieres and B. Dieny, *Nature Mater.* **3**, 877 (2004).
63. K. J. Lee and B. Dieny, *Appl. Phys. Lett.* **88**, 132506 (2006).
64. Y. Acremann, J. P. Strachan, V. Chembrolu, S. D. Andrews, T. Tyliszczak, J. A. Katine, M. J. Carey, B. M. Clemens, H. C. Siegmann and J. Stöhr, *Phys. Rev. Lett.* 96, 217202 (2006).
65. M. Hosomi, H. Yamagishi, T. Yamamoto, K. Bessho, Y. Higo, K. Yamane, H. Yamada, M. Shoji, H. Hachino, C. Fukumoto, H. Nagao and H. Kano, IEDM *Technical Digest*, 459 (2005).
66. Y. A. Bychkov and E. I. Rashba, *JETP Lett.* **39**, 78 (1984).

67.	P. R. Hammar, B. R. Bennett, M. J. Yang and M. Johnson, *J. Appl. Phys.* **87**, 4665 (2000).
68.	P. R. Hammar and M. Johnson, *Appl. Phys. Lett.* **79**, 2591 (2001).
69.	P. R. Hammar and M. Johnson, *Phys. Rev. Lett.* **88**, 066806 (2002).
70.	G. L. Chen, *Phys. Rev. B* **47**, 4084 (1993).
71.	M. I. D'yakanov, *Sov. Phys. JETP* **33**, 1053 (1971).
72.	KIST report on a Development of the Technology of Spintronic Devices, December, 2004 (UC-E1802-7634-9).
73.	R. M. Stroud, A. T. Hanbicki, Y. D. Park, G. Kioseoglou, A. G. Petukhov and B. T. Jonker, *Phys. Rev. Lett.* **89**, 166602 (2002).
74.	J. Hong, J. Lee, S. Joo, K. Rhie, B. C. Lee, J. Lee, S.-Y. An, J. Kim and K.-H. Shin, *J. Kor. Phys. Soc.* **45**, 197 (2004).
75.	R. Dornhaus and G. Nimtz, *Narrow-Gap Semiconductors, Springer Tracts in Modern Physics*, Vol. 98, Springer-Verlag, New York, 1983.
76.	P. Pfeffer and W. Zawadzki, *Phys. Rev. B* **59**, R5321 (1999).
77.	P. Murzyn, C. R. Pidgeon, P. J. Phillips, J.-P. Wells, N. T. Gordon, T. Ashley, J. H. Jefferson, T. M. Burke, J. Giess, M. Merrick, B. N. Murdin and C. D. Maxey, *Phys. Rev. B* **67**, 235202 (2003).
78.	T. Koga, J. Nitta, T. Akazaki and H. Takayanagi, *Phys. Rev. Lett.* **89**, 046801 (2002).
79.	D. L. Smith and R. N. Silver, *Phys. Rev. B* **64**, 045323 (2001).
80.	J. D. Albrecht and D. L. Smith, *Phys. Rev. B* **66**, 113303 (2002).
81.	J. D. Albrecht and D. L. Smith, *Phys. Rev. B* **68**, 035340 (2003).
82.	H. Ohno, H. Munekata, T. Penney, S. von Molnar and L. L. Chang, *Phys. Rev. Lett.* **68**, 2864 (1992).
83.	H. Ohno, A. Shen, F. Matsukura, A. Oiwa, A. Endo, S. Katsumoto and Y. Iye, *Appl. Phys. Lett.* **69**, 363 (1996).
84.	T. Kasuya and A. Yanase, *Rev. Mod. Phys.* **40**, 684 (1968).
85.	J. K. Furdina, *J. Appl. Phys.* **64**, R29 (1988).
86.	A. H. MacDonald, P. Schiffer and N. Samarth, *Nature Mater.* **4**, 195 (2005).
87.	A. M. Nazmul, S. Sugahara and M. Tanaka, *Phys. Rev. B* **67**, 241308 (2003).
88.	T. Dietl, H. Ohno, F. Matsukura, J. Cibert and D. Ferrand, *Science* **287**, 1019 (2000).
89.	Y. D. Park, A. T. Hanbicki, S. C. Erwin, C. S. Hellberg, J. M. Sullivan, J. E. Mattson, T. Ambrose, A. Wilson, G. Spanos and B. T. Jonker, *Science* **295**, 651 (2002).

90.   S. Cho, S. Choi, S. C. Hong, Y. Kim, J. B. Ketterson, B. J. Kim, Y. C. Kim and J. H. Jung, *Phys. Rev. B* **66**, 033303 (2002).
91.   G. Kioseoglou, A. T. Hanbicki, C. H. Li, S. C. Erwin, R. Goswami and B. T. Jonker, *Appl. Phys. Lett.* **84**, 1725 (2004).
92.   Y. J. Zhao, T. Shishidou and A. J. Freeman, *Phys. Rev. Lett.* **90**, 047204 (2003).
93.   C.-M. Hu and T. Matsuyama, *Phys. Rev. Lett.* **87**, 066803 (2001).
94.   R. A. de Groot, F. M. Muller, P. G. Van Engen and K. H. J. Buschow, *Phys. Rev. Lett.* **50**, 2024 (1983).
95.   J.-H. Park, E. Vescovo, H.–J. Kim, C. Kwon, R. Ramesh and T. Venkatesan, *Nature (London)* **392**, 794 (1998).
96.   Y. Sakuraba, J. Nakata, M. Oogane, Y. Ando, H. Kato, A. Sakuma, T. Miyazaki and H. Kubota, *Appl. Phys. Lett.* **88**, 022503 (2006).
97.   T. Marukame, T. Ishikawa, S. Hakamata, K. Matsuda, T. Uemura and M. Yamamoto, *Appl. Phys. Lett.* **90**, 012508 (2007).

# 14

## The Longer Term: Quantum Information Processing and Communication

Philippe Jorrand

CNRS, Laboratoire d'Informatique de Grenoble,
46 avenue Félix Viallet, 38000 Grenoble,
France.

Philippe.Jorrand@imag.fr

..................................

Information is physical. Today's information processing and communication are classical: they are based upon the laws of Newton's and Maxwell's classical physics. This assertion holds all the way, from commercial computers and networks, up to their most abstract models, e.g. Turing machines. Research in quantum information was born some twenty five years ago, with the encounter of two major scientific achievements of the 20th century, namely quantum physics and information sciences. A technological motivation for that is an extrapolation of Moore's law which seems to indicate that the amount of matter needed for one bit will be reduced to one particle sometimes before year 2020. A deeper, scientific driving force of this interdisciplinary research is that of looking for the consequences of having computation and communication based directly upon the laws of quantum physics, i.e. our current ultimate knowledge of the world of elementary particles, as described by quantum mechanics. Breakthroughs in cryptography, communications, information theory and algorithmics have shown that this transplantation from classical to quantum has far reaching consequences, both quantitative and qualitative, and opens new avenues for research within the foundations of computer science and physics. The principles and most striking results and hot topics of this promising way of encoding, processing and communicating information are briefly introduced in this chapter from a mostly computer science point of view, with detailed examples in algorithmics and in cryptography. At the end of the chapter, a short list of some significant articles, textbooks and reports on quantum information are suggested for further reading.

## 1.    Introduction

For the last 25 years, properties of elementary particles which had been identified by quantum physics, formalized by quantum mechanics and confirmed again and again by experiments all along the 20th century, properties which go sometimes against our classical intuition of what the world is around us, are being considered as potential resources for encoding, processing and communicating information. In 1982, Richard Feynmann suggested that using quantum physics instead of classical physics as the physical layer for carrying information and performing computations would render feasible information processing tasks are out of reach of today's computers because of the complexity of these tasks. In 1985, David Deutsch, a theoretician physicist from Oxford University, formalizes this intuition by defining a quantum Turing machine, i.e. a quantum analogue of the abstract device defined by Alan Turing in 1936 and which remains the fundamental model of what all classical computations are. Deutsch provides a theoretical confirmation of Feynmann's intuition, by proving indeed that his quantum Turing machine can perform tasks that are only reproducible at an exponential cost by the classical Turing machine.

It took about 10 years before less abstract, more realistic, but outstanding and really surprising algorithmic results, theoretical at first, then experimental, where found, that confirmed Feynmann's original intuition supported by Deutsch's abstract device. The first result was seemingly strange and totally unexpected. In 1993, Charles Bennett, from IBM Research Yorktown, Gilles Brassard, from the University of Montreal, and a few others, elaborate the theoretical principles of a quantum teleportation protocol, which relies on the use of a feature of quantum objects which has no classical counterpart, namely entangled states: the state of a quantum system $a$ localized at point $A$ can, after having been destroyed at point $A$, become the state of another quantum system $b$, localized at a distant point $B$, without the state of $a$ being known neither at point $A$ nor at point $B$, and without any quantum system carrying the state of $a$ being transported on a trajectory from $A$ to $B$.

One year later, in 1994, Peter Shor, from AT&T, shows that finding the prime factors of an integer number in a time which is a polynomial function of the number of digits needed for writing that integer, is possible with quantum computing, whereas the best classical algorithm known for solving this problem takes an exponential time. This exponential drop of complexity was immediately noticed as a threat to the most widely used cryptographic systems. This was the real trigger for a very wide and very diverse expansion of research activities in quantum information processing and communication. In 1996, Lov Grover, from Lucent Technologies, designs a quantum algorithm for finding an item in unordered database that takes a time which is the square root of the time needed by classical computers for the same problem: this also came as a big surprise to the algorithmic research community. In 1997, Anton Zeilinger, from the University of Vienna, realizes the first experimental teleportation of the state of a photon, an experiment which has now been repeated in many other places, over distances much beyond the size of a laboratory, like a recent experiment over 144 km between two Canary Islands. Then, from 1999 to 2002, Isaac Chuang, from IBM Research Almaden, designs and builds the first quantum computer, based on NMR technology, which, although of a very modest size with only 7 quantum bits, has permitted to show experimentally that the new quantum algorithmic ingredients imagined and applied in theory by Shor and Grover in their algorithms, could indeed be implemented in practice within a physical quantum system.

These splendid theoretical results on quantum information and its processing, followed by their experimental confirmations, have become evidence that problems that are out of reach from classical information processing become feasible when the new quantum paradigms for computation and communication are used. This opens technological perspectives that are still quite remote in the future, but scientifically fascinating, and probably with immense consequences.

The principles and most striking results and hot topics of the quantum way of encoding, processing and communicating information are briefly presented in this chapter, from a computer science point of view. Section 2 introduces the strict minimum of quantum mechanics required for understanding how quantum objects and properties can be exploited within information processing tasks. Section 3 deals with quantum algorithms, with some details on Shor's and Grover's algorithms. Section 4, on quantum

cryptography, explains step by step a quantum secret key distribution protocol. And Section 5 provides an overview of other hot topics in this rapidly
expanding domain of investigations. Given the rather informal style of this
chapter, instead of inserting references to publications within the text, the
last section suggests a few articles, books and reports for further reading.

## 2.     From Quantum Physics to Information

Quantum mechanics is the mathematical formulation of laws for the physics
of elementary particles. It has been elaborated in the first half of the 20th
century and can be considered as relying on four postulates:

(i) the state of a quantum system (e.g. a photon, an electron, an ion, or a
collection of those) is a vector of norm (length) 1 in a $d$-dimensional
complex vector space, i.e. a column vector with $d$ components which
are complex numbers such that the sum of the squares of their moduli
$|z_j|^2$, for $j$ in $\{1, 2, ...d\}$, is 1 (the square of the modulus of a complex
number $z = x+iy$ is $|z|^2 = x^2 + y^2$); within the scope of this chapter,
except in the description of Shor's algorithm, instead of complex components, it will be all right to assume that state vectors always have
real components $x_j$ such that their squares $x_j^2$ add to 1;

(ii) the evolution of the state of an isolated quantum system (i.e. not inter-
acting with a neighbouring physical system) is deterministic, linear,
and characterized by a unitary operator, that is by a $d$x$d$ unitary matrix
applied to the state vector, where unitary means that the new state
vector after applying the matrix has the same norm, i.e. 1, as the state
vector before; the unitarity property implies that this evolution is also
reversible;

(iii) the measurement of a quantum system (i.e. the observation of the
state of a quantum system by the classical world) is an interaction of
that system with another system comprising a measuring device; the
measurement operation irreversibly modifies the state of the measured
system by performing a projection of its state vector before measurement, onto a probabilistically chosen basis vector among the $d$ vectors
of a basis of the vector space, with renormalization to norm 1 of the
resulting projection; the probability to be projected onto the $j$th basis
vector is $|z_j|^2$ (or $x_j^2$, if real) and the measurement operation returns an
information (e.g. the integer number $j$) to the classical world, which

tells which basis vector was chosen for the projection, but tells nothing about the state before the measurement, except the fact that its projection onto the *j*th basis vector was not zero;

(iv) the state space of a quantum system composed of several quantum subsystems is the tensor product of the state spaces of its components (given two vector spaces $P$ and $Q$ of dimensions $p$ and $q$ respectively, their tensor product is a larger vector space of dimension $p \times q$), i.e. given two quantum systems with state vectors having $p$ and $q$ components respectively, the state vector of the larger quantum system composed of these two subsystems has $p \times q$ components.

The question is then: how to take advantage of these postulates to the benefits of information processing and communication?

## 2.1. Making a quantum system compute

The most widely developed approach to quantum computation exploits all four postulates in a rather straightforward manner. The elementary physical carrier of information is a qubit (quantum bit), i.e. a quantum system (photon, electron, ion, …) with a 2-dimensional state space (postulate (i), e.g. the polarization of a photon or the spin of an electron; the state of a *n*-qubit memory register is a vector in a $2^n$-dimensional vector space, i.e. the tensor product of $n$ 2-dimensional vector spaces (postulate (iv)). Then, by imitating in the quantum world the most traditional organization of classical computation, quantum computations are considered as comprising three steps in sequence:

- first, preparation of the initial state of a *n*-qubit quantum register (postulate (iii) can be used for that, possibly with postulate (ii));
- second, computation, by means of a deterministic unitary transformation of the *n*-qubit register state (postulate (ii)), i.e. by applying a $2^n \times 2^n$ unitary matrix to it. Such a matrix can always be obtained or approximated by means of matrix and tensor products of $2 \times 2$ and $4 \times 4$ unitary matrices, i.e. by applying elementary operators on 1 and 2 qubits respectively. An adequate set of such basic operators could constitute an instruction set for a quantum computer;
- third, output of a classical result (e.g. an integer number) by measuring part or all of the register (postulate (iii)). Since measurement is probabilistic, the main rule of the game for quantum algorithmics is to get a final quantum state such that the probability to obtain a result relevant

for the intended computation should be as close to 1 as possible, while using a minimal number of operators to reach that state.

## 2.2.    *Informational and computational consequences*

The postulates of quantum mechanics can be given an informational and computational interpretation, thus providing the elementary quantum ingredients which are at the basis of quantum algorithm and quantum communication protocol design. Section 3 shows ways in which these ingredients have indeed far reaching quantitative consequences in terms of algorithmics and how they allow very significant drops of complexity for some classes of problems. Section 4 shows how they can be used to achieve communication security in ways that are qualitatively out of reach with classical information only.

### 2.2.1.    *Superposition*

At any given moment, the state of a quantum register of $n$ qubits is a vector in a $2^n$-dimensional complex vector space, i.e. a vector with $2^n$ complex components (in fact, most of the time real within this chapter), one for each of the $2^n$ different values on $n$ bits. The standard basis of this vector space comprises the $2^n$ vectors $|i\rangle$, for $i$ in $\{0,1\}^n$, where $\{0,1\}^n$ is the set of the $2^n$ integers on $n$ bits, and $|i\rangle$ is Dirac's notation for quantum vector states. This fact is exploited computationally by considering that a register of $n$ qubits can actually contain at any given moment a superposition of part (some vector components may be zero) or all of the $2^n$ different values on $n$ bits, whereas a classical register of $n$ bits may contain only one of these values at any given moment.

### 2.2.2.    *Quantum parallelism and deterministic computation*

Let $f$ be a function from integers on $n$ bits to integers on $m$ bits (from $\{0,1\}^n$ to $\{0,1\}^m$), and $x$ be a quantum register of $n$ qubits initialized in a state which is a superposition of all values in $\{0,1\}^n$ (this initialization can be done by one very simple quantum computation step). Then, computing $f(x)$ is achieved by a deterministic, linear and unitary operation $U_f$ on the state of $x$: because of the linearity of quantum mechanics, a single application of operation $U_f$ will distribute over all $2^n$ basis states $|i\rangle$, for $i$ in $\{0,1\}^n$, that are superposed in $x$, i.e. will produce all $2^n$ values of $f$ in a single computation step. Performing such an operation $U_f$ for any, possibly non linear and non reversible $f$ while obeying the linearity and unitarity laws of the quantum world, requires a register of $n+m$ qubits formed of the register $x$, augmented with a register $y$ of $m$ qubits. Initially, $y$ is in any arbitrary state $|s\rangle$ on $m$ qubits: before the application of $U_f$, the larger register of $n+m$ qubits contains a superposition of all pairs $|i,s\rangle$ for $i$ in $\{0,1\}^n$. After the application of $U_f$, this $n+m$-qubit register contains a superposition of all pairs $|i, s{+}{+}f(i) >$ for $i$ in $\{0,1\}^n$, where $++$ is bitwise addition modulo 2 (exclusive or). It is easy to verify that, for any $f$, this operation $U_f$, on a register of $n+m$ qubits, is its own inverse and is unitary, i.e. quantum mechanically legitimate. In many instances, it will be applied with $s{=}0$, which results in a superposition of all simpler pairs $|i, f(i)\rangle$ for $i$ in the domain $\{0,1\}^n$ of $f$.

### 2.2.3. *Probabilistic measurement and output of a result*

After $f$ has been computed in this way, i.e. in a single step for all values in its domain of definition, all possible $f(i)$'s, for $i$ in $\{0,1\}^n$, are superposed in the $y$ part ($m$ qubits) of the register of $n+m$ qubits, each of these values facing (in the pair $|i,f(i)\rangle$) their corresponding $i$ which is still stored in the $x$ part ($n$ qubits) of that register. Observing the contents of $y$ will project the state of the $y$ part on one of the $2^m$ basis vectors $\{|j\rangle\}$, for $j$ in $\{0,1\}^m$, and will return only one classical value, the probabilistically chosen $j$, among all possible values of $f$. This value is chosen with a probability which depends on $f$ since, e.g. if $f(i) = j$ for more than one values of $i$, the probability of obtaining $j$ as a result will be higher than that of obtaining $k$ if $f(i) = k$ for only one value of $i$ (and the probability of obtaining $l$ if there is no $i$ such that $f(i) = l$ will of course be 0). Since this measurement also causes the state of the $y$ part to collapses to $|j\rangle$, all other values of $f$ which were previously in $y$ are irreversibly lost.

### 2.2.4. *Interference*

Using appropriate unitary operations, the $2^n$ computations of $f$ can be made to interfere with each other. Destructive interference will lower the probabilities of observing some values, whereas additive interference will increase the probabilities of observing other values and bring them closer to 1. Because of probabilistic measurement, a major aim of quantum algorithmics will be to assemble the unitary operations for a given computation in such a way that, when a final measurement is applied, a relevant result has a high probability to be obtained.

### 2.2.5.  Entangled states

Measuring $y$ after the computation of $f$ is in fact measuring only $m$ qubits (the $y$ part) among the $n + m$ qubits of a register. The state of this larger register is a superposition of all pairs $|i, f(i)\rangle$ for $i$ in $\{0,1\}^n$ (e.g., in this superposition, there is no pair like $|2, f(3)\rangle$): this superposition is not a free cross-product of the domain of definition $\{0,1\}^n$ of $f$ by its co-domain $\{0,1\}^m$, i.e. there is a strong correlation between the contents of the $x$ and $y$ parts of the register. As a consequence, if measuring the $y$ part returns a value $j$, with the state of that part thus collapsing to the basis state $|j\rangle$, the state of the larger register will itself collapse to a superposition of all remaining pairs $|i,j\rangle$ such that $f(i) = j$. This means that, in addition to producing a value $j$, the measurement of the $y$ part also causes the state of the $x$ part to collapse to a superposition of all elements of the $f^{-1}(j)$ set of predecessors of $j$ in the domain of $f$. This correlation between the $x$ and $y$ parts of the register is called entanglement: the state of a quantum system composed of $p$ sub-systems is not, in general, reducible to an $p$-tuple of the states of the sub-system. Entanglement has no equivalent in classical physics and it constitutes the most powerful resource for quantum information processing and communication.

### 2.2.6.  No-cloning

A simple two line proof shows a major consequence of the linearity of all operations that can be applied to quantum states: the state of a qubit $a$ (this state is in general an arbitrary superposition, i.e. a vector made of a linear combination of the two basis state vectors $|0\rangle$ and $|1\rangle$), cannot be duplicated and made the state of another qubit $b$, unless the state of $a$ is simply either $|0\rangle$ or $|1\rangle$ (i.e. not an arbitrary superposition). This no-cloning theorem holds more generally for the state of any quantum system, including of course registers of $n$ qubits used during a quantum computation. In programming terms, this means that the "value" (the state) of a quantum variable cannot be used twice nor copied into another quantum variable, which would come as a shock to most programmers.

# 3.     Quantum Algorithms

Richard Feynman launched in 1982 the idea that computation based upon quantum physics could be exponentially more efficient than based upon classical physics. Then, after the pioneering insight of David Deutsch in the mid eighties, who showed, by means of a quantum Turing machine, that quantum computing could indeed not, in general, be simulated in polynomial time by classical computing, it was ten years before the potential power of quantum computing was demonstrated on actual computational problems.

The first major breakthrough was in 1994, when Peter Shor, from AT&T, published a quantum algorithm operating in polynomial time $O(p^3)$ for factoring a $p$-bit long integer $P$ ($p = \log P$), whereas the best classical algorithm currently known for this problem is exponential in $p$. In 1996, Lov Grover, from Lucent Technologies, published a quantum algorithm for searching an unordered database of size $N$, which achieves a quadratic speedup (it operates in $N^{1/2}$ steps) when compared with classical algorithms which, for the same problem, need up to $N$ steps.

## 3.1.     *Shor's algorithm: Exponential speedup of integer factoring*

Factoring is the problem of finding the prime factors of an integer. Although there is no known proof that this cannot be solved in a reasonable time, i.e. in a number of operations which is a polynomial function of the size of the data (here, the size is the number of bits or digits required to write down the integer to be factorized), there is no known algorithm achieving that in polynomial time: the most efficient factoring algorithm known today takes an exponential time. However, this algorithmic obstacle, although not backed by mathematical evidence, thus not fully trustable, is currently the main resource upon which the security of RSA, the most widely used cryptographic protocol, is based.

Shor's quantum algorithm factorizes integers in polynomial time. It relies on a known polynomial cost reduction of the problem of factoring to the problem of finding the period of a function. Then, since period finding can be achieved by a Fourier Transform, the key of Shor's algorithm is a Quantum Fourier Transform (QFT), which is indeed exponentially more efficient than the classical Fourier Transform, thanks to quantum parallelism, entanglement and tensor product. Shor's result implies that once a quantum computer with a sufficiently large number of qubits is available, most currently used cryptographic protocols will be broken in few minutes. However, as shown in section 4 of this chapter, quantum information provides also means for building the security of communications upon definitely trustable physical principles, rather than on the not proved exponentiality of integer factoring.

Given a $p$-bit long integer $P$, Shor's algorithm goes as follows:

**Step 1:** Choose at random an integer $a$ between 1 and $P$: $1 < a < P$.
**Step 2:** If $GCD(a,P) = 1$, continue. Otherwise, the problem is solved!
**Step 3:** Using a unitary quantum operator $U_{fa}$, compute the function $f_a(k) = a^k \mod P$, with $k$ an integer modulo $N$, where $N$ is the power of 2 such that $P^2 \leq N \leq 2P^2$. Then, using QFT, find the period $r$ of this function. Group theory tells that this function is indeed periodic, and that its period $r$ is such that $a^r = 1 \mod P$, that is $a^r - 1 = 0 \mod P$.
**Step 4:** If $r$ is even, the equation $a^r - 1 = 0 \mod P$ can be rewritten as $(a^{r/2} - 1)(a^{r/2} + 1) = 0 \mod P$. Furthermore, if $r$ is such that $a^{r/2} \neq \pm 1 \mod P$ (i.e. $a^{r/2}$ is not a trivial solution of the equation), then $GCD(a^{r/2} - 1, P)$ or $GCD(a^{r/2} + 1, P)$, or both, are factors of $P$. Otherwise if $r$ is odd, or if $a^{r/2}$ is a trivial solution of the equation, then return to step 1.

Steps 1, 2 and 4 are classical computation and are all polynomial in $p = \log P$, the number of bits that are used for encoding $P$. The quantum algorithmic breakthrough made by Shor lies entirely within step 3, where the issue is finding the period $r$ of the function $f_a(k) = a^k \mod P$.

In this explanation of Shor's algorithm, some important and non trivial technicalities will be ignored for concentrating on Shor's main quantum algorithmic ideas. We assume that the unitary operation $U_{fa}$ is applied to a register made of $n+p$ qubits, with $n = \log N$, composed a $n$-qubit part $x$ for the arguments to the function, and a $p$-qubit part $y$ for storing the results (see Section 2.2.2 in this chapter). With all values in the domain of definition $\{0,1\}^n$ of the function $f_a$ initially superposed in the $x$ part, and 0 initially stored the $y$ part, applying $U_{fa}$ produces, in one step, all pairs $|i, f_a(i)\rangle$ for $i$ in the domain $\{0,1\}^n$ of $f_a$, with $i$ in part $x$ and $f_a(i)$ in part $y$ of the register. Once $U_{fa}$ has been applied, the $y$ part of the register is measured. Since $f_a$ is periodic, with a yet unknown period $r$, there is a probability $1/r$ to get any one of the $r$ different values spanned by $f_a$ within one period (these values are all distinct). Let $j$ be that value: measuring the $y$ part projects the state of that part of the register onto the basis vector $|j\rangle$. Then, because the $x$ and $y$ parts of the register have been put into an entangled state by $U_{fa}$ (see Section 2.2.5), measuring the $y$ part also projects the $x$ part onto a superposition of all predecessors $f_a^{-1}(j)$ of $j$ by $f_a$. Since $f_a$ is periodic, if $i_0$ is the smallest among these predecessors of $j$, the state of the $x$ part thus collapses to a superposition of states $\{|i_0\rangle, |i_0 + r\rangle, |i_0+2r\rangle, ...\}$.

At this point, the contents of the $x$ part of the register provide exactly the information needed by a Discrete Fourier Transform (DFT) to get the period of the function. Furthermore, DFT is a unitary operation, thus quantumly legitimate. Applying DFT to the $x$ part would replace the superposition of states it contains by another superposition of states $\{|0\rangle, |N/r\rangle, |2N/r\rangle, |3N/r\rangle, ...\}$. Then, measuring the $x$ part would produce a value $q=kN/r$ which, after an expected average of $n$ trials, should allow to compute $r$. However, nothing has been gained yet in terms of complexity, since the DFT unitary operator applied to the x part is a $2^n \times 2^n$ matrix : it is exponential in $n$, hence in $p$!

This is where Shor has designed a very clever way of decomposing the DFT matrix into matrix and tensor products of $n(n+1)/2$ elementary unitary matrices operating on 1 and 2 qubits. This Quantum Fourier Transform computes the same operation as DFT, but in the order of $n^2$ steps instead of $2^{2n}$ for DFT, and instead of $n2^n$ for FFT, the Fast Fourier Transform. Finally, since an average of $n$ iterations of QFT are to be expected for finding a satisfactory $r$ (this is probabilistic because of measurement) the complexity of Shor's quantum factoring algorithm is $O(n^3)$, which achieves an exponential speedup compared with today's best classical algorithm for the same problem.

## 3.2. Grover's algorithm: Quadratic speedup of unordered search

A simple example shows what is achieved by Grover's quantum algorithm. Consider a telephone directory which contains the names and phone numbers of $10^6$ people. Since it is organised in alphabetical order of names, telephone numbers are unordered. Now, given a phone number, finding the unique name of the person who has that number "costs", in the worst case, answering $10^6$ times the query "does the person whose I am currently reading the name in the directory have the phone number I have been given?". This is the best that classical computing can achieve when faced with the problem of searching an unordered database: if the size of the database is $N$, the classical "query complexity" of the problem of finding the unique element which satisfies a given "oracle" $f$ (able to answer the query "is this the person with that number?") is $O(N)$, considering that "time" goes one step ahead each time a query is made to the oracle $f$ (the use of the term "oracle" is due to the fact that we don't care how the answer to a query is found, we don't even care about the cost of finding the answer to one query, we only take into account the number of queries to the oracle).

Grover's algorithm relies upon a very subtle use of interference, now known as amplitude amplification, which performs a stepwise increase of the probability of obtaining the relevant item in the database by means of a measurement, and which brings this probability as close to 1 as possible after $N^{1/2}$ steps: the quantum query complexity of unordered database search is $\Theta(N^{1/2})$ ($\Theta(h(N))$ denotes the fact that the exact complexity of a problem is of the order of $h(N)$, if $N$ is the size of the input, whereas $O(h(N))$ tells that $h(N)$ is an upper bound of the complexity). In the case of our telephone directory, Grover's algorithm finds the correct answer after exactly $10^3$ queries to the quantum oracle $U_f$, instead of up to $10^6$ queries to the classical oracle $f$ when classical means are used, which represents a quadratic speedup.

For reasons of pedagogical simplification, we take $N = 2^n$ as the size of the database, for some $n$. The problem of unordered database search can then be simply formalized by means of a function $f$ (the oracle), which takes its argument in $\{0,1\}^n$ (which encode the persons' names); and returns 1 or 0, depending on whether or not the argument given to it has the unique number we are looking for. This means that $f$ returns 1 for only one value $i_0$ in $\{0,1\}^n$, and 0 for all other values. The problem of unordered data base search is the problem of finding this unique $i_0$. Classically, this may require up to $N$ queries to the oracle $f$. We don't care about how the oracle is implemented.

Corresponding to the classical function $f$, there exists a unitary quantum operation $U_f$ which operates on $n + 1$ qubits: $n$ qubits for the arguments, and, since $f$ returns 1 or 0, 1 qubit for the results (see Section 2.2.2). If the $n$-qubit argument part initially contains a superposition of all $2^n$ basis states $|i\rangle$, for $i$ in $\{0,1\}^n$, and if the result qubit state is initially $|0\rangle$, applying $U_f$ results in these $n + 1$ qubits to contain a superposition of all pairs $|i, f(i)\rangle$ for $i$ in the domain $\{0,1\}^n$ of $f$. Only one of these pairs is of the form $|i,1\rangle$, the pair where $i = i_0$. All other pairs are of the form $|i, 0\rangle$.

The state of the $n$-qubit part before applying $U_f$ is a vector in a $2^n$-dimensional vector space, where all $2^n$ components (i.e. the lengths of the projections of this vector on each of the $2^n$ vectors of the standard basis of the vector space) have the same real value $1/2^{n/2}$. This vector can be given the following graphical representation, where, facing each of the $2^n$ dimensions listed along the horizontal line, a stick of length $1/2^{n/2}$ represents the corresponding vector component:

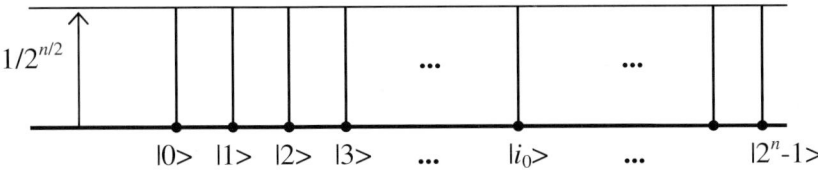

In quantum mechanics, the values of the components of a state vector are called amplitudes. Here, the amplitudes are all real, positive, and equal to $1/2^{n/2}$. The state pictured here is a uniform superposition of all standard basis states of the $2^n$-dimensional vector space. One can easily verify that the sum of the squares of the amplitudes is 1, and if one measures that state, there is a uniform probability $1/2^n$ (square of the amplitude) to get any of the $2^n$ values in $\{0,1\}^n$.

Now, using $U_f$ as a building block, it is possible to design another unitary quantum operation $V_f$ (this construction is very simple), which takes as input any superposition of the $2^n$ basis states (i.e. possibly with non uniform amplitudes), which makes a query to $U_f$, and transforms its input state by simply inverting the amplitude corresponding to basis state $|i_0\rangle$, the unique state for which $U_f$ produces the pair $|i_0,1\rangle$: if this amplitude was positive, $V_f$ makes it negative, and vice-versa. For example, applying $V_f$ to the initial uniform superposition produces a state with a negative amplitude corresponding to basis state $|i_0\rangle$:

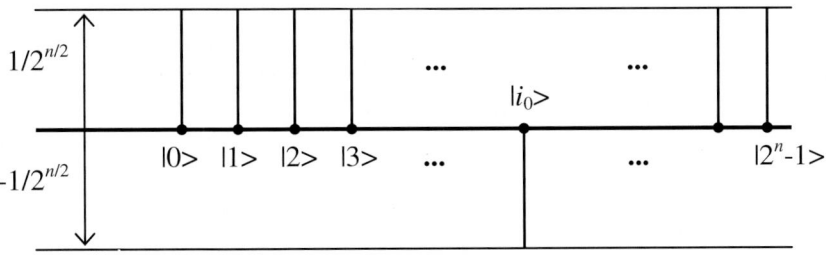

In addition to $V_f$, Grover's algorithm uses another unitary operation $R$, which can be implemented with matrix and tensor products of elementary operators on 1 and 2 qubits, and which performs a reflection of the amplitudes in its input state with respect to the average of these amplitudes. For example, in the state obtained above after applying $V_f$, the average $a$ of the amplitudes can be visualized by a dotted line slightly below the positive amplitudes:

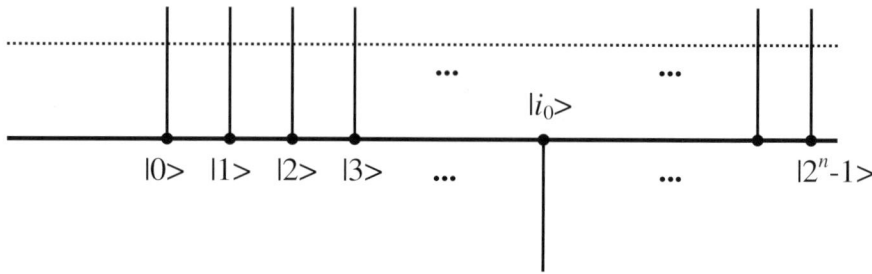

If this state is given as input to $R$, all the positive amplitudes $1/2^{n/2}$ will be replaced by their reflection with respect to that line, that is by smaller amplitudes $a - (1/2^{n/2} - a) = 2a - 1/2^{n/2}$, and the negative amplitude $-1/2^{n/2}$ will be replaced by a positive amplitude with a much larger absolute value $2a + 1/2^{n/2}$:

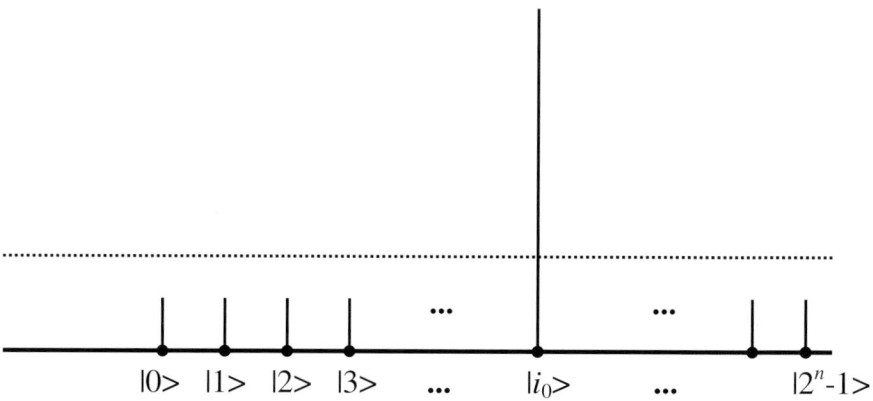

Initially, all basis states had the same amplitude in the uniform super-position. Now, after having applied $V_f$, then $R$ (i.e. the matrix product, or composition $R \cdot V_f$) the amplitude of state $|i_0\rangle$ in the new superposition of all basis states has been amplified, while the amplitudes of all other basis states have decreased. A measurement performed at that point would return $i_0$ with a higher probability than any other value, but if the database is very large, the odds are still high that a wrong result be still produced (the larger the database, the smaller the changes in the amplitudes after applying $V_f$ and $R$).

Grover's idea is to take the composition $G = R.\ V_f$ as an elementary building block, and to repeat it a sufficient number of times, so that the amplitude of $|i_0\rangle$ is amplified in a stepwise fashion and brought as close to 1 as possible, while all other amplitudes are reduced and brought close to zero. The question is then: starting with the initial uniform superposition, how many times should $G$ be repeated to reach this optimum situation before measurement? Since each time $G$ is applied, $V_f$, therefore $U_f$, is also applied, the number of times $G$ should be applied is also the number of queries that will be made to the quantum oracle $U_f$. An elegant geometric proof provides the answer to this question.

Consider a 2-dimensional space where the vector $|i_0\rangle$ is on the horizontal axis in the unit circle, and a renormalized projection of the vector sum of all other basis vectors of the $2^n$-dimesional space is on the vertical axis. The projection onto this 2-dimensional space of the state vector corresponding to the initial uniform superposition has a component of length $1/2^{n/2}$ along the $|i_0\rangle$ axis, and can be pictured as a vector separated by an angle $\theta$ from the vertical axis, with $\sin \theta = 1/2^{n/2}$:

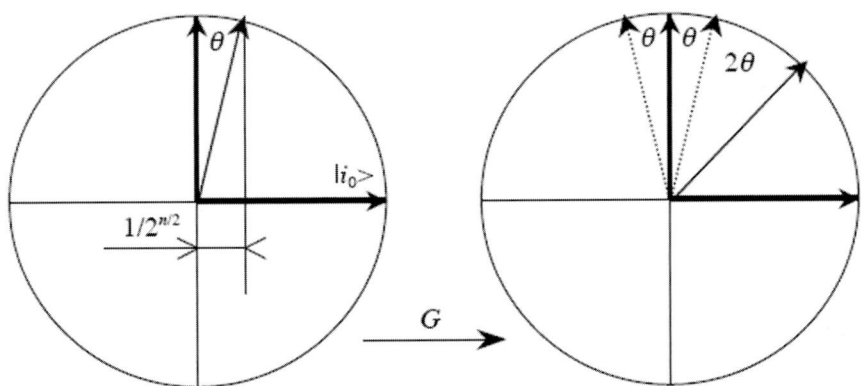

Applying $G$ means applying $V_f$ first, then $R$. Applying $V_f$ to the 2-dimensional projection of the state vector is negating its component along the $|i_0\rangle$ axis, i.e. reflecting it with respect to the vertical axis. Then, applying

$R$ is reflecting the new vector with respect to the 2-dimensional projection of the uniform superposition, which is the initial state vector projection. Thus, as a whole, applying $G$ is making two successive reflections with respect to two axes which are separated by an angle $\theta$: straightforward geometric reasoning shows that applying $G$ in this 2-dimensional space amounts to a rotation by an angle $2\theta$.

The goal is to arrive as close as possible to the vector $|i_0\rangle$, since at that point, the amplitude of $|i_0\rangle$ in the state vector will be as close to 1 as possible, hence also the probability to get the value $i_0$ (the name of the person with the given telephone number) out of a measurement. Since each application of $G$, i.e. each query to the oracle $U_f$, moves the state vector by angle $2\theta$ toward $|i_0\rangle$, the best that can be done is arriving within and angle of $\theta$ before $|i_0\rangle$ and $\theta$ after $|i_0\rangle$.

If reaching this optimum situation needs $k$ queries to $U_f$, the angle spanned from the vertical axis to that final position of the 2-dimensional state vector will be $\theta + k2\theta$, and such that:

$$\pi/2 - \theta \le \theta + k2\theta \le \pi/2 + \theta$$

Assuming that the size $N = 2^n$ of the database is very large, the angle $\theta$ is small and can be replaced by its sinus $1/N^{1/2}$ in the above relation. This leads immediately to $k = \pi/4 N^{1/2}$. Therefore, of the order of $N^{1/2}$ queries to the oracle provide the answer to the question with very high probability. There exist a proof that $N^{1/2}$ is also the lower bound for the number of these quantum queries. Hence the quantum query complexity of unordered database search is $\Theta(N^{1/2})$. Notice that checking in the telephone book whether the answer is correct has a logarithmic cost. If it is not correct, the whole game has to be played again. A known result in probability theory (Chernoff bound) tells that the probability to have a majority of wrong answers, after $p$ trials of an algorithm which produces correct answers with a probability higher that $1/2$ (which is clearly the case here), decreases exponentially with $p$. This means that the probability to get a correct answer with Grover's algorithm can be quickly brought as close to 1 as we want.

## 3.3.  *A note about quantum algorithmic techniques*

Integer factoring is a special case of a larger class of problems (called the "Hidden Subgroup Problem") which can also benefit from an exponential speedup when an adequate form of Shor's Quantum Fourier Transform exists. Such a QFT is known to exist when the problem domain is a commutative group (e.g. integers modulo $P$, for factoring). Research is currently very active for extending Shor's approach to non commutative groups. A few very specific problems in this larger class have been solved polynomially, but no general solution is known yet. One of the challenges is finding a polynomial algorithm for graph isomorphism, which is a problem of very high interest in many domains.

The technique due to Grover has also been extended to the more general quantum algorithmic principle of amplitude amplification. This has been used for finding the quantum query complexity of some problems on graphs. For example, given a $n$-vertex graph specified by its adjacency matrix (a $nxn$ matrix with $a_{i,j}=1$ if there is an edge between vertices $i$ and $j$, 0 otherwise), the problems of finding the minimum weight spanning tree of that graph, of checking whether the graph is fully connected, or of checking if there is a path between any two vertices if the graph is directed, all have a classical query complexity $\Theta(n^2)$. With the help of amplitude amplification and a few other quantum algorithmic techniques, these problems have been found to have a quantum query complexity $\Theta(n^{3/2})$.

Other algorithmic techniques than QFT and amplitude amplification are also being developed, like quantum random walks, which already appear to be promising for a number of other classes of problems.

## 4.    Quantum Cryptography

No communication channel can be guaranteed 100% safe: a message sent by person A (Alice) to person B (Bob) can always be observed by an eavesdropper (Eve). Cryptographic techniques have been designed for improving this situation by hiding the actual contents of a confidential message inside what is sent on the channel, in such a way that it is easy for Alice to encrypt the message, easy for Bob to decrypt what he has received and recover the original contents, but very difficult for Eve to discover information about the confidential contents from what is sent on the channel. For this to be possible, Alice and Bob must have previously reached an agreement about the encrypting-decrypting process. There are two broad classes of techniques in classical cryptography for achieving that: secret key cryptography and public key cryptography.

## 4.1.    Secret key, public key

With secret key cryptography, Alice encrypts the message with a key and Bob decrypts what he has received with the same key. This implies that the key is known in advance by both Alice and Bob, and by no one else. Under some further conditions, this method can be made 100% secure. But it has a severe drawback. The key has to be distributed in advance among Alice and Bob: this requires an absolute security of the channel used for that purpose, which is unachievable since passive observation of a channel is always possible. In spite of that, this method has actually been used in at least one crucial situation: the red telephone which linked the White House with the Kremlin during the Cold War was encrypted with a secret key that both ends had to agree upon. The key was physically transported by a trusted courier between Washington and Moscow.

With public key cryptography, Alice encrypts with Bob's public key, which can be known to every one, and Bob decrypts with his private key, known to him only. The two keys must be related by a mathematical property which, while it must be easy for Alice to encrypt and for Bob to decrypt, guarantees that it is very difficult for Eve, who knows Bob's public as anyone else does, to find Bob's private key. The most widely used encryption technique used on internet, RSA, uses this method and its security is based on the unproved exponentiality of classical integer factoring. A proof which would invalidate this conjecture would also retroactively destroy the security of all messages encrypted with this method and, when a quantum computer with a sufficiently large number of qubits is available, Shor's polynomial quantum factoring algorithm will definitely invalidate this method.

## 4.2.    Secure quantum Key Distribution (QKD)

Quantum cryptography is actually not cryptography at all: no message is encrypted by Alice nor decrypted by Bob. It is more accurate to talk about secure Quantum Key Distribution (QKD). QKD allows Alice and Bob to agree safely on a common key that they intend to use later for encrypting a confidential message with a classical secret key cryptographic method. QKD achieves that while making no assumption about the security of the channels used by Alice and Bob during the QKD protocol. QKD relies upon the properties of quantum measurement: quantum measurement is probabilistic and it irreversibly modifies the state of the qubit or quantum system that is measured. There exist several QKD protocols. The principles of one of them are explained in the following paragraphs. In all these protocols, the main idea is that the properties of quantum measurement allow Alice and Bob to detect the presence of an eavesdropper, and to estimate and eliminate the amount of information that may have been obtained by Eve.

## 4.2.1. *Back to quantum measurement*

Consider four specific states, traditionally called $|0\rangle$, $|1\rangle$, $|+\rangle$ and $|-\rangle$, among the infinity of states that a qubit can take. In the 2-dimensional state space of qubits, these four states can be visualized as four vectors respectively horizontal, vertical, at 45° and at 135° in the unit circle:

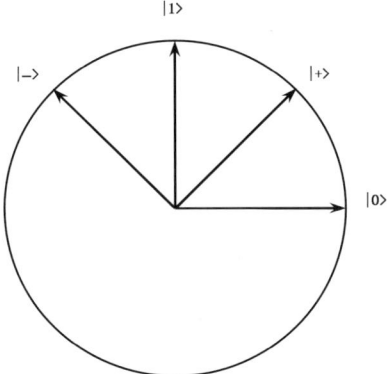

As described thus far in this chapter, quantum measurement is performed in the standard basis $\{|0\rangle, |1\rangle\}$. Consider indeed the measurement of state $|0\rangle$ in that basis: with probability 1, it will be projected onto itself and the classical value 0 will be produced. Same story for state $|1\rangle$ measured in that basis: with probability 1, it is projected onto itself, and the value 1 is produced. Consider now the measurement of state $|+\rangle$ in the standard basis. Its amplitudes on $|0\rangle$ and $|1\rangle$ are both equal to $1/2^{1/2}$, which means that with probability 1/2 (square of the amplitude) its measurement in the standard basis will project $|+\rangle$ onto $|0\rangle$, the state will become $|0\rangle$ and the value 0 will be produced; and with probability 1/2, the projection will be onto $|1\rangle$, the state will become $|1\rangle$ and the value 1 will be produced. Exactly the same story can be told for a measurement of $|-\rangle$ in the standard basis.

But any other basis than the standard basis could be chosen to perform a quantum measurement. For example, the diagonal basis $\{|+\rangle, |-\rangle\}$. When measuring $|0\rangle$, $|1\rangle$, $|+\rangle$ and $|-\rangle$ in the diagonal basis, the situation will be exactly the dual of their respective measurements in the standard basis. Measuring $|+\rangle$ ($|-\rangle$) in the diagonal basis will, with probability 1, project $|+\rangle$ ($|-\rangle$) onto itself, the state will remain $|+\rangle$ ($|-\rangle$), and the value 0 (1) will be produced. Measuring $|0\rangle$ ($|1\rangle$) in the diagonal basis will, with probability 1/2, project $|0\rangle$ ($|1\rangle$) onto $|+\rangle$, the state will become $|+\rangle$ and the value 0 will be produced; with probability 1/2, the projection of $|0\rangle$ ($|1\rangle$) will be onto $|-\rangle$, the state will become $|-\rangle$ and the value 1 will be produced.

Qubit in states $|0\rangle$, $|1\rangle$, $|+\rangle$ and $|-\rangle$ are used by the QKD protocol presented in the next paragraph to encode bits of a sequence of bits that will constitute the key on which Alice and Bob want to agree: a bit 0 will be encoded by a qubit in either of the two states $|0\rangle$ or $|+\rangle$, whereas a bit 1 by a qubit in either of the two states $|1\rangle$ or $|-\rangle$. Then, exploiting the probabilities explained above of measuring these states in either the standard or the diagonal bases, Alice and Bob will be able to detect the undesired observations made by Eve on the quantum channel along which these qubits have been transported from Alice to Bob.

### 4.2.2. The BB84 Quantum Key Distribution protocol

The BB84 protocol for quantum key distribution is one of the oldest achievements in the domain of quantum information. Its theoretical principles were discovered and published in 1984 by Charles Bennett, from IBM Research Yorktown, and Gilles Brassard, from the University of Montreal. BB84 is also the first theoretical result in quantum information which is currently giving rise to the design and marketing of commercial products for secure information transmission, where the security is based on the measurement postulate of quantum mechanics.

The BB84 QKD protocol proceeds in three steps:

**Step 1:** Alice sends qubits to Bob through a public quantum channel. Initially, Alice builds on her side a random sequence of $4n$ bits (0's and 1's), where $n$ is the length of the key that Alice and Bob want to agree upon. Alice sends these bits, one by one, to Bob, encoded by qubits. For each 0 and 1, Alice flips a coin for choosing at random between two possible encoding bases: a qubit for encoding a 0 will be, at random, either in state $|0\rangle$ or in state $|+\rangle$. Similarly, a qubit for encoding a 1 will be, at random, either in state $|1\rangle$ or in state $|-\rangle$. For each bit, she remembers which encoding basis, standard or diagonal, she has used.

At the other end of the quantum channel, Bob receives qubits. For each qubit he receives, he does not know whether this qubit encodes a 0 or a 1, and he does not even know in which basis, standard or diagonal, Alice has encoded this 0 or this 1. Thus, for each qubit, Bob flips a coin for choosing at random which measurement basis he will use to get a 0 or a 1 from this qubit. For each qubit, he remembers in which basis he has made the measurement.

When Alice has sent all $4n$ qubits for encoding the $4n$ bits of her initial sequence, Bob has also built on his side a sequence of $4n$ bits produced by the measurements of the $4n$ qubits he has received. Assuming that there is no eavesdropper, at any given common position in these two sequences of bits, the probability to have identical bits is 1 if Alice and Bob have used the same bases, Alice for encoding, Bob for measuring; this probability if 1/2 if they have used different bases.

**Step 2:** Alice and Bob now communicate through a public classical channel, e.g. a telephone line. Alice tells to Bob the sequence of her encoding bases, without revealing any of the 0's and 1's that she had encoded, and Bob tells to Alice the sequence of his measurement bases, without revealing any of the 0's and 1's that he had obtained. In their respective sequences of $4n$ bits, they keep, each on her/his side, only the bits which are at the positions where they have used the same basis. Given the probabilities due to coin flipping, this amounts approximately to half of the 0's and 1's in Alice's original random sequence of bits: both Alice and Bob now have sequences of 2n bits.

These two sequences should be identical, up to acceptable errors of transmission, if Eve was not intercepting and observing the qubits on the quantum channel during step 1.

**Step 3:** Alice and Bob detect the presence of Eve, using again a classical channel. We assume that Eve has access to the quantum channel, that she intercepts all qubits, she measures them in a basis that, like Bob, she chooses each time at random, and she forwards each qubit, once measured, to Bob. Bob has of course no means of knowing whether the qubits he receives come directly from Alice of have been measured by Eve on their way to him.

Each time Eve chooses a measurement basis which is the same as the encoding basis that Alice had used, the qubit forwarded by Eve to Bob is in the same state as the qubit initially sent by Alice: in that case, there is no way of detecting that Eve knows the 0 or the 1 which was initially encoded by Alice. This is the same situation as passive observation of a classical channel, except that Eve does not know (yet) that she has used the correct basis, nor that Bob may use a different basis to measure that qubit, which implies that the corresponding bit will eventually be discarded in step 2.

However, each time Eve chooses a measurement basis which is not the same as the encoding basis that Alice had used, the qubit forwarded by Eve to Bob will not be in the same state as the qubit sent by Alice. In this case, Eve's observations leave traces that Alice and Bob will be able to detect. Because of Eve's coin flipping, this is the case for half of the qubits which correspond to the bits that Alice and Bob will have kept after step 2.

Consider a qubit in this situation: if it was encoding a bit in the standard basis, i.e. it was initially put in state $|0\rangle$ or $|1\rangle$ by Alice, Eve has measured it in the diagonal basis, which means that she forwards to Bob a qubit in state $|+\rangle$ or $|-\rangle$. Since this is one of the positions where Alice and Bob had used the same basis, Bob measures this qubit in the standard basis, and gets a 0 or a 1 with a probability of 1/2. The dual situation holds if the state of the qubit initially sent by Alice was $|+\rangle$ or $|-\rangle$. The net result is that 25% of the bits kept by Alice and Bob at step 2 are different, although they had used the same basis, and the reason is that Eve was observing the qubits on the quantum channel.

In order to detect that, Alice and Bob choose at random 50% of the positions that they had kept at step 2, i.e. $n$ positions. On the classical channel, they compare the corresponding 0's and 1's in their respective sequences, position by position, and they discard these $n$ bits since Eve may be listening to their conversation. The probability that Alice's and Bob's bits are identical at all the $n$ compared positions in spite of eavesdropping is $(3/4)^n$, i.e. it decreases exponentially with $n$ (e.g. it is of the order of $3.10^{-13}$ for $n = 100$).

Finally, if the error rate is acceptable for a normally noisy channel, the remaining n bits will constitute a secret key, after error recovery, and privacy amplification if needed. Otherwise, Alice and Bob start over the whole protocol.

### 4.2.3. *Other issues, experiments, and QKD on the market*

The BB84 quantum key distribution protocol uses four different qubit states for encoding the 0's and 1's of a secret key, in such a way that the properties of quantum measurement allow Alice and Bob to detect the presence of an eavesdropper. There exist other QKD protocols, e.g. using only two non orthogonal states, or using entangled states of two qubits shared by Alice and Bob, but the security of all of them relies upon the unavoidable perturbation of quantum states due to measurement, for detecting the undesirable observations made by Eve. All these protocols also comprise procedures of reconciliation (for correcting errors due to channel noise), and of privacy amplification (for transforming the key so as to decrease the amount of information that Eve may have obtained about it, e.g. when she has used the same basis as Alice; this is possible if this leakage of information is estimated below a predefined threshold). All these protocols and procedures must remain security effective for more subtle classes of attacks than simply intercepting, measuring and forwarding all qubits. Other security issues are also investigated, like authentication, signature, secret sharing, with the aim of discovering improvements that can be brought by the use of quantum resources.

Most experimental implementations of QKD protocols are based on the BB84 protocol, and use photons for implementing qubits. Photons are transmitted either on optical fibres or in open air. The latest experiments, on distances of over 140 km in open air, together with theoretical studies, indicate that high fidelity ground-satellite transmission of individual photons should soon become feasible, thus enabling QKD over arbitrary distances. Several companies (e.g. id Quantique in Geneva, MagiQ Technologies in New York) are now manufacturing and marketing plug and play QKD systems.

## 5.    Hot Topics and Perspectives of Quantum Information

A rapid survey of quantum algorithmics and QKD has been done in Sections 3 and 4. These have been historically the first main topics in quantum information processing and communication. They have triggered the expansion of this new territory of scientific exploration, because they have clearly shown that the passage to the quantum scale comes with new computational opportunities which have no classical counterparts. But research in quantum information processing and communication is expanding much beyond algorithms and protocols. While facing sometimes a number of formidable and stimulating obstacles, most notably in physics, this research now explores a large number of topics, where new and promising advances are being made.

## 5.1. Quantum computation models and foundational structures

Much of the quantum informatics research to date has focussed on a quest for new quantum algorithms and new kinds of quantum protocols, and great advances have been made. However, many important basic questions which are fundamental to the whole quantum informatics endeavour still remain to be answered, such as: what are the true origins of quantum computational algorithmic speedup? How do quantum and classical information logically interact during a computation? What are the limits of quantum computation? These are all questions which explore the foundational structures and boundaries of quantum information and computation.

In the mid-eighties, David Deutsch, from Oxford University, has pointed out that one of the most fundamental abstract models of what a computation is, the original, classical Turing machine designed in the thirties by Alan Turing, was entirely relying on the untold hypothesis that computations are performed by devices which obey the laws of classical physics. Deutsch developed a more abstract, but physically grounded view of what a computation is, namely the simulation of a physical system by another physical system. Based on that, he designed a quantum analogue of the Turing machine, with which he showed two major results, some ten years before the discovery of Shor's and Grover's algorithms: (i) the set of functions that can be computed by a quantum Turing machine is the same as the set of functions that can be computed by a classical Turing machine; and (ii) quantum computations can perform tasks that cannot be simulated classically (i.e. by a classical Turing machine), better than with an exponential complexity cost for the simulation. This means that the Turing machine, be it classical or quantum, sets the limits: in terms of computability, quantum computation and classical computation are proved equivalent. The promises of quantum computation are elsewhere: enlarge as far as possible the boundaries of what is reasonably computable.

Until the end of the nineties, it seemed that Deutsch's quantum Turing machine, and a computationally equivalent but simpler and more practical model, the quantum circuit model, could supply canonical quantum analogues of the classical computational models. Both models are in fact formalized descriptions of the three step approach to quantum computation sketched at the beginning of Section 2.1 in this chapter, where the computation is performed by means of unitary transformations applied to a register of qubits. This is now considered as the traditional model of quantumcomputation. But other, very different models have emerged in the last few years since year 2000.

One of the most developed among these models is measurement-based quantum computation: since quantum measurement modifies the state of the measured quantum system, and since a computation is always implemented physically as a modification of the state of a physical system, why not take measurement as the main operation for driving quantum computations? But measurement is probabilistic: since each measurement step during such a computation tells to the classical world which probabilistic choice has been taken, it is always possible to adapt consequently what has to be done at the next or future steps, thus giving rise to a notion of classically controlled quantum computation. It has been proved that the measurement-based model of quantum computation has the same computability power as the traditional, unitary-based model, with no significant loss in terms of complexity.

The most extreme form of measurement-based quantum computation, but probably the most promising in terms of computational properties and of physical implementability, is the so-called one-way quantum computer designed by Hans Briegel at the University of Innsbruck: a grid of qubits is initially set in a globally entangled state, with some of the qubits containing the initial input data for the computation, and some others being designated as the output qubits. Then, each computation step consists in measuring only one qubit at a time, in an adequately chosen basis (standard, diagonal, or other). Each measurement separates the measured qubit from the global entangled state, and modifies the global and still entangled state of all the others. This modification of the remaining global state is driven and propagated in a stepwise fashion, until the result is stored in the state of the output qubits. The choice of which qubit to measure at each step and of which basis to use may of course depend on classical values produced by previous measurements. This gives rise to a whole new collection of extremely interesting algorithmic possibilities for optimising and parallelising quantum computations, which were absent in the traditional model. The one-way quantum computation model may well be the first, and unexpected way to physically implement a quantum computer of significant size, were the resource that is consumed along a computation is the global entangled state initially established over the grid of qubits.

Other, yet again different models have also appeared. With adiabatic quantum computation, information is encoded in the Hamiltonian of a quantum system and a computation is a slow transformation from an initial to a final Hamiltonian, while staying at the minimum energy level along the path from initial to final. This can be roughly viewed as a quantum analogue of simulated annealing. With topological quantum computation, information is encoded in the topological properties of a set of particles, and computation exploits these properties with techniques inspired by the mathematics of knot and braid theory.

All these models of quantum computation have features which are both theoretically and experimentally of great interest, and the methods developed to date for the traditional quantum circuit model do not carry over straightforwardly to them. In this situation, there is no confidence that a comprehensive paradigm has yet been found. It is even more than likely that many new ways of letting a quantum system compute have been overlooked until now, and wait to be discovered.

## 5.2. Quantum information theory

The information contents of a system of $n$ qubits is paradoxical. Although, according to the postulates of quantum mechanics, $2^n$ complex numbers are necessary for specifying the state of these $n$ qubits (the vector state of their system has $2^n$ components), a theorem proved in 1973 by a Russian theoretician physicist, Alexander Holevo, has the consequence that $n$ qubits can be used for encoding $n$ classical bits of information, and not more than $n$. This results is a clear bound that limits the amount of information that can be transmitted by sending qubits on a quantum channel.

However, with the assistance of entangled states and classical communication, the transmission of 2 classical bits is sufficient for the teleportation of arbitrary qubit states, as discovered by Charles Bennett and five other scientists in 1993: the unknown state $|q\rangle$ of a qubit $a$ located at point $A$ can become the state of a qubit $b$ located at a distant point $B$, after having been measured at point $A$, and without any qubit in state $|q\rangle$, nor any qubit in a state related to $|q\rangle$, being transported along a trajectory from $A$ to $B$. For achieving that, the qubit $b$ and another qubit, $c$, are initially set in an entangled state, $c$ is placed at point $A$, and $b$ is sent to a distant point $B$: although spatially separated, the 2-qubit system $\{b,c\}$ is entangled. The qubit $a$, which is in an unknown state $|q\rangle$, is also placed at point $A$. Then; two operations are applied at point $A$ to the 2-qubit system $\{a,c\}$: a unitary operation to entangle them, thus also entangle the 3-qubit system $\{a,b,c\}$, and a measurement: this measurement of two qubits produces two bits of information at point $A$. These two classical bits are sent to point $B$ where, because of the measurement performed at point $A$ and thanks to the global entanglement of $\{a,b,c\}$, they constitute enough information for choosing which among four unitary operators (the so-called Pauli operators) has to be applied locally to qubit $b$ so that it is finally set in state $|q\rangle$. The "magic" there is that two classical bits are sufficient for recovering at point $B$ the two complex components of state $|q\rangle$. A dual protocol to teleportation, dense coding, shows that with the assistance of entanglement, sending one qubit on a quantum channel from point $A$ to point $B$ is enough for communicating two classical bits of information from $A$ to $B$, which is an information compression unachievable by classical means.

As a further strangeness of quantum information, the no-cloning theorem, which is a straightforward consequence of the linearity of quantum mechanics, proved in 1982, tells that it is impossible to duplicate an unknown quantum state, i.e. there exists no quantum copy machine that would take as input an original qubit $a$ in state $|q\rangle$ and a "blank" qubit $b$ in state e.g. $|0\rangle$, and produce as output the qubit $a$ still in state $|q\rangle$ and the qubit $b$ also in state $|q\rangle$. It should noticed that teleportation does not contradict the no-cloning theorem: in the teleportation protocol, the qubit $a$ is measured at point $A$, which implies that its state $|q\rangle$ collapses onto $|0\rangle$ or $|1\rangle$, hence the original is destroyed and there remains only one qubit in state $|q\rangle$, the qubit $b$ at point $B$.

Quantum information theory reconsiders in the quantum setting the whole set of questions that are part of classical information theory, along similar lines to those initially established in 1948 by Claude Shannon. Both quantum and classical bits can now be taken as elementary carriers of information, and both quantum and classical channels can be used for transmission. Example of some questions studied in quantum information theory: how is classical or quantum information transmitted along a quantum channel, noisy or not? How and to what extent do entangled states facilitate the transmission of information? Besides the known notion of capacity of a classical channel, several related notions appear when quantum information and quantum channels enter the picture, like simple quantum capacity, for the transmission of qubits on quantum channels, or simple classical capacity, for the transmission of classical bits on quantum channels, or classically assisted quantum capacity, or again entanglement assisted classical capacity.

## 5.3.    EPR and entangled quantum states

What entangled quantum states are is suggested in Section 2.2.5, where it is shown that in a quantum system composed of two subsystems (parts $x$ and $y$ of a quantum register, in Section 2.2.5), the respective states of the subsystems are strongly correlated and therefore cannot be considered independently of one another. This is a general situation in quantum mechanics: the state of a quantum system composed of $n$ subsystems is not, in general, reducible to a $n$-tuple of the states of its components.

Such a situation, where the state of a part is not a part of the state of the whole, has no equivalent in classical physics, and does not fit with our intuition of what the world is around us. In 1935, Einstein, Podolsky and Rosen already understood that the mathematics of quantum mechanics implied that such strange states would be part of the quantum world. They expressed their discontent in a famous paper entitled "Can quantum-mechanical description of physical reality be considered complete." Briefly stated, they pointed at the fact that, according to them, there was some information hidden in quantum states that was not told by quantum mechanics. In their understanding, this had very severe and far reaching consequences on physics, since they thought that not considering this hidden information as corresponding to some physical reality would imply that instantaneous transmission of information is possible, hence contradict relativity theory. One can understand that Einstein had reasons to worry. Known as the EPR paradox, their question quickly became a centrepiece in the debate over the interpretation of the quantum theory. This debate continues in some circles, despite the now widely accepted fact that EPR is not a paradox at all. But it took about 50 years to arrive at a convincing evidence that the observable consequences of such states, as they were predicted in theory by the physicist John Bell in 1964, can indeed be confirmed experimentally, as this has been achieved convincingly for the first time by Alain Aspect's Bell experiment in 1982 in Orsay. As a result, it is now well understood that measuring parts of an entangled system does not transmit information at all. Einstein, Podolsky and Rosen would not have worried if they had known that in the 30s.

Entangled states have become the key quantum resource in quantum algorithmics and in other quantum feats like one-way quantum computation, teleportation and quantum key distribution: although the correlation that entanglement establishes among parts of a quantum system does not permit, alone, the transmission of information, this correlation itself contains an amount of information that can be exploited computationally: as an example, see in Section 2.2.5 how a function can be inverted, for free, for a randomly chosen value in its co-domain. This is why entangled states are a topic of research, for a better understanding of what they are and how they can be taken advantage of.

One of the questions is central: given the mathematical description of the state of a system (a vector state, in trivial cases, but more generally a so-called density matrix, when the knowledge about the state of the system is a probability distribution over a set of possible states), how to decide whether that state is entangled? Other questions are related to quantifying the amount of entanglement contained is a state: given two systems, how to associate measures to the entanglements of their respective states and decide whether one of them is more entangled than the other? Which operations and measurements, applied locally by partners who have distributed among them the components of an entangled quantum system, will allow them to evolve the state of that system toward another specified entangled state? Except for small systems and for specific classes of entangled states (so-called graph states), general answers are not known yet, and there is still a long way to go to reach a satisfactory understanding of what entangled states are.

## 5.4. Distributed quantum algorithms

The paradigm of distributed computation is the situation where two partners, Alice and Bob, have to compute a function $f(x,y)$, while $x$ is given to Alice only and $y$ to Bob only. The rule of the game is for Alice and Bob to achieve that in such a way that the number of bits that they exchange among them is as small as possible before they get to the result. This scenario can be generalized to any number of partners. The minimal number of bits required for computing $f$ in that way is a lower bound of the communication complexity of the distributed computation of function $f$. The communication complexity, which is a function of the size of the inputs $x$ and $y$, is not the same for all functions, and it has been evaluated in the classical setting for some classes of functions.

In the quantum setting, is has been found that, for some classes of functions, the number of exchanged qubits (the quantum communication complexity) is significantly lower than the number of bits for the same functions. There are even classes of function with an exponential drop of communication complexity. There are still many open questions in quantum communication complexity, even for some classes of very simple functions. An intriguing question, among others, is the analysis of situations where the partners share a set of qubits previously established in a globally entangled state. It has been found that, for some classes of functions, this allows a significant drop in communication complexity. Understanding precisely how and why entangled states become a computational resource that can improve communication complexity still needs a deeper analysis of such situations for distributed computations.

## 5.5. *Quantum error correcting codes*

Quantum information is fragile. It is carried by elementary particles which have to be operated upon and observed within some limited region of space, which implies that there are other particles inside that same region of space. The unavoidable consequence of this obvious physical fact is that the particles which are supposed to carry information that is relevant for the execution of some information processing task, interact with other particles which have nothing to do with that task, but just happen to be sitting there also. Because of these interactions, and after some time, usually very short, the useful and the undesirable particles will constitute a entangled quantum system, which means that the state of the useful particle is no longer relevant for the intended information processing task. This is the unavoidable physical phenomenon of quantum state decoherence which, as briefly mentioned in the next paragraph, is the main obstacle attacked by physicists who wish to find a practically usable physical implementation for qubits.

The question is then: how to process and communicate information in a reliable manner, in spite of the perturbations due to decoherence? A part of the answer is hoped to be in the hands of physicists, as told in the next paragraph. But this is not enough: even if physicists succeed in finding a physical qubit which stays coherent during a very long time, perturbations of the qubit state cannot be avoided. Quantum error correcting codes have been designed for taking care of such perturbations and for recovering, whenever possible, the original quantum state.

Like in the classical case, quantum error correcting codes rely upon redundancy. But the difficulty is much higher in the quantum case: it is not possible to maintain multiple copies of the same state, because of the no-cloning theorem, there is a continuum of possible perturbations, because of complex amplitudes, and, last but not least, the observation of the state destroys the state. Several systems of quantum error correcting codes have been designed. The idea is always to identify a set of possible errors, e.g. $|1\rangle$'s changed into $|0\rangle$'s and vice versa, positive amplitude changed to negative, etc., each type of error being associated with a corresponding correcting unitary operator, and to have logical qubits, i.e. the qubits as viewed by the information processing task, implemented by several physical qubits. For example, one such scheme uses five physical qubits for one logical qubit, another one 9 physical qubits for one logical qubit. A general theory of quantum error correcting codes has also been elaborated. But several major questions still remain unanswered.

One of the major results is the threshold theorem, according to which a quantum algorithm, however complicated, can be made fault tolerant as long as the error rate due to physical perturbations at each computation step is below a constant threshold, now estimated at $10^{-4}$. The idea is to perform the computation on the logical qubits, and to have each computation step followed by a correcting step. But this theorem makes simplifying assumptions on the type of errors that can happen, and on the independence among errors on distinct qubits: what would more realistic assumptions look like is still an open question and, more generally, what the limits of quantum error correcting codes are remains unknown.

## 5.6. Implementing quantum computers

Last but not least, the abstract qubits used without much metaphysical hesitations by the theoretician designers of quantum algorithms and protocols must, some day, be inscribed on a physical layer, in such a way that these algorithms and protocols can actually run. This is a formidable scientific and technological challenge. Pessimists even claim that the physical implementation of a quantum computer with a number of qubits large enough to perform practically useful information processing tasks (e.g. factorize very large integers with Shor's algorithm) is unfeasible. But it is interesting to notice that this does not prevent some of the most renown among these pessimists to work, very hard and with outstanding results, toward the physical implementation of qubits. There are indications that success lies indeed somewhere, far ahead on the road.

Although the threshold theorem tells that there is no physical principles that would definitely prevent such an implementation, decoherence is still there and it is, by far, the main obstacle thrown by Nature across the road. Two contradictory requirements must be satisfied: (i) the qubit must be as isolated as possible from other particles in its environment, so that its state remains coherent as long as possible, and stays in any case within the predefined class of error states that are recoverable by quantum error correcting codes; and (ii) the qubit must be manipulatable by its environment since all operations that have to be applied to it (unitaries and measurements) during the execution of an algorithm are necessarily controlled by the classical environment. The physical problem is thus to find a physical layer which would provide a satisfactory compromise between these two opposite requirements.

Five criteria have been identified at the end of the 90s by David DiVincenzo, from IBM Research, that are now widely agreed upon, and that any candidate qubit implementation must satisfy in order to be considered as a viable qubit to build a usable quantum computer:

(i) It must be possible to initialize the qubits in some predefined standard state, e.g. the state $|0\rangle$.

(ii) A universal set of elementary unitary operators must be applicable to the qubits, i.e. a quantum instruction set such that all computable functions can be realized on the computer.

(iii) It must be possible to measure qubit in at least one basis, e.g. the standard basis $\{|0\rangle, |1\rangle\}$.

(iv) The qubit implementation must be scalable, i.e. it must allow the coexistence and individual accessibility of a large number of qubits.

(v) The coherence time of the qubits must be significantly larger than the time required for applying any of the elementary unitary operators, so as to allow sufficient time for error recovery at each step. The current estimate is $10^4$ times the duration of an elementary operation.

The key challenge is to combine the necessary access to qubits, for initialization, control of operations and measurement, with a high degree of isolation, so that a long coherence time is guaranteed, within a scalable system.

Different candidate implementations for qubits are under study and many experiments and evaluations are being conducted all over the world. Six approaches are mentioned here, among many others:

(a) Nuclear magnetic resonance
(b) Trapped ions
(c) Trapped neutral atoms
(d) Photons
(e) Electronic spins
(f) Josephson junctions

An evaluation by the Advanced Research and Development Activity, of how each of these approaches satisfies the DiVincenzo criteria shows that some of them are rather promising, like trapped ions, whereas others already seem behind, like RMN which does not scale up properly.

This evaluation is summarized in the following table, where grey positions mean that the experiments conducted so far indicate that the approaches satisfy the criteria, blank position mean that not enough experiments have been conducted yet to get a reliable evaluation, and black means that the approach will most probably not satisfy the criteria:

|  | (i) | (ii) | (iii) | (iv) | (v) |
|---|---|---|---|---|---|
| (a) |  | ▨ |  | ■ |  |
| (b) | ▨ | ▨ | ▨ |  |  |
| (c) | ▨ |  |  |  |  |
| (d) |  |  |  |  | ▨ |
| (e) |  |  |  |  |  |
| (f) | ▨ |  |  |  |  |

There is still a long way to go before a satisfactory physical implementation of qubits is found. Then, higher level architectural considerations will have to be addressed with, in addition to many purely quantum issues, the necessary cooperation between quantum and classical processors. Fifteen to twenty years before a quantum computer is available on the market is considered an optimistic estimate.

## 6.    Further Reading

I. Articles on the main foundational results mentioned in this chapter:

- Bennett, C. H. and Brassard, G., Quantum cryptography: Public key distribution and coin tossing. In *Proceedings of IEEE International Conference on Computers Systems and Signal Processing*, Bangalore, India, 175–179, 1984.
- Bennett, C. H., Brassard, G., Crepeau, C., Jozsa, R., Peres, A. and Wootters, W. K., Teleporting an unknown quantum state via dual

classical and Einstein-Podolski-Rosen channels, *Physical Review Letters*, **70** 1895–1899, 1993.
- Shor, P. W., Algorithms for Quantum computation: Discrete logarithms and factoring. In *Proceedings 35th Annual Symposium on Foundations of Computer Science, IEEE Proceedings*, 1994.
- Grover, L. K., A fast quantum mechanical algorithm for database search. In *Proceedings 28th ACM Symposium on Theory of Computing (STOC'96)* 212–219, 1996.
- Bouwmeester, D., Pan J. W., Mattle, K., Eibl, M., Weinfurter, H. and Zeilinger, A., Experimental quantum teleportation, *Nature*, **390**, 575, 1997.
- Vandersypen, L. M. K., Steffen, M., Breyta, G., Yannoni, C. S., Sherwood, M. H. and Chuang, I. L., Experimental realization of Shor's quantum factoring algorithm using nuclear magnetic resonance, *Nature*, **414**, 883, 2001.

II. A short, well written and easy to read introduction:

- E. G. Rieffel and W. Polak, An introduction to quantum computing for non-physicists. Los Alamos ArXiv e-print, http://arxiv.org/abs/quant-ph/9809016, 1998. Also published in *ACM Computing Surveys*, **32**(3), pp 300–335, 2000.

III. An excellent textbook, well organised for a course, covers most topics:

- M. A. Nielsen and I. L. Chuang, *Quantum Computation and Quantum Information*, Cambridge University Press, 2000.

IV. Another textbook, with a deeper approach and a more theoretical style:

- A. Y. Kitaev, A. H. Shen and M. N. Vyalyi, *Classical and Quantum Computation*, American Mathematical Society, Graduate Studies in Mathematics, **47**, 2002.

V. Two reports and roadmaps on quantum information processing:

- *A Quantum Information Science and Technology Roadmap*, ARDA, http://qist.lanl.gov, 2004.
- *QIPC (Quantum Information Processing and Communication) — Strategic report on current status, visions and goals for research in Europe*. EU document, http://qist.ect.it/Reports/reports.htm, 2005.

# Index